Tijuana River Valley
HISTORICAL ECOLOGY INVESTIGATION

PREPARED FOR

THE CALIFORNIA STATE COASTAL CONSERVANCY
JANUARY 2017

Samuel Safran[1]

Sean Baumgarten[1]

Erin Beller[1]

Jeff Crooks[2]

Robin Grossinger[1]

Julio Lorda[3]

Travis Longcore[4]

Danielle Bram[5]

Shawna Dark[5]

Eric Stein[6]

Tyler McIntosh[1]

DESIGN AND PRODUCTION • RUTH ASKEVOLD[1] & SAMUEL SAFRAN[1]

[1] San Francisco Estuary Institute
[2] Tijuana River National Estuarine Research Reserve
[3] Facultad de Ciencias, Universidad Autónoma de Baja California
[4] University of Southern California
[5] California State University at Northridge
[6] Southern California Coastal Water Research Project

PREPARED BY

San Francisco Estuary Institute-Aquatic Science Center
Publication #760

FUNDED BY

California State Coastal Conservancy

ADDITIONAL FUNDING FROM

National Estuarine Research Reserve System Science Collaborative

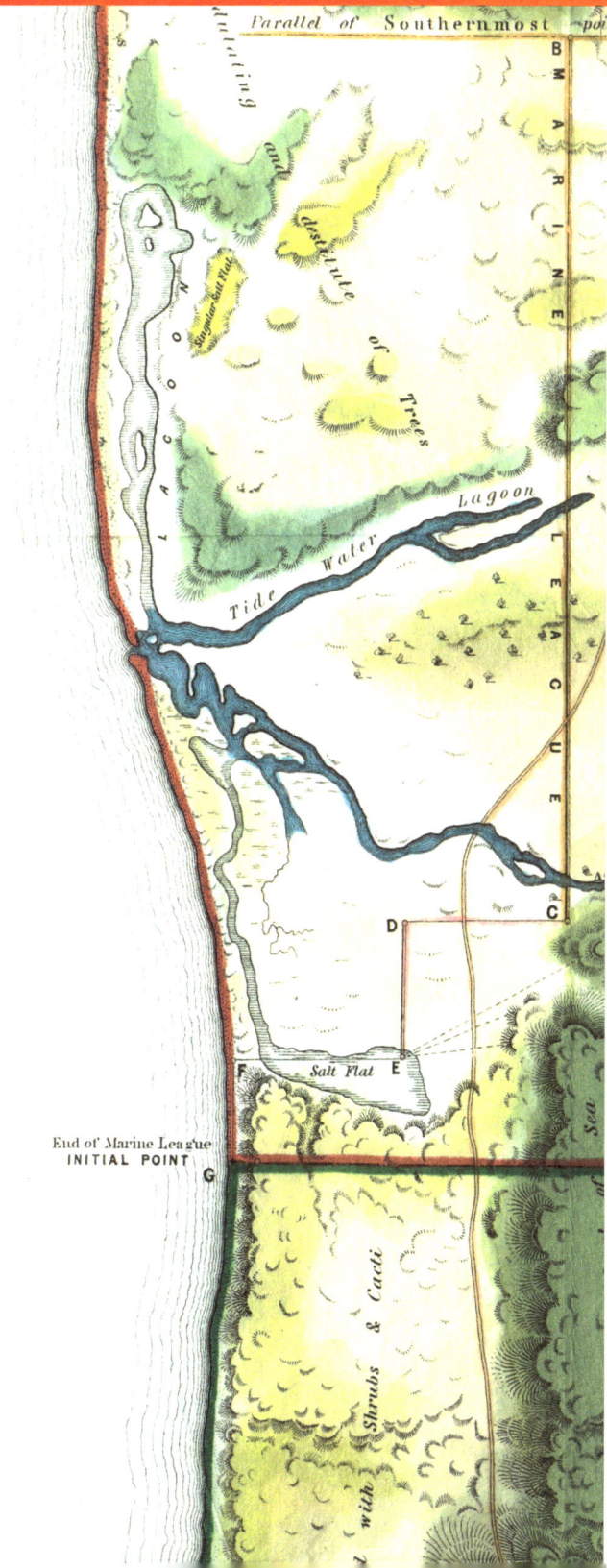

to San Diego Bay

Oneonta Slough

Tijuana River Slough

Tijuana Estuary

Mid-valley Slough

Old River Slough

North Slough

Tijuana River

Yogurt Canyon

Goat Canyon

Smuggler's Gulch

HISTORICAL CONDITIONS | Tijuana River valley, circa 1850

This map reconstructs the habitat types of the lower Tijuana River valley as they appeared ca. 1850, prior to major landscape modifications. Historically, the Tijuana River was the dominant feature of the valley and a critical driver of numerous physical and ecological processes. While the river was characterized by prolonged dry periods with little to mark its course but a sandy channel, major storms periodically transformed the river into a powerful force that abruptly shifted its course and flooded vast areas of the valley floor. These flood events redistributed tremendous amounts of sediment and uprooted riparian vegetation, maintaining a heterogeneous mosaic of floodplain habitats that included sandy river wash, dense riparian scrub, and groundwater-fed ponds. Wetlands, ranging from perennial freshwater wetlands to vernal pools and alkali meadows, occupied extensive areas outside of the river corridor, while terrestrial habitat types such as grassland and coastal sage scrub occupied higher and drier soils on a low mesa north of the valley floor. At the mouth of the Tijuana River, the daily ebb and flow of the tides maintained a broad estuary with a diverse array of habitat types and associated species.

Additional details about this map and the sources used to develop it can be found in this report and associated metadata. Note that the Mexican portion of the study area only encompassed the historical river corridor, so habitat types outside of corridor are not depicted as they are in the U.S.

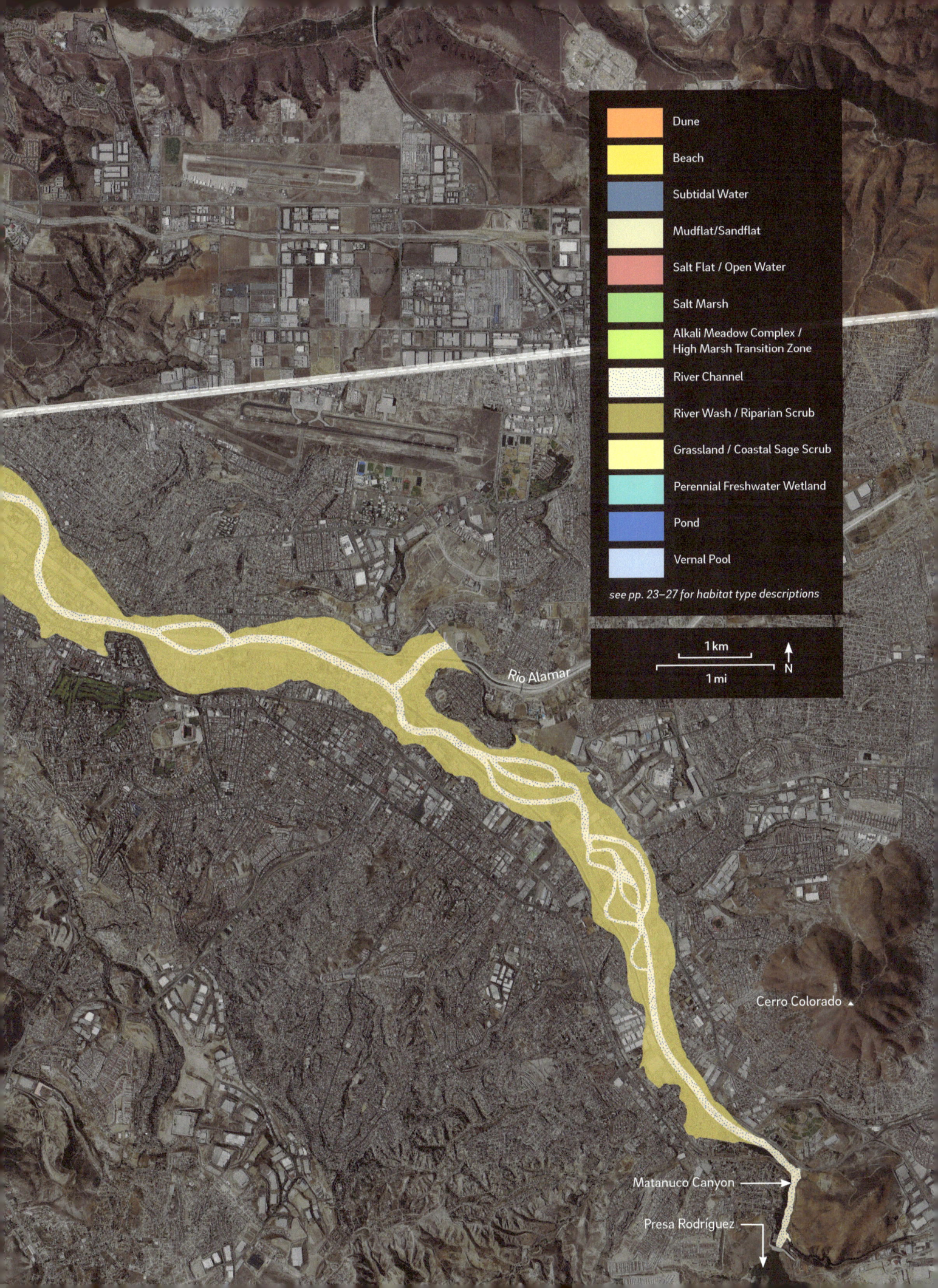

Dune

Beach

Subtidal Water

Mudflat/Sandflat

Salt Flat / Open Water

Salt Marsh

Alkali Meadow Complex /
High Marsh Transition Zone

River Channel

River Wash / Riparian Scrub

Grassland / Coastal Sage Scrub

Perennial Freshwater Wetland

Pond

Vernal Pool

see pp. 23–27 for habitat type descriptions

1 km

1 mi

N

Río Alamar

Cerro Colorado ▲

Matanuco Canyon →

Presa Rodríguez →

SUGGESTED CITATION

Safran SM, Baumgarten SA, Beller EE, Crooks JA, Grossinger RM, Lorda J, Longcore TR, Bram D, Dark SJ, Stein ED, McIntosh TL. *Tijuana River Valley Historical Ecology Investigation.* 2017. Prepared for the State Coastal Conservancy. A Report of SFEI-ASC's Resilient Landscapes Program, Publication # 760, San Francisco Estuary Institute-Aquatic Science Center, Richmond, CA.

REPORT AVAILABILITY

Report and GIS layers are available on the SFEI website at *www.sfei.org/projects/tijuana*.

COVER CREDITS

Front | *Boundary survey map:* Gray 1849, courtesy Coronado Public Library

Back | *[1] Car in dunes photo (ca. 1929):* Photo #UT 1982, courtesy San Diego History Center; *[2] Historical river courses graphic:* p. 114 of this report; *[3] Locational probability analysis graphic:* p. 116 of this report; *[4] Habitat change analysis:* p. 177 of this report; *[5] Pronghorn photo:* Bob Wick [BLM], June 2015, CC BY 2.0; *[6] Historical aerial photo:* San Diego County 1928; *[7] Border monument photo (ca. 1885):* BANC PIC 1982.078—ALB:7, courtesy The Bancroft Library; *[8] Northern harrier nest in marsh photo (1924):* Photo #L5722, courtesy San Diego Natural History Museum; *[9] Modern aerial photo:* courtesy Google Earth; *[10] Boundary survey map:* International Boundary Commission 1901, courtesy of the David Rumsey Historical Map Collection; *[11] Flood photo (ca. 1918):* Photo #89:17358-6, courtesy San Diego History Center.

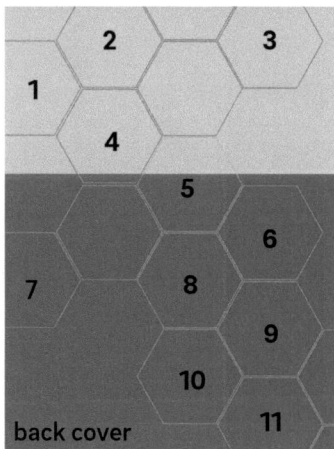

CONTENTS

ACKNOWLEDGMENTS

This project was primarily funded by the California State Coastal Conservancy, with additional funding from the National Estuarine Research Reserve System Science Collaborative. We would like to extend our thanks to Greg Gauthier and Megan Cooper of the Coastal Conservancy, who helped develop the vision for the project and provided the support and resources that allowed for its successful completion.

The report benefited greatly from the input of our technical review team. Technical reviewers Brian Bledsoe, Jeff Haltiner, David Jacobs, John Largier, Bruce Orr, Ellen Wohl, Richard Wright, and Joy Zedler each provided insightful guidance and comments on project mapping, analyses, and reporting.

We would like to thank SFEI staff, including Julie Beagle, Joanne Cabling, Emily Clark, Scott Dusterhoff, Steve Hagerty, Jeremy Lowe, and Katie McKnight, Micha Salomon for assisting with analyses, interpretation, graphical development, and the report-development process. Two interns from the Bill Lane Center for the American West at Stanford – Alexandra Peers and Rachel Powell – contributed significantly to the project. Staff from the Tijuana River Estuarine Research Reserve also assisted with research efforts; thank you in particular to Monica Almeida, Dani Boudreau, Michelle Cordrey, and Kristen Goodrich.

Many individuals assisted with the collection and compilation of historical data. These include Cristina Bourassa and Lora Robbins from the Tijuana River National Estuarine Research Reserve, as well as Dominic Kovacs, Marianne Jara, Joel Osuna, Andrea Rodriguez, and Randy Wolter from CSU Northridge. Outside researchers, including Exequiel Ezcurra, Paula Rebert, and John Cloud, helpfully responded to questions on historical data sources.

We are indebted to the staff and volunteers at the local regional, state, and national archives and institutions that we visited over the course of the project (see p. 17). In particular we would like to thank Carlos Vidali Rebolledo of the Mapoteca Manuel Orozco y Berra, Israel Sandré Osorio of the Archivo Histórico del Agua, Margaret Dykens of the San Diego Natural History Museum, Christian Esquevin of the Coronado Public Library, Tiffany Chow of the Water Resources Collections and Archive, Holly Reed of the National Archives and Records Administration, Jane Kenealy of the San Diego History Center, David Kessler of The Bancroft Library, and Mattie Taormina of Stanford University for their extraordinary help and patience.

1
INTRODUCTION

The lower Tijuana River valley, straddling the border between southern California and northern Baja California, is a dichotomous landscape. In Mexico, the Tijuana River – confined to a concrete flood channel – is embedded within a highly modified urban area. As the river crosses the border into the United States, the high-rises and freeways give way to farms and fields, and a dense riparian forest marks the course of the river as it flows towards its outlet at the Pacific Ocean. Despite these differences, the valley on both sides of the border has been dramatically altered over the past centuries, albeit in different ways. Understanding what this landscape looked like in the past, and how it has changed over the time, is key to effectively managing it in the future.

In the relatively recent past (within the last 165 years), the valley supported a remarkable diversity of plants and animals adapted to a wide range of habitat types. Some, like coastal California gnatcatcher and San Diego pocket mouse, thrived in the dry scrublands on the edges of the valley. Others, like sandhill crane and wandering skipper, flourished in the extensive wetlands on the valley floor. Still others, like the California black rail, made their home in the tidal sloughs and marshes of the Tijuana Estuary. The river valley was truly a dynamic landscape: tides pumped water in and out of the estuary twice daily, while massive floods periodically swept through, scouring away vegetation and reshaping the valley floor.

(BANC PIC 1982.078–ALB:7 "Monument, on boundary line between United States and Mexico," ca. 1885, courtesy The Bancroft Library)

Over the past several centuries, development and other land uses have resulted in dramatic physical and ecological changes in the valley. In the U.S., the extent of every major mapped historical habitat type (occupying >60 ha in the mid-19th century) has decreased by ~40–80%, while in Mexico the loss has exceeded 90%. Some habitat types, such as Vernal Pools, have completely vanished from the valley. In addition to this overall loss, habitat conversion has altered the structure of the estuary and the riparian corridor, further contributing to the transformation of the valley. Nevertheless, native plants and animals still survive and flourish in many areas. The nationally significant Tijuana Estuary continues to be an immensely productive ecosystem that supports a diverse array of birds, fish, and invertebrates. Upstream of the estuary, the Tijuana River supports an extensive riparian forest, one of the largest in coastal southern California, that provides habitat for a wide range of wildlife, including the federally endangered Least Bell's Vireo and Southwestern Willow Flycatcher. Though regulated by dams and partially channelized, the river continues to be a dynamic force capable of re-configuring the landscape.

With its binational jurisdiction and complex mosaic of urban and suburban development, agriculture, and open space, the Tijuana River valley is a challenging environment for land managers and conservation practitioners. In addition to habitat loss and conversion, key management

challenges include the presence of threatened and endangered species, invasion of non-native species, changes in streamflow and sediment dynamics, loss of tidal prism, and water quality degradation. The landscape also has unique restoration opportunities – for example, much of the lower river valley is relatively undeveloped and publicly owned – though determining appropriate restoration goals is itself a challenge. What types of habitats should be restored, and where? What physical processes are needed to maintain those habitats? How can critical disturbance events such as flooding and channel movement be accommodated?

The goal of this project is to help guide thinking around these and other questions by exploring the historical ecological and physical characteristics of the Tijuana River valley. While a historical perspective does not prescribe specific restoration goals or strategies, it can provide insights that help us to understand how the system functions today and to identify appropriate restoration targets. The absence of historical information limits our ability to see the future potential of the landscape. Conversely, an understanding of the valley's historical trajectory – of the drivers of change as well as the fundamental characteristics and processes that have remained unchanged – helps us to envision and create a landscape that supports the needs of its human residents while also optimizing biodiversity and ecological resilience.

(Photo by Samuel Safran, April 2015)

Project background & objectives

The *Tijuana River Valley Historical Ecology Investigation* addresses a regional data gap by reconstructing the landscape and ecosystem characteristics of the river valley prior to the major modifications of the late 19th and 20th centuries. The research presented here, funded by the California State Coastal Conservancy, supplies foundational information at the regional and system scale about how the Tijuana Estuary, River, and Valley looked and functioned in the recent past, as well as how they have changed over time. The ultimate goal of this study is to provide a new tool and framework that, in combination with contemporary research and future projections, can support and guide ongoing restoration design, planning, and management efforts in the valley.

The study draws on hundreds of historical documents to interpret and reconstruct the ecological and hydrogeomorphic characteristics of the valley circa the late 1700s to late 1800s, shortly after the arrival of Europeans (when written documents about the area were first produced) but prior to major subsequent landscape modifications. Data used in this report extend from 1769 through the 21st century, and range from travel diaries and family photographs to technical reports and government surveys. The resulting report describes the distribution of historical habitat types, analyzes hydrogeomorphic processes such as inlet dynamics and river movement, discusses driving physical processes, and quantifies change over time.

One of the primary products of this investigation is a map documenting historical habitat type patterns across the river valley, from the estuary to the present-day site of the Rodríguez Dam in Mexico (a total area of 4,250 ha [10,500 ac]). Information used to make this map was compiled and synthesized in an accompanying Geographic Information System (GIS) database, which includes historical sources and our level of certainty for each mapped feature. (See p. 16 for mapping methodology; the geodatabase may be downloaded at www.sfei.org/he.) This report complements the mapping with additional detail, context, and analysis.

The report is organized into eight chapters: this chapter provides an overview of study goals and objectives and summarizes major findings. Chapter 2 (p. 14) provides a review of mapping and analytical methodology, and Chapter 3 (p. 31) summarizes the physical and historical environmental context for the region. Chapter 4 (p. 66) describes the historical characteristics of the river valley outside of the river corridor and estuary; the river corridor and estuary are addressed separately in Chapter 5 (p. 87) and Chapter 6 (p. 135), respectively. Chapter 7 (p. 167) analyzes how historical habitats in the valley have changed over time. The report closes with Chapter 8 (p. 192), which summarizes the study's management implications and recommended future research.

Looking across the international border, towards the town of Tijuana, ca. 1890. *(Photo #FEP 1188, courtesy San Diego History Center)*

Why historical ecology?

The use of historical data to study past ecosystem characteristics is an interdisciplinary field referred to as "historical ecology" (Swetnam et al. 1999, Rhemtulla and Mladenoff 2007). Historical ecology is a powerful tool to reconstruct the form and function of past landscapes, enhancing our understanding of contemporary landscapes and helping us envision their future potential.

It can be tempting to see historical ecological research as an exercise in nostalgia, or as a restoration panacea that provides a prescriptive template from which to recreate the past. It is neither. Today's systems operate under different contexts than yesterday's, and we could not turn back the clock even if we wanted to. At the same time, many physical controls – from topography to geology – have remained relatively stable, and history can provide relevant clues about how natural, resilient systems persisted in a particular place in the recent past. Historical ecology is not just about the "way things were," but also the way they worked, providing invaluable insight into system dynamics today (Safford et al. 2012a). Historical ecological research is valuable for supporting current planning and restoration efforts in a number of ways:

- **Archival documents are a rich dataset of relevant ecological information**, with the potential to change assumptions about the past landscapes and to reveal ecological functions it previously provided.

- Historical ecology provides an opportunity to examine **system patterns, processes, and drivers** at broad spatial and temporal scales, describing the conditions to which native species are adapted and revealing fundamental characteristics and dynamics often difficult to discern in the contemporary landscape.

- Historical research can help foster a **shared understanding of local landscape history and habitat values**, establishing a common reference point and collective sense of place across diverse stakeholders. It is also an effective educational and communication tool shown to make stakeholders more receptive to future changes in management (Hanley et al. 2009).

- Historical ecology is a critical component in identifying **locally appropriate restoration targets** (Jackson and Hobbs 2009). It provides the context needed to document change over time, using this understanding to **recognize both the constraints and opportunities** posed by the contemporary landscape (Higgs 2012). This can ultimately translate into project cost savings by revealing restoration strategies that are realistic for the site and would require minimal maintenance. Conversely, ignoring historical context can lead to inappropriate restoration targets (e.g., Kondolf et al. 2001). It is a core principal of scenario planning that potential futures should emerge as logical trajectories from the past through the present (Peterson et al. 2003).

- Similarly, historical ecology can help us design and manage more flexible, **resilient future ecosystems** (Safford et al. 2012b). The study of historical landscapes can provide clues to how ecosystems were adapted to a highly variable climate regime, buffering the effects of environmental extremes. As a result, historical ecology has particular relevance in the context of global climate change: as we anticipate a more variable future climate, we can learn from the ways in which intact dynamic ecosystems were able to respond and adapt to extreme, variable conditions in the recent past (Harris et al. 2006).

Major findings

The following pages summarize some of the key findings of the *Tijuana River Valley Historical Ecology Investigation*. This section is meant to serve as a both a broad overview of the report, as well as a guide to its content; follow the page references to explore each theme in more detail. The images are drawn from the body of the report, where they can be seen with their supporting materials (legends, captions, labels, etc.).

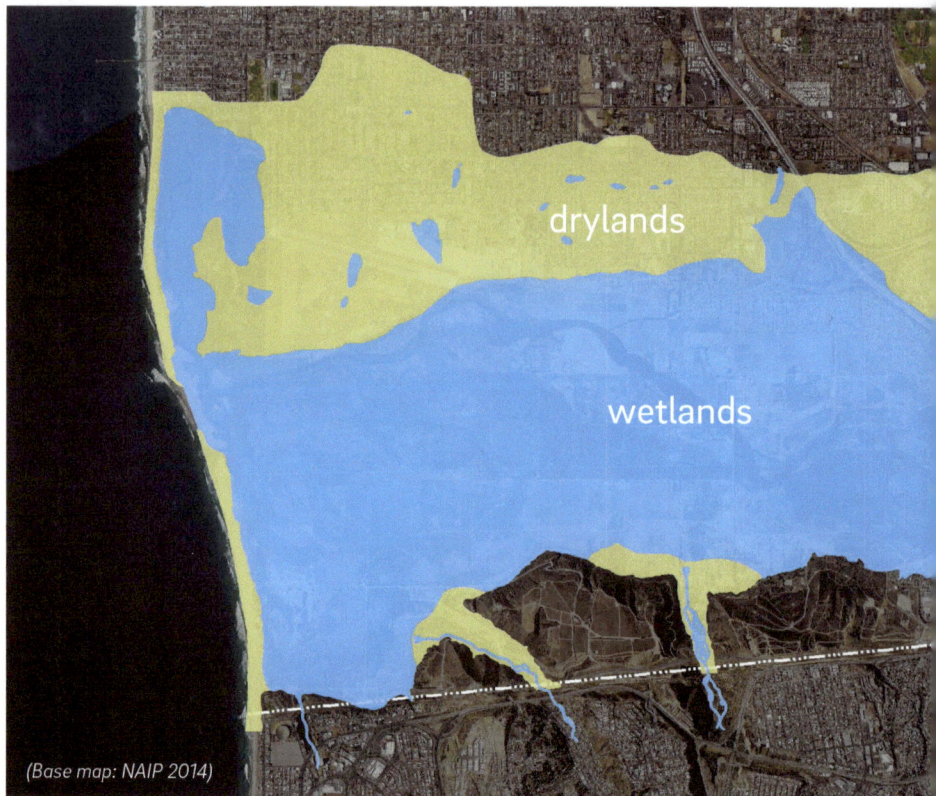

drylands

wetlands

(Base map: NAIP 2014)

Historical habitat types; pp. ii–iii

② A COMPLEX MOSAIC OF HABITATS | As it flowed northwest towards the ocean, the Tijuana River entered a broad valley characterized by a diversity of wetland and terrestrial vegetation types arranged along topographic, hydrologic, and salinity gradients. An extensive alkali meadow complex, supported by seasonally high groundwater levels and subject to seasonal flooding, occupied much of the valley floor adjacent to the river corridor (**p. 70**). Drier areas on low mesas to north and east of the river corridor were covered by grassland and coastal sage scrub dotted with freshwater wetlands, ponds, and vernal pools in depressional areas, supporting a unique flora and fauna (**p. 78**). The river itself supported a range of aquatic and riparian habitats, home to a variety of both xeric and obligate wetland species (**p. 126**). At the western end of the valley as the river entered the Pacific Ocean, was a mosaic of sandy beaches and dunes (**p. 156**) and estuarine wetlands such as salt marshes, mudflats, and seasonally flooded salt flats (**pp. 146–155**).

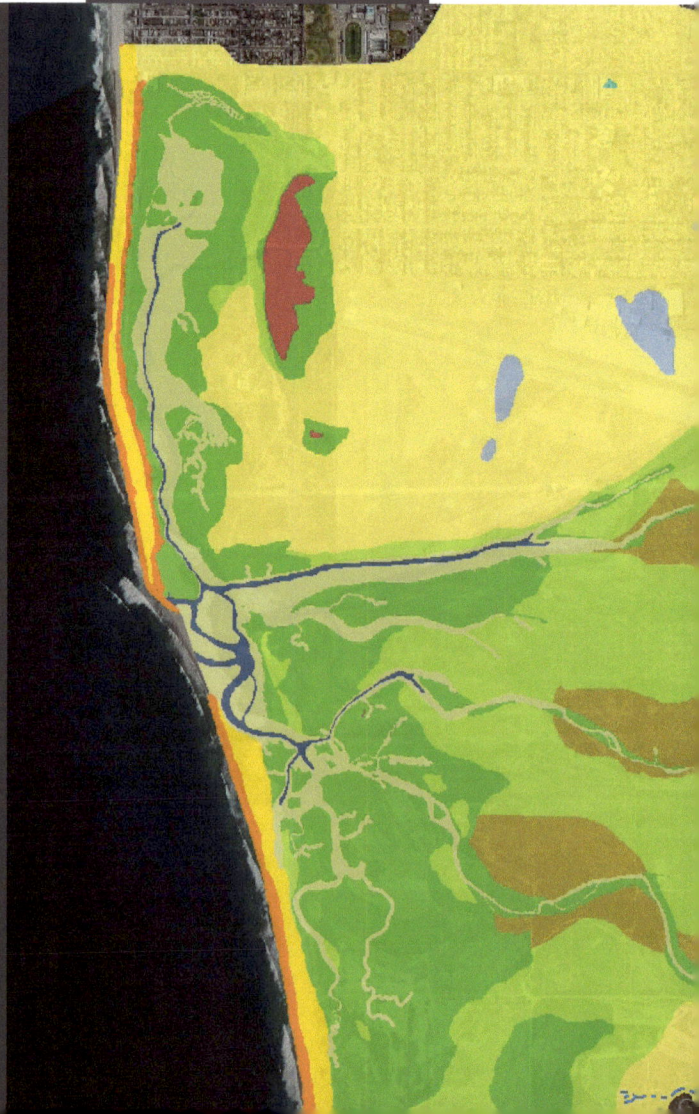

1 **WETLANDS IN A DRY LANDSCAPE** The Tijuana River valley's semi-arid climate is the driest found in coastal California (**p. 34**). Despite this dry climate, the valley floor was dominated by an array of wetland habitat types, which together covered more than 75% of the mapped area. Wetland habitat types were quite diverse, including estuarine (e.g., salt marsh; **p. 152**), palustrine (e.g. riparian scrub; **p. 122**), lacustrine (e.g., pond; **p. 104**), and riverine features (e.g., river wash; **p. 125**). Some of these wetlands were perennial, while others were seasonal or ephemeral. Many of these wetlands were supported by seasonally high groundwater (**p. 52**), including extensive alkali meadows (**p. 70**) and a few stream reaches with year-round flow (**p. 100**).

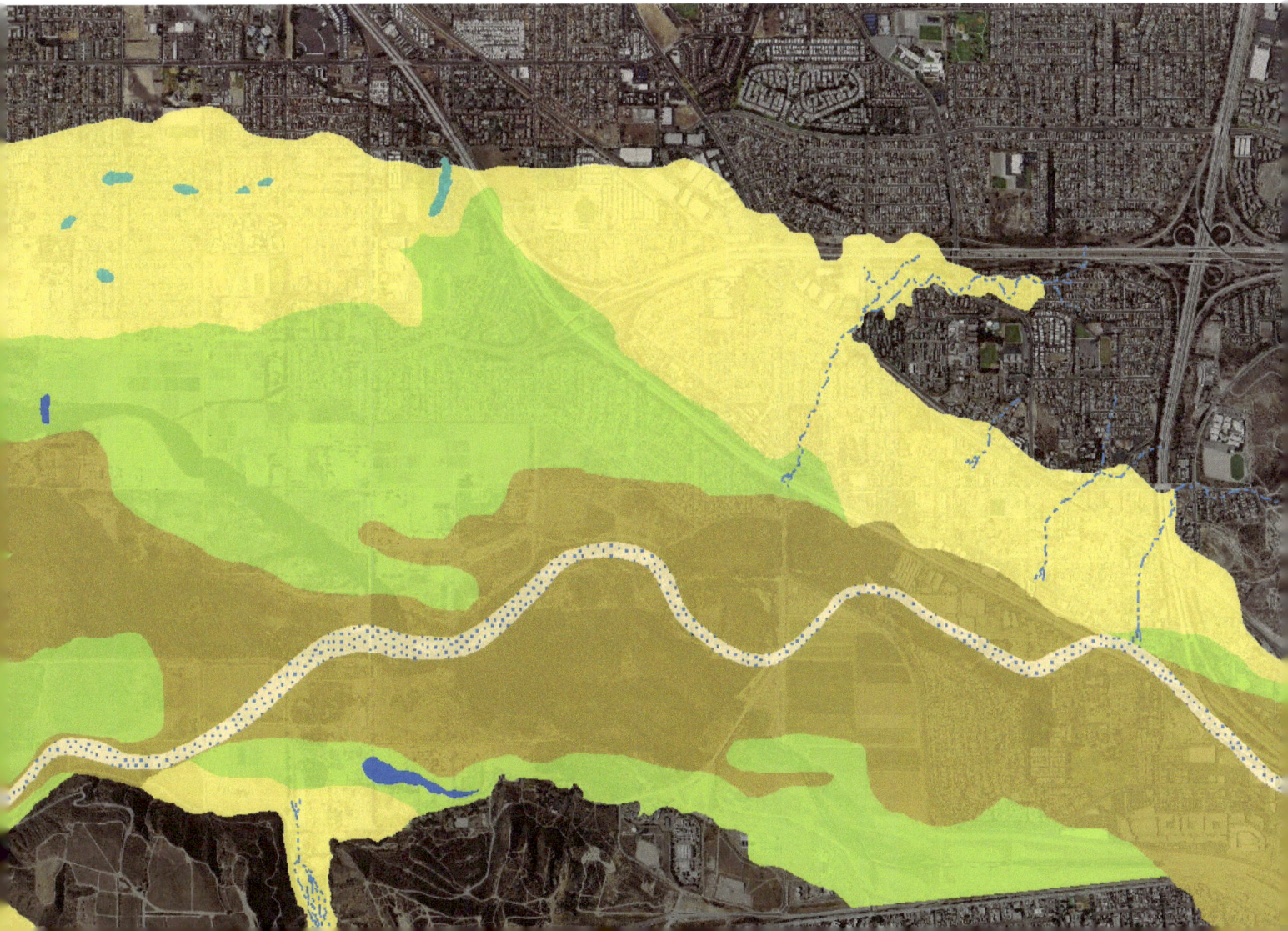

A BIODIVERSE LANDSCAPE

3 At the continental scale, the Tijuana River Valley's coastal semi-arid climate, which straddled the Mediterranean climate to the north and the desert climate to the south, is relatively rare (**p. 34**), a factor that likely contributed to the area's native species diversity. Historical records (pre-1950) document the presence of at least 280 taxa of plants, 125 birds, 30 mammals, 29 fish, 19 reptiles, and 4 amphibians. At least 14 of these animal species have not recently been observed in the lower valley and have possibly been extirpated: Bell's sparrow (*Artemisiospiza belli*; **p. 75**), fulvous whistling-duck (*Dendrocygna bicolor*), sandhill crane (*Grus canadensis*; **p. 71**), California condor (*Gymnogyps californianus*), black rail (*Laterallus jamaicensis*), pronghorn (*Antilocapra americana*), Pacific pocket mouse (*Perognathus longimembris pacificus*; **p. 150**), Western pond turtle (*Actinemys marmorata*; **p. 105**), Western banded gecko (*Coleonyx variegatus*), California glossy snake (*Arizona occidentalis*; **p. 75**), red diamond rattlesnake (*Crotalus ruber*), ring-necked snake (*Diadophis punctatus*), long-nosed snake (*Rhinocheilus lecontei*), and Southern California toad (*Anaxyrus boreas halophilus*). Also likely present historically, though no specific records were located, were the grizzly bear (*Ursus arctos;* Laliberte and Ripple 2004) and sea otter (*Enhydra lutris;* Riedman and Estes 1988). Some of these extirpations might be attributable to the near-complete losses of certain historical habitat types (**p. 175**). The lower valley has supported at least nine animal taxa and eight plant taxa that are currently listed as threatened or endangered at the state or federal level.

(Plant voucher: #112588, courtesy The Herbarium of the San Diego Natural History Museum; Pronghorn photo: Bob Wick [BLM], June 2015, CC BY 2.0)

(4) **A DYNAMIC LANDSCAPE WITH STABLE ELEMENTS** | The lower valley was a dynamic environment. Streamflow was highly variable both within and between years; some years brought tremendous floods, while others had no flow at all (**p. 96**). The inlet to the estuary migrated hundreds of meters up and down the coast (**p. 143**). During large floods, the river moved even further, frequently shifting laterally by more than 1,000 m (**p. 112**). Major floods uprooted large swaths of vegetation, creating a riparian environment that varied considerably over space (**p. 130**) and time (**p. 132**). But within the dramatic variability were aspects of stability. Inlet migration, for instance, was ultimately limited to a relatively small portion of the shoreline by coastal processes (**p. 143**). River movement was bounded by geologic controls and confined to a defined river corridor (**p. 116**). And the river reliably entered the estuary at one of three primary tidal sloughs (**p. 119**).

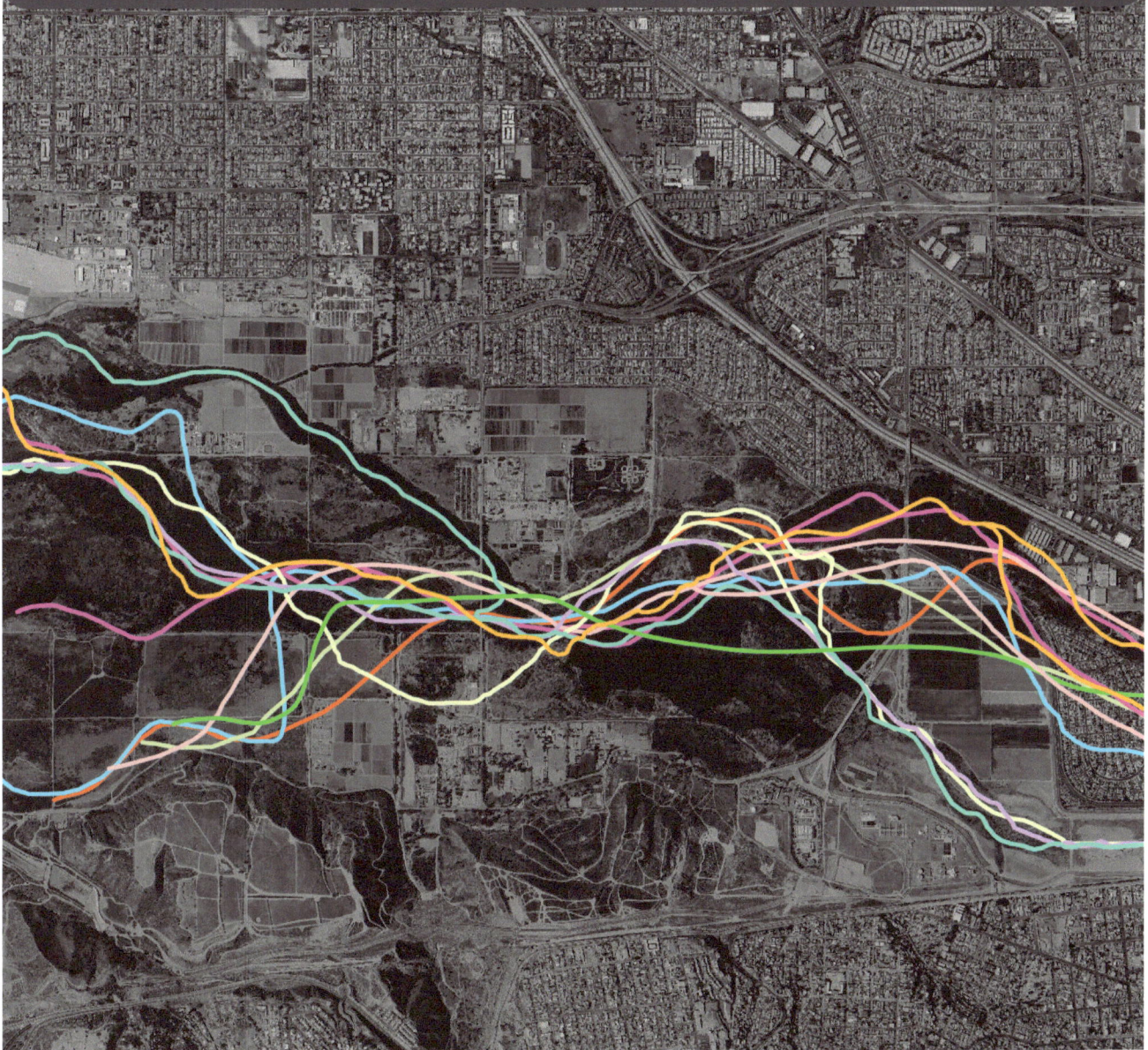

River movement over time; pp. 114–15

5 **A HETEROGENEOUS RIVER CORRIDOR** | The Tijuana River occupied a broad corridor, generally more than a kilometer wide, with a range of dynamic geomorphic features (**p. 90**). Shaped by relatively infrequent but high-magnitude flood events that had the power to dramatically shift the position of the river, the river corridor featured one or more sparsely vegetated, meandering low-flow channels; a wider, shifting, often braided high-flow channel; and a more densely vegetated floodplain. Unlike the tall willow forests found today, the corridor was dominated by shorter willow scrub interspersed with less densely vegetated areas of sandy river wash (**p. 122**) and perennial pools carved out by floods and filled with groundwater (**p. 104**). The river corridor hosted dozens of riparian plant species, ranging from obligate wetland species to more xeric alluvial scrub species, which together suggest that there was a range of hydrological conditions and a high degree of historical habitat heterogeneity (**p. 126**).

1852 "T-sheet" showing historical estuary; Ch. 6

(Harrison 1852, courtesy NOAA)

(Box 111, 3/3, photo #S-47, courtesy San Diego History Center)

6 **A PICKLEWEED PLAIN** | The Tijuana Estuary occupies the western edge of the valley, where the Tijuana River meets the Pacific Ocean. Historically, the estuary extended along 4.5 km (2.8 mi) of coastline and reached nearly 2.3 km (1.4 mi) inland (**p. 138**). Aside from a single inlet, which historical sources suggest was predominantly open in at least the intertidal range (**p. 142**), the estuary was separated from the ocean by a narrow strip of beach and dune (**p. 156**). Tidal channels and flats branched out to the north, east, and south, transporting water, sediment, and nutrients in and out of the estuary twice daily (**p. 146**). The vast majority of the estuary was vegetated with salt marsh; several large, unvegetated salt flats were found in higher elevation areas of the marsh plain (**p. 155**). Along the inland margins of the estuary was an extensive transition zone from high marshes to a variety of terrestrial habitat types (including non-tidal wetlands; **p. 160**).

8 **MAJOR POTENTIAL TO RECOVER LOST HABITATS AND FUNCTIONS** | In addition to documenting many of the ways in which the valley has been altered over time, this report also highlights opportunities to recover lost habitats and desired ecological functions (**p. 195**). In the U.S., most of the valley is undeveloped and in public ownership, which may help support landscape-scale restoration and recovery of key ecological processes supported by the historical landscape. Large floods still have the potential to drive river movement and associated riparian processes, given the space. Groundwater levels have recovered, raising the possibility that the valley could once again support groundwater-dependent seasonal wetlands. Ongoing efforts in the estuary seek to recover tidal prism, restore habitats that have been lost, and improve tidal wetlands' resilience to climate change. Though urban development limits some opportunities in Mexico, there are actions and best management practice (including native landscaping, green infrastructure, and low impact development) that would help to reestablish some desired ecological processes and services and provide co-benefits to those who live in the valley. Finally, there are a number of possible remnant habitat patches, which could have unique value for restoration efforts (**p. 179**).

7 **A TRANSFORMED VALLEY** | Urban and agricultural development, hydrologic changes, and invasions of non-native species, have resulted in widespread habitat loss and conversion in the lower Tijuana River valley over the past 150 years (**p. 175**). The largest single driver of habitat loss has been land conversion for urban development and agriculture, which now cover more than two-thirds of the valley. In the estuary, sediment accumulation, changes in the hydrology of the Tijuana River, and other factors have led to major shifts in habitat types (e.g. from unvegetated intertidal flats to vegetated salt marsh). A seaward shift in the inland extent of salt marsh and a landward movement of the dune system has resulted in an overall "compression" of the estuary (**p. 181**). The river corridor, which has been developed and channelized in Mexico (**p. 188**), has also been altered in the United States. The transition to perennial streamflow in the 1980s, accompanied by changes in groundwater levels, were likely key factors enabling the conversion from a riparian corridor dominated by riparian scrub to one dominated by riparian forest (**p. 184–87**). Additionally, sparsely vegetated areas of sandy river wash, which contributed to historical habitat heterogeneity, are no longer present. On the low mesa north of the valley, nearly all of the historical Grasslands / Coastal Sage Scrub matrix and associated wetlands have been lost to urban development (**p. 190**). Alkali meadow complexes (a facultative wetland type that once dominated the valley floor) have been replaced by farms, grasslands, and coastal sage scrub (drier terrestrial habitat types **p. 189**).

2 METHODOLOGY

The reconstruction of the past landscape of the lower Tijuana River valley presented in this report is a synthesis of data gleaned from hundreds of historical maps, texts, and photographs. When carefully scrutinized, compared across space and time, and analyzed with a modern understanding of environmental science, these disparate historical sources reveal a great deal about how the landscape functioned in the recent past. This chapter describes the methodology we used to accomplish this, including our process for data collection, compilation, and interpretation. It includes the methods we used to synthesize the historical data into a single map of historical (ca. 1850) habitat types, as well as a description of the other analyses we carried out to assess landscape processes and change over time.

For additional details on the methodology used to reconstruct historical landscape characteristics, please refer to Grossinger (2005), Grossinger et al. (2007), Stein et al. (2010), and Beller et al. (2016). This work has also benefitted from other historical ecology studies of coastal California, including those carried out for the Ballona Creek Watershed (Dark et al. 2011), Los Angeles Coastal Prairie (Mattoni and Longcore 1997), Northern San Diego County Lagoons (Beller et al. 2014), lower Santa Clara and Ventura rivers (Beller et al. 2011), and Sacramento-San Joaquin Delta (Whipple et al. 2012).

Data collection & compilation

Because a single dataset rarely provides sufficient information for accurate interpretation of complex systems, reconstructing historical landscape characteristics requires a broad range of historical data sources (Grossinger and Askevold 2005). We visited 40 institutions in California and Mexico to collect relevant data for this study, including local and regional historical archives, county offices, and public and private libraries and museums (Table 2.1). We also conducted searches of approximately 30 websites and electronic databases to obtain publicly available digital material. In total, we reviewed thousands of documents, and ultimately collected hundreds of unique sources, including approximately 180 maps, 200 landscape photographs, and 400 textual documents. The most important datasets collected for this study are summarized on pages 18–19.

Data collection efforts focused on the period from early Spanish explorers in 1769 (which represent the first available written records) to the time of the first aerial photography in the late 1920s. While this time period represents only a short time in the natural history of the lower Tijuana River valley, it is a relevant span for understanding how habitats were formed and maintained within a large-scale geomorphic and climatic context relatively similar to today's. This snapshot provides an opportunity not just to reconstruct landscape patterns during the late 18th and 19th centuries, but also to understand the natural processes that shaped the distribution, diversity and abundance of habitats during this period – processes that in many cases may still be active. We also collected and compiled contemporary sources, including geologic maps, soil surveys, elevation datasets, and modern aerial photography that supported our interpretation and mapping of historical sources. While these datasets clearly depict an altered landscape, when used in conjunction with earlier sources they aid the interpretation of the historical landscape.

Once collected, data were compiled into more accessible formats for mapping and interpretation. We used a geographic information system (Esri's ArcGIS 10 software) to synthesize and compare many types of spatial data. We georeferenced more than 50 high-priority maps, including U.S. Coastal [and Geodetic] Survey (USC[G]S) T-sheets, U.S. Geological Survey (USGS) topographic quads, General Land Office (GLO) survey plats, soils maps, and boundary surveys. Additionally, we orthorectified, georeferenced, and mosaicked the earliest available aerial imagery (about 200 images, taken in 1928, 1946, and 1955) into a continuous coverage of the study area. GLO survey data (over 1,400 individual data points) were also transcribed and georeferenced using software adapted from the Forest Landscape Ecology Lab at the University of Wisconsin-Madison (Manies 1997, Radeloff et al. 1998, Sickley et al. 2000). Sources not compiled within the GIS (e.g., textual data, landscape and oblique photography, and maps too spatially imprecise to be georeferenced) were transcribed and/or organized by topic and date to allow for use of these data during interpretation and mapping.

Though the data collection process was extensive, it was inevitably not exhaustive. Undoubtedly, additional sources of information will surface in the future that will refine and enrich the understanding of the historical landscape presented in this report.

Table 2.1. (opposite page) Source institutions from which data were collected for this study. In addition to these archives, numerous online repositories were also consulted. (*Archive photos by Erin Beller and Julio Lorda, 2014*)

United States	
San Diego region	
	Coronado Public Library
	International Boundary and Water Commission Records Office
	San Diego History Center
	San Diego Natural History Museum
	San Diego Public Library
	San Diego State University- Malcolm A. Love Library
	San Diego State University- Special Collections
	Scripps Institution of Oceanography Archives
	University of California (UC) San Diego- Geisel Library
	UC San Diego- Mandeville Department of Special Collections
Los Angeles region	
	California State University Northridge Library
	Huntington Library
	National Archives- Riverside
	Seaver Center for Western History Research at the Los Angeles Museum of Natural History
	UC Los Angeles- Spence/Fairchild Collection
	UC Riverside- Water Resources Center Archives
San Francisco Bay Area	
	California Historical Society
	Society of California Pioneers
	Stanford Library & Special Collections
	UC Berkeley- Doe Library
	UC Berkeley- Earth Sciences and Map Library
	UC Berkeley- Hearst Anthropology Museum
	UC Berkeley- Marian Koshland Bioscience and Natural Resources Library
	UC Berkeley- The Bancroft Library
Other areas	
	California State Railroad Museum (Sacramento, CA)
	National Archives (Arlington, VA)

Mexico	
Tijuana	
	Instituto Municipal de Arte y Cultura- Archivo Historico de Tijuana
	Biblioteca Alberto Limón Padilla
	Casa de la Cultura Juridica
	Centro Cultural Tijuana
	Comisión Nacional del Agua
	Direccion de Catastro, Ayuntamiento de Tijuana
	Universidad Autónoma de Baja California- Instituto de Investigaciones Históricas
	Instituto Nacional de Estadística y Geografía
	Sociedad de Historia de Tijuana
Mexico City	
	Archivo General de la Nación
	Archivo Histórico del Agua
	Fundación ICA
	Universidad Nacional Autónoma de México- Instituto de Investigaciones Históricas
	Mapoteca Manuel Orozco y Berra

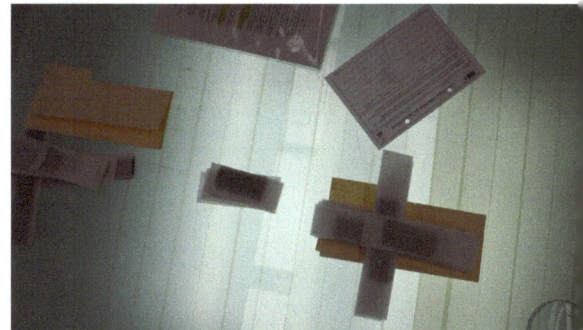

Primary historical data sources for the Tijuana River valley

This study required the collection and compilation of a wide variety of sources that spanned multiple centuries, languages, and formats. Careful review of these historical documents served as the foundation for the mapping and interpretation of the Tijuana River valley's recent ecological past. Below, we summarize the primary cartographic, textual, and pictorial sources used in this study.

1 **Mexican land grant sketches and court testimony (1820s-1850s).** As the Mission system disintegrated in the 1830s, influential Mexican citizens submitted claims to the government for land grants. A *diseño*, or rough sketch of the solicited property, was included with each claim. Diseños often show notable physical landmarks that would have served as boundaries or provided natural resources, such as creeks, wetlands, springs, and forests. While diseños are not as spatially accurate as subsequent surveys, they provide extremely early glimpses of former landscape features and patterns. A diseño drawn of Rancho Tijuana (U.S. District Court California, Southern District ca. 1840; shown here) is the earliest known map of the river valley.

2 **U.S. and Mexico Boundary Commission surveys (1849–1901).** The Treaty of Guadalupe Hidalgo, which ended the Mexican-American War in 1848, defined the boundary between the U.S. and Mexico and called for the creation of government commissions from both nations that would be jointly responsible for locating, marking, and mapping the border (Rebert 2001). The newly formed commissions met in the port of San Diego on July 3rd, 1849 and soon set off to establish the initial point of the boundary at the Pacific Ocean and begin the survey eastward. From this effort, the U.S. and Mexican commissions each produced a series of maps of the border, including the Tijuana River valley. These maps depict natural features such as marshes, rivers, and springs, as well as cultural features such as Indian villages, Mexican ranches, fields, and roads. Photographs, sketches, and reports from the expedition also provide rich additional detail on the fauna and flora of the borderlands. The border was resurveyed at the turn of the century, generating additional materials. Taken together, the early and accurate data of the Boundary Surveys are a unique and invaluable resource for understanding the historical Tijuana River valley.

3 **U.S. Coast Survey maps (1852–1933).** The U.S. Coast Survey (USCS; later the U.S. Coast and Geodetic Survey [USCGS]) was established in 1807 by Thomas Jefferson to survey and map the American coastline. USC[G]S maps covering the landward portion of the coastline, known as "topographic sheets" or "T-sheets," are a highly valuable source for understanding the physical and ecological characteristics of the coastline prior to extensive Euro-American modification. Because of their relatively early survey dates, high scientific rigor, and impressive detail, T-sheets are widely used by researchers studying the historical U.S. shoreline (Grossinger et al. 2011). The earliest T-sheet depicting the Tijuana River Estuary, surveyed by A.M. Harrison in the winter of 1851–2, was one of the first T-sheets produced in the young state of California. Mapped at a scale of 1:10,000 (more than twice the resolution of modern USGS topographic quadrangles), the T-sheet offers a detailed look at the estuary's beaches, dunes, sloughs, flats, ponds, and marshes. Seventy years later, the T-sheet was resurveyed and remapped by USCGS from 1933 aerial photographs. Because this later T-sheet is heavily annotated and covers the entire U.S. portion of the study extent, it serves as a useful "ground truth" for features visible in the historical aerials that we compiled from the same period.

4 **General Land Office (GLO) Public Land surveys (1854–1880).** Established in 1812, the GLO was charged with surveying and overseeing the sale of public lands in the western U.S. In areas not claimed through the land grant system, the U.S. Public Land Survey divided the land into a grid of 1x1 mile squares (known as "sections"). Surveyors systematically walked section boundaries, keeping detailed field notes on the natural and cultural features encountered along the way. Notes and plat maps from these surveys are useful for their ecological information and have been extensively utilized in historical landscape reconstruction and land cover change research (Buordo 1956, Radeloff et al. 1999, Collins and Montgomery 2001, Brown 2005, Whipple et al. 2011). The U.S. portion of our study extent was covered by the Public Land Survey and surveyed during the mid-19th century, providing early and spatially accurate data.

5 **U.S. Geological Survey (USGS) topographic quadrangles (1904-present).** The USGS (established in 1879) began producing topographic quadrangles ("quads") of the San Diego area in 1904. The maps provide complete coverage of the U.S. portion of the study extent and offer information on the region's topography and hydrography. Because they were regularly reproduced and utilize standardized symbologies, the quads are also particularly useful for assessing change over time.

6 **Historical aerial photography (1928–1955).** The earliest historical aerial imagery—with complete coverage for the U.S. portion of the study extent—was flown by San Diego County during the winter of 1928–1929. The Mexican portion of the study extent is covered by vertical aerials from 1946 and 1955. While the photographs were taken after substantial modifications to the river valley, they nevertheless reveal many remnant and relict ecological features, traces of which are often still evident on the landscape. Historical aerial photos are useful for interpreting and mapping features depicted on earlier, less spatially accurate sources.

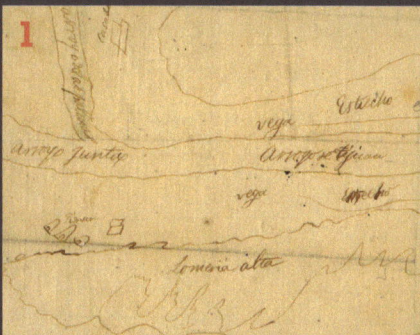

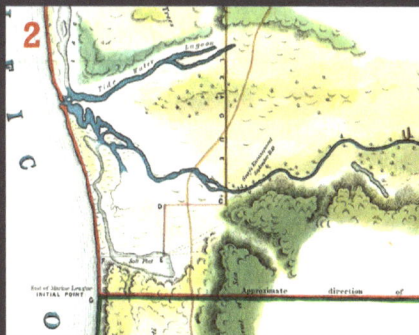

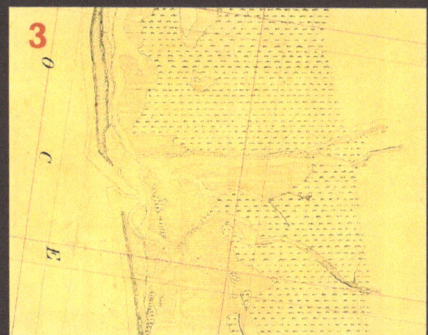

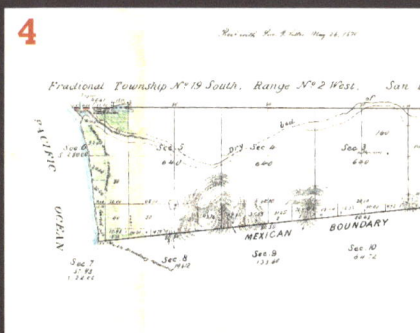

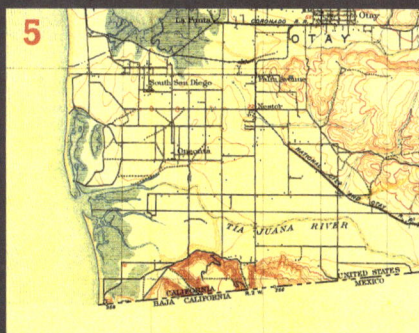

(1: U.S. District Court California, Southern District ca. 1840, courtesy The Bancroft Library; 2: Gray 1849, courtesy Coronado Public Library; 3: Harrison 1852, courtesy NOAA; 4: Day 1870, courtesy Bureau of Land Management; 5: USGS 1904; 6: San Diego County 1928)

Data interpretation

Accurately interpreting documents produced during different eras, using different techniques, for differing purposes, and by different authors can be challenging (Grossinger and Askevold 2005). Only when compared can individual historical sources reveal prevailing landscape patterns and processes (Harley 1989, Swetnam et al. 1999). With this in mind, we utilized an iterative process of source comparison and inter-calibration to interpret the historical dataset and to yield evidence. The inter-calibration of multiple independent data sources helps to uncover (and often resolve) inconsistencies between individual sources and reveal persistent landscape features and patterns. Ultimately, this process allowed us to take a large body of often subjective information (e.g., a traveler's description of the Tijuana River) and form a reliable, comprehensive, and coherent body of data (Grossinger 2005, Grossinger et al. 2007).

An important component of the source intercalibration process is interpreting data in the context of daily-, annual-, and decadal-scale variation in climate. Knowing the season in which a source originated, or whether it reflects conditions before or after a major flood, for example, influences the interpretation of that source. A photograph of the Tijuana River running with water in July tells us more about the flow permanence of that reach than the same photograph taken during January. Similarly, the potential effects of various land use changes must be taken into account when evaluating any particular data source (see Chapter 3).

Once collected, compiled, and interpreted, historical data were synthesized into a map of the lower Tijuana River valley's historical habitat types (printed on the inside cover of this report). Rather than portray conditions at a specific point in time, we endeavored to map the general diversity and distribution of habitat types during average dry-season conditions just prior to significant Euro-American landscape modification (referred to as "ca. 1850" throughout this report). The map is meant to integrate a large body of information into a single image and to serve as a tool for landscape interpretation that helps enhance our understanding of regional ecological patterns and processes.

The primary tool for developing the historical habitat type map was the project's GIS. We used the GIS to store and organize historical information and to evaluate and compare landscape features over space and time (Figure 2.1). Through this process, we determined how best to classify and delineate features in the historical habitat type map. The GIS was also used to draw (or "digitize") each feature and to record the sources that aided their interpretation and mapping.

To fully document the provenance of the historical habitat type map, we recorded three kinds of data sources on a feature-by-feature basis: digitizing sources (those used to draw each mapped feature), primary interpretation sources (those used to guide the classification and interpretation of each mapped feature), and supporting interpretation sources (those used to enhance our understanding of each mapped feature). We did not attempt to document every piece of evidence, but only those that contributed most to each feature's interpretation and delineation. Additionally, we assigned each feature an estimated certainty levels to indicate our confidence in its historical presence and classification (interpretation), size, and location (Table 2.2) following standards discussed in Grossinger et al. (2007). Certainty levels were determined based on a combination of source date, accuracy of the digitizing source, and the diversity and quality of supporting evidence. Attributing features in this way allows users to assess the accuracy of specific map elements and to identify the full suite of original information sources (Grossinger 2005, Stein et al. 2010).

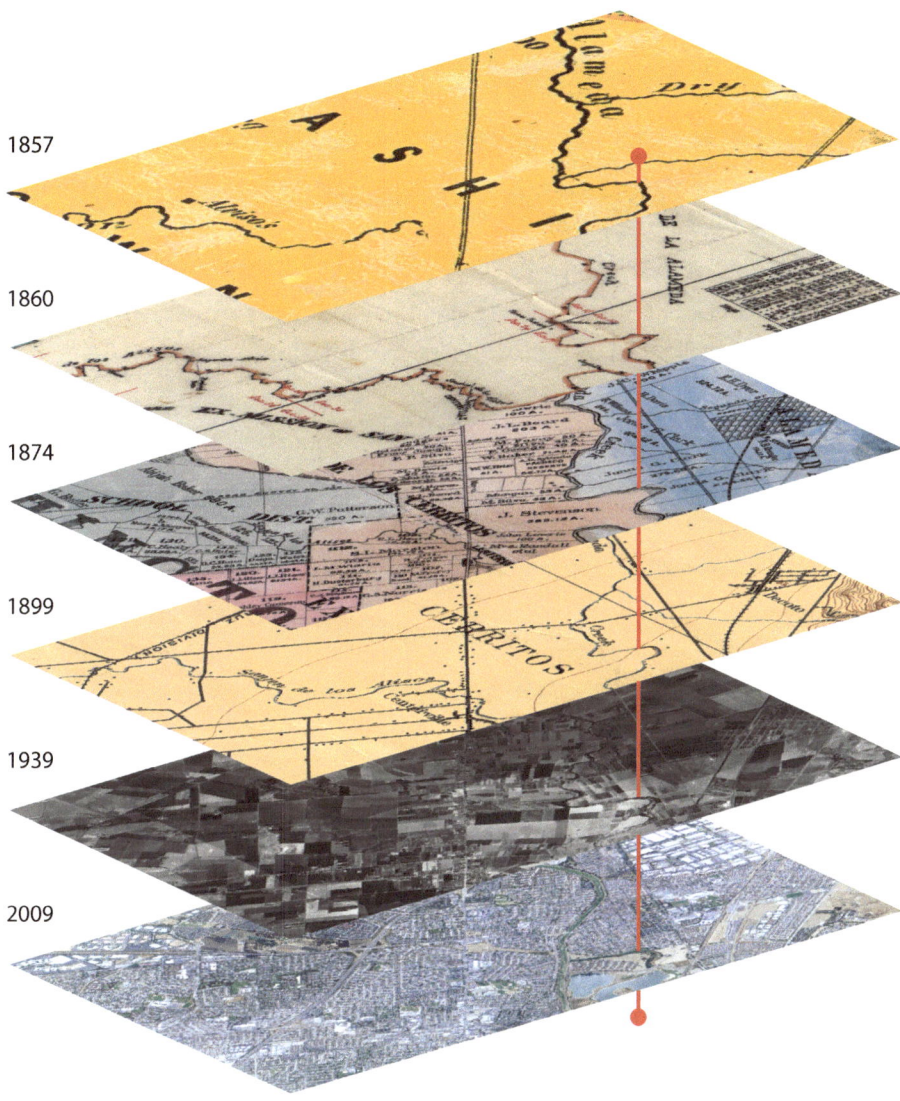

Figure 2.1. Assembling maps from different time periods in geographic information system allows for comparison of features across space and time (adapted from Stanford et al. 2013).

Table 2.2. Mapping certainty levels. Each mapped feature was assigned a certainty level of high, medium, or low for each of three characteristics (interpretation, size, and location). Interpretation certainty describes our confidence that the habitat type assigned to the feature is accurate and that the feature is representative of the historical period. Size certainty describes our confidence that the feature's spatial extent is accurately depicted. Location certainty describes our confidence that the feature existed at the mapped location (adapted from Grossinger et al. 2007).

Certainty Level	Interpretation	Size	Location
High/ "Definite"	Feature definitely present before Euro-American modification	Mapped feature expected to be 90%-110% of actual feature size	Expected maximum horizontal displacement less than 50 meters (150 ft)
Medium/ "Probable"	Feature probably present before Euro-American modification	Mapped feature expected to be 50%-200% of actual feature size	Expected maximum horizontal displacement less than 150 meters (500 ft)
Low/ "Possible"	Feature possibly present before Euro-American modification	Mapped feature expected to be 25%-400% of actual feature size	Expected maximum horizontal displacement less than 500 meters (1,600 ft)

There are necessarily limitations to the process of mapping historical habitat types. First, habitat types that were characterized by particularly small or ephemeral features are unlikely to be well-represented in historical records and maps. As a result, some historical habitat types, such as Perennial Freshwater Wetlands and Vernal Pools, were likely under-mapped. While there was no minimum mapping unit for mapping polygonal features, we aimed to illustrate features and landscape characteristics that could be mapped consistently across the study area. Second, the mapping exercise often required us to draw sharp boundary lines where the true boundaries were likely broader transitional gradients. On a number of occasions, when potentially distinct habitat types could not be reliably distinguished across the full study extent, we opted instead to lump them into combined classes (e.g. River Wash / Riparian Scrub, Alkali Meadow Complex / High Marsh Transition Zone, and Grassland / Coastal Sage Scrub). And finally, the highly dynamic nature of the Tijuana River made it particularly challenging to represent historical conditions in a single, static, two-dimensional map. For example, though the active river channel shifted with relative frequency, we opted to represent a single well-documented course from ca. 1850. Other analyses were used to illustrate temporal variability and capture historical ecosystem dynamics (variation in the location of the river, for instance, is shown in a separate map on pp. 114–15). Many of the complexities uncovered in the historical data that could not be represented in the habitat type map are described more fully in this report.

Historical habitat types classification

Reliable evidence was found for mapping 14 historical habitat types. These classes are intended to capture broad-scale landscape patterns and to be comparable with contemporary classification systems. Although they represent the greatest level of detail that could be consistently mapped across the study extent, complex fine-scale patterns and considerable variation in species assemblages would have existed within each habitat type. Brief descriptions of each historical habitat type are provided below, along with a short account of the sources and methods used to map their historical extent and distribution (the map can be found on the inside cover of this report). For more specific information on the sources and accuracy of any particular mapped feature, please refer to the GIS metadata (available online at www.sfei.org/he).

BEACH

DESCRIPTION	For the purposes of this study, Beaches are the relatively flat zones of sandy sub-strate along the shore that exist above the high water line either up to, between, or behind elevated dune ridges. It excludes portions of the larger beach-dune system that lie within the littoral zone (like the swash zone) and instead includes areas such as the "upper beach" (Pickart and Barbour 2007), "coastal strand" (Dugan and Hubbard 2010), and "deflation plain" (Pickart and Barbour 2007). Although the upper beach supports lower species richness and plant cover than the higher foredunes, this habitat type likely supported some areas of ephemeral beach vegetation (probably dominated by sticky sand verbena *[Abronia maritime]*) on relatively short-lived shadow dunes and beach mounds (Pickart and Barbour 2007).
MAPPING METHODS	Digitized exclusively from the earliest available USCS T-sheet (Harrison 1852), as indicated by the regular, low-density stipple pattern used to indicate sandy beaches along the coast (Shalowitz 1964, Grossinger et al. 2011) and supported by other 20th century historical sources.

DUNE

DESCRIPTION	Dunes are coastal upland habitats formed along the shore that develop from the vertical accumulation of wind-blown sand. Many dune systems support extensive areas of vegetation, which, in the relatively narrow dunes associated with the historical Tijuana Estuary, most likely consisted of low pioneering herbs and subshrubs such as sticky sand verbena (*Abronia maritima*), pink sand verbena (*Abronia umbellata*), dune ragweed (*Ambrosia chamissonis*), and salt bush (*Atriplex leucophylla*) (Purer 1936, Zedler et al. 1999). Grasses, such as salt grass (*Distichlis spicata*), and larger shrubs, such as lemonade berry (*Rhus integrifolia*), were likely also present in some areas. Dunes are closely associated with beaches, and the two habitat types can be difficult to distinguish. Since the historical sources from which our mapping of Dune and Beach is derived simply distinguished higher topographic features from lower areas, what is categorized as "Dune" here likely includes both what contemporary authors would consider "foredune" and "dune ridge" (Pickart and Barbour 2007). Low-lying swales or deflation plains between ridges are included with the Beach habitat type.
MAPPING METHODS	Digitized exclusively from the earliest available USCS T-sheet (Harrison 1852), as indicated by the dense stipple pattern used to show raised topography at the beach margin (Shalowitz 1964, Grossinger et al. 2011) and supported by other 20th century sources.

(Beach photo: Samuel Safran, April 2015; Dune photo: Michael Hedin, December 2012, CC BY-SA 2.0)

SUBTIDAL WATER

DESCRIPTION

Subtidal Water is the area within an estuary that remains submerged during the lowest tides, where the bottom lies below the elevation of Mean Lower Low Water (MLLW) and where the substrate is therefore permanently inundated.

MAPPING METHODS

Digitized exclusively from the earliest available USCS T-sheet (Harrison 1852), as indicated by the area below the dotted line of MLLW (Grossinger et al. 2011) and supported by other early maps and photographs.

MUDFLAT / SANDFLAT

DESCRIPTION

Mudflats and Sandflats are soft-bottom intertidal habitats with less than 10% vascular plant cover that are exposed between low and high tides. They occur approximately from the elevation of MLLW to Mean Tide Level (Goals Project 1999). In the Tijuana Estuary, the majority of Mudflat and Sandflat habitat occurs along the margins of intertidal channels. Mudflats and Sandflats are distinguished by sediment composition and grain size, with differences generally attributable to the relative strength of prevailing water currents; Mudflats are associated with lower-energy areas generally higher in the intertidal zone and farther from the estuary mouth, while Sandflats are associated with higher energy areas generally lower in the intertidal zone and closer to the estuary mouth (Elliot et al. 1998, Zedler et al. 1999).

MAPPING METHODS

Digitized primarily from the earliest available USCS T-sheet (Harrison 1852), as indicated by the unvegetated area between MLLW and the lower limit of marsh vegetation or land (Grossinger et al. 2011) and supported by other early maps and landscape photographs. Intertidal channels drawn as polygons on the T-sheet were represented as polygons on our map, while channels drawn as lines on the T-sheet were represented as lines. Historical aerials (San Diego County 1928) were used to digitize additional intertidal channels thought to be representative of the historical period and to improve the spatial accuracy of some areas of intertidal flat shown by the T-sheet.

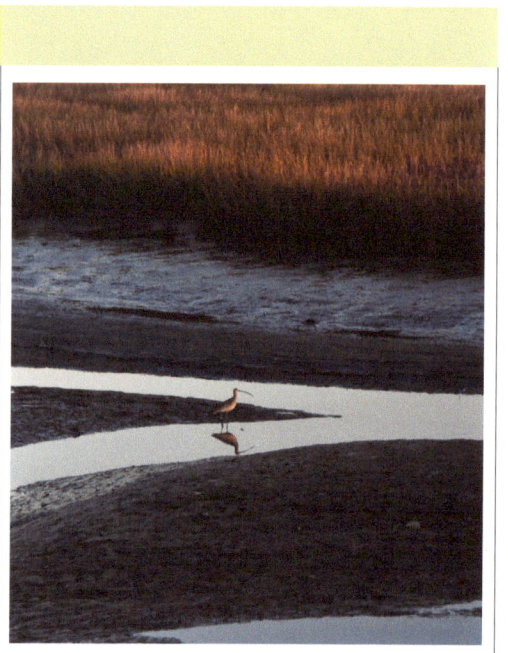

SALT MARSH

DESCRIPTION

Salt Marshes are emergent wetlands dominated by salt-tolerant vegetation. Although this habitat type is often subject to tidal inundation, the frequency of tidal inundation varies widely depending on elevation, inlet closure dynamics, and climate, causing spatial heterogeneity as well as wide temporal fluctuations in salt marsh salinity. As mapped, this habitat type includes areas of salt marsh that were largely non-tidal. Common salt marsh plant species in the Tijuana River Estuary likely included California cordgrass (*Spartina foliosa*), pickleweed (*Sarcocornia pacifica*), saltgrass (*Distichlis spicata*), shoregrass (*Monanthochloe littoralis*), Parish's glasswort (*Arthrocnemum subterminale*), and alkali heath (*Frankenia salina*) (Purer 1942, Grewell et al. 2007).

MAPPING METHODS

Digitized primarily from the earliest available USCS T-sheet (Harrison 1852), as indicated by closely spaced parallel lines (Shalowitz 1964, Grossinger et al. 2011). Supported by other 19th century maps, photographs, and texts. To increase spatial accuracy, some areas mapped by Harrison (1852) were adjusted to align marshland edges with major stable topographic features.

(Subtidal Water photo: Samuel Safran, April 2015; Mudflat/Sandflat photo: Lars Dugalczyk, December 2008, CC BY-ND 2.0; Salt Marsh photo: Samuel Safran, April 2015; Salt Flat photo: Samuel Safran, April 2015; Alkali meadow photo: Tony Frates, November 2013, CC BY-NC-SA 2.0; River Channel photo: LIPP, Box 85, photo #3070, ca. 1920, courtesy Water Resources Collections and Archives, UC Riverside)

SALT FLAT / OPEN WATER

DESCRIPTION	Salt Flats are unvegetated or sparsely vegetated areas where high soil salinities largely preclude the growth of vegetation (Pennings and Bertness 1999). Salt Flats trap water during the rainy season or extreme high tides and are temporarily transformed into areas of open water. During the dry season, high evaporation rates cause the flats to once again dry out, concentrating salts in the process. In some places, areas mapped as Salt Flat / Open Water would have supported small patches of marsh vegetation that are not represented in the historical synthesis mapping.
MAPPING METHODS	Digitized exclusively from the earliest available USCS T-sheet (Harrison 1852), as indicated by unvegetated areas at the upland margins of the estuary disconnected from tidal channels. Supported by mid-19th century Boundary Commission and GLO surveys.

ALKALI MEADOW COMPLEX / HIGH MARSH TRANSITION ZONE

DESCRIPTION	The High Marsh Transition Zone is the irregularly flooded ecotone between estuarine and terrestrial communities. In the historical Tijuana River valley, the area of High Marsh Transition Zone often graded into non-tidal alkali meadows. Alkali meadows are characterized by fine-grained alkaline soils that have a high residual salt content and support a distinctive, salt-tolerant herbaceous plant community (Holland 1986). These habitats are restricted to zones of shallow groundwater (Elmore et al. 2006) and are subject to seasonal to intermittent flooding, with subsequent drying through the summer. Salt grass (*Distichlis spicata*) is the dominant vegetation of alkali meadows, though other salt-tolerant species, such as yerba mansa (*Anemopsis californica*), alkali sacaton (*Sporobolus airoides*), Coulter goldfields (*Lathsenia glabrata* ssp. *coulteri*), and California croton (*Croton californicus*), were likely present.
MAPPING METHODS	Digitized primarily from historical soil surveys (Storie and Carpenter 1930), as indicated by the alkali soil types classified as "Foster very fine sandy loam" or "Alviso very fine sandy loam" (outside of areas mapped as Salt Marsh). Supported by other 19th and 20th century texts. Some vegetated upland areas mapped by Harrison (1852) within the estuary were also mapped as Alkali Meadow Complex / High Marsh Transition Zone based on their landscape position. To increase spatial accuracy, the upland edges of some features were aligned with sudden slope breaks evident in modern topographic maps (USGS 1967).

RIVER CHANNEL

DESCRIPTION	The River Channel classification corresponds to the Tijuana River's low-flow channel; that is, the smaller, generally meandering channel inset within a wider braided high flood zone (Graf 1988). As the most frequently inundated portion of the river corridor, it featured a sparsely vegetated sandy bed.
MAPPING METHODS	In the U.S., the River Channel was digitized from a variety of data sources, primarily 19th century maps (e.g., Ilarregui and de Chavero 1850, Harrison 1852, USGS 1904, Ervast 1921) and GLO surveys (Freeman 1854, Pascoe 1869). The most spatially accurate sources served as anchors for the location of the channel, while other, more generalized sources were used to connect these points and capture the general shape of the channel. Segments of the mapped River Channel were derived from courses shown by later (post-1850) sources where they were determined to correspond well to the approximate mid-century location of the channel.

In Mexico, the low-flow channel was digitized primarily from an early railroad survey map (Rankin 1909), as supported by other 19th and 20th century maps and aerial photographs. Because the railroad survey map could not be georeferenced with a high degree of precision, the channels it depicts were mapped by roughly digitizing the shapes and then repositioning these shapes based on the major topographical features shown on the original document. The mapped River Channel should be considered representative of river planform in Mexico prior to major development and not the precise location of the low-flow channel ca. 1850. |

RIVER WASH / RIPARIAN SCRUB

DESCRIPTION

The character of the River Wash / Riparian Scrub habitat type varies significantly over space and time along gradients of water availability, sediment supply, and disturbance regimes. At one extreme, it includes sparsely vegetated areas of river wash found in recently flood-scoured zones of the active channel. At the other extreme, dense patches of riparian scrub vegetation support some small trees, primarily willow (*Salix* spp.) and cottonwood (*Populus* spp.). Along gradients of water availability, this riparian habitat type can include both xeric alluvial scrub communities and riverine marshes where shallow-rooted herbaceous wetland species grow in saturated soils. This habitat type is primarily characterized by riparian scrub vegetation with few to no tall trees.

MAPPING METHODS

In the U.S., digitized primarily from historical soil surveys, as indicated by the "Cajon fine sand" and "River wash" soil types (Storie and Carpenter 1923) and supported by 19th century maps, surveys, and photographs. In Mexico, digitized primarily from sources indicating the extent of the river corridor, including historical photographs (Unknown 1946, Unknown 1955) and a Department of Public Works map (CDWR ca. 1942). Where the river bed was shown to abut the valley's low terraces, modern topographic maps and historical orthophotos were used to increase the spatial accuracy of the mapped features.

GRASSLAND / COASTAL SAGE SCRUB

DESCRIPTION

A broad category encompassing herbaceous and shrub cover. Vegetation communities included in this category range from treeless herbaceous cover and coastal prairie (which may have included native bunchgrasses and annual grasses, in addition to annual forbs, wildflowers, and shrubs) to coastal sage scrub (including coyote brush [*Baccharis pilularis*] and California sagebrush [*Artemisia californica*]). Where it is mapped in the valley's tributary canyons, it also includes alluvial scrub, considered a type of coastal sage scrub distinguished by both its physiographic position on alluvial fans/ephemeral floodplains and its physiognomy as a shrubland dominated by woody shrubs and small tress (including large evergreen shrubs like laurel sumac [*Malosma laurina*] and lemonade berry [*Rhus integrifolia*]; Smith 1980).

MAPPING METHODS

Digitized primarily from historical soil surveys (Storie and Carpenter 1930), as supported by other 19th century historical sources (e.g., Hardcastle and Gray 1850 and Harrison 1852).

PERENNIAL FRESHWATER WETLAND

DESCRIPTION

Permanently flooded to intermittently exposed, permanently saturated palustrine wetlands. Common plants within the freshwater wetlands likely included cattails (*Typha* spp.), sedges (e.g., *Cyperus* spp.), rushes (*Juncus* spp.), and tules (*Schoenoplectus acutus*).

MAPPING METHODS

Digitized exclusively from early USGS topographic quads (USGS 1943), as indicated by the standard symbol for "marsh or swamp" (USGS 2003) and supported by historical soil surveys.

(River Wash photo: #1113, 1910, courtesy San Diego History Center; Grassland / Coastal Sage Scrub photo: Samuel Safran, April 2015; Perennial Freshwater Wetland photo: USGS, May 1994; Vernal Pool photo: Sally Brown/USFWS, July 2016, CC BY 2.0; Pond photo: Allan Ferguson, March 2015, CC BY 2.0; Temporary stream photo: nicklafrance, 2013, CC BY-NC 4.0)

VERNAL POOL

DESCRIPTION	Seasonally or intermittently flooded depressions, characterized by a relatively imper-meable subsurface soil layer and distinctive vernal pool flora. Vernal pools of coastal terraces in southern California are described by Zedler (1987).	
MAPPING METHODS	Mapped exclusively from early USGS topographic quads (USGS 1943), as indicated by the standard symbol for "intermittent lake/pond symbol" (USGS 2003) and sup-ported by historical aerial photographs and soil surveys. Since we only mapped Vernal Pools where indicated in the USGS quad, this habitat type is likely under-represented. Additional pools were likely embedded within the Grassland / Coastal Sage Scrub matrix on compact clay subsoils.	

POND

DESCRIPTION	Permanently flooded depressions, largely devoid of emergent palustrine vegetation.	
MAPPING METHODS	Digitized from historical aerial photographs (San Diego County 1928) and historical boundary surveys (Hardcastle and Gray 1850). Supported by early USGS topographic quads and other general 19th century texts.	

TEMPORARY STREAM

DESCRIPTION	Temporary Streams are channels that lack continuous year-round flow and instead feature either intermittent or ephemeral flows (Meinzer 1923; McDonough et al. 2011). For our purposes, this habitat type is limited to tributaries of the larger Tijuana River (which is mapped separately; see "River Channel" above) and includes a range of channel morphologies and settings (including braided channels in dry washes, con-fined channels at the base of steep ravines, and channels where the streams enter onto the valley floor). Due to their small drainage basins, most of these features were likely ephemeral streams (those that receive no water from springs or lack continued supply from other surface sources and flow only in direct response to precipitation), but a lack of specific historical information necessitates the broader "temporary streams" classification.	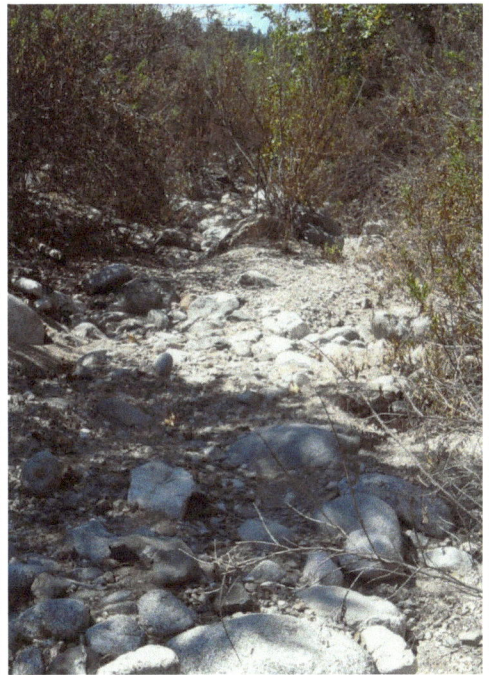
MAPPING METHODS	We digitized temporary streams primarily from historical aerial photographs (San Diego County 1928) and the earliest USGS topographic quad (USGS 1904). To facili-tate analyses of change over time, we used the contemporary courses of intermittent streams (USGS 2014) to represent the historical position when the historical and contemporary sources depicted a similar channel shape and showed <15 m horizontal displacement. The termini, or distributaries, of intermittent streams were primarily mapped from the same sources, as indicated by the forked "disappearing stream" symbol in USGS quads (USGS 2003) and visible in historical aerials. All temporary stream features were mapped as one-dimensional line features. Although individual braided channel segments are highly dynamic features, we mapped braids visible in the early 20th century sources as representative of the general expected planform ca. 1850.	

Additional analyses

In addition to the historical habitat type map, we performed a number of supplemental analyses to further investigate landscape dynamics and change over time. These analyses are listed below by chapter. Instead of describing the methods for each of these analyses here, we embed them throughout this report alongside the corresponding results.

River (Chapter 5)

- River courses map (pp. 114–15, 120–21)

- River locational probability analysis (pp. 116–17, 120–21)

Estuary (Chapter 6)

- Tidal prism volume calculation (p. 140)

- Inlet condition analysis (p. 145)

- Historical estuarine-terrestrial transition zone analysis (p. 161)

Habitat change over time (Chapter 7)

- Habitat type change analysis (p. 172–74)

- Persistent habitats (p. 178)

- Habitat type change matrix (p. 178)

(Photo by Samuel Safran, October 2015)

RANCHO NACIONAL

SAN DIEGO BAY.

Limit of One League from San Antonio Hills.

Due East

S. F. Argüello's Ranch.
House.

VALLEY O

Road to Tijuana

P A C I F I C   O C E A N.

VALLEY   O F   T

Average length
Average width.

Well of San Antonio.

Due East

UNITED

Initial Point
of Boundary.

SAN   ANTONIO   HILLS.

PLAN OF THE

RANCHO OF ME

COUNTY OF SAN DIEG

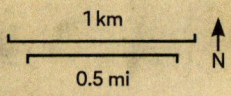

1 km

N

0.5 mi

3
ENVIRONMENTAL SETTING & HISTORY

The Tijuana River valley is a dynamic landscape that has evolved in response to both natural processes and human influences. An understanding of the physical and anthropogenic drivers that have influenced the landscape is critical to interpreting historical data and reconstructing historical ecological conditions in the valley. This chapter first provides an overview of the physical setting of the Tijuana River valley and watershed, with a focus on regional climate, geology, hydrology, and coastal dynamics. This overview is followed by a discussion of the primary land and water uses that have affected ecological conditions within the valley, including native land management, grazing and agriculture, groundwater extraction, dams and surface water diversion, wastewater discharge and urban runoff, urban development and population growth, and conservation measures.

(Poole 1854, courtesy The Bancroft Library)

WATERSHED OVERVIEW

The Tijuana River watershed encompasses approximately 4,530 km² (1,750 mi² in southern San Diego County and northern Baja California.

The Tijuana River watershed encompasses approximately 4,530 km² (1,750 mi²) in southern San Diego County and northern Baja California (SDSU Dept. of Geography 2005; Fig. 3.1). Seventy-three percent of the watershed area lies within Mexico. Most of the Tijuana River itself is located in Mexico; the river crosses back into the United States approximately 10 km (6 mi) from its mouth and empties into the Pacific Ocean just south of Imperial Beach, CA. Elevations within the watershed range from sea level at the mouth of the Tijuana River to over 1,900 m (6,200 ft) in the mountains further inland (Wright 2005; Fig. 3.1). The Tijuana Estuary, one of the largest remaining coastal wetlands in southern California, lies at the downstream end of the watershed adjacent to the international border.

A broad diversity of vegetation types occurs within the watershed, distributed along gradients in elevation, precipitation, temperature, and geology. Coastal sage scrub is the dominant vegetation type in low-elevation areas in the western portion of the watershed where it has not been replaced with urban development. The eastern regions are dominated by chaparral, which covers 56% of the total watershed area. The highest-elevation areas to the east support stands of Jeffrey pines and other coniferous forests. Riparian vegetation, grasslands, and other vegetation types cover a comparatively small fraction of the watershed. Salt Marsh and other estuarine habitat types are found along the coast within the Tijuana Estuary (O'Leary 2005).

Though much of the watershed is sparsely inhabited, several large urban areas exist in the western and central regions. The vast majority of the population within the watershed is concentrated in the city of Tijuana; other population centers include the city of Tecate in Mexico and the cities of San Ysidro and Imperial Beach in the United States (Wright 2005).

Figure. 3.1. The Tijuana River watershed (opposite page) encompasses approximately 4,530 km² (1,750 mi²) in southern San Diego County and northern Baja California. The lower Tijuana River valley (our study extent) covers just a small portion of the watershed. A longitudinal profile of the river **(opposite page, bottom)** shows the elevation of the river from the estuary to the Sierra de Juárez mountain range. *(Watershed and stream geodata: Tijuana River Watershed Atlas, San Diego State University; Shaded relief: courtesy ESRI, USGS, NGA, NASA, CGIR; Locator map: courtesy ESRI)*

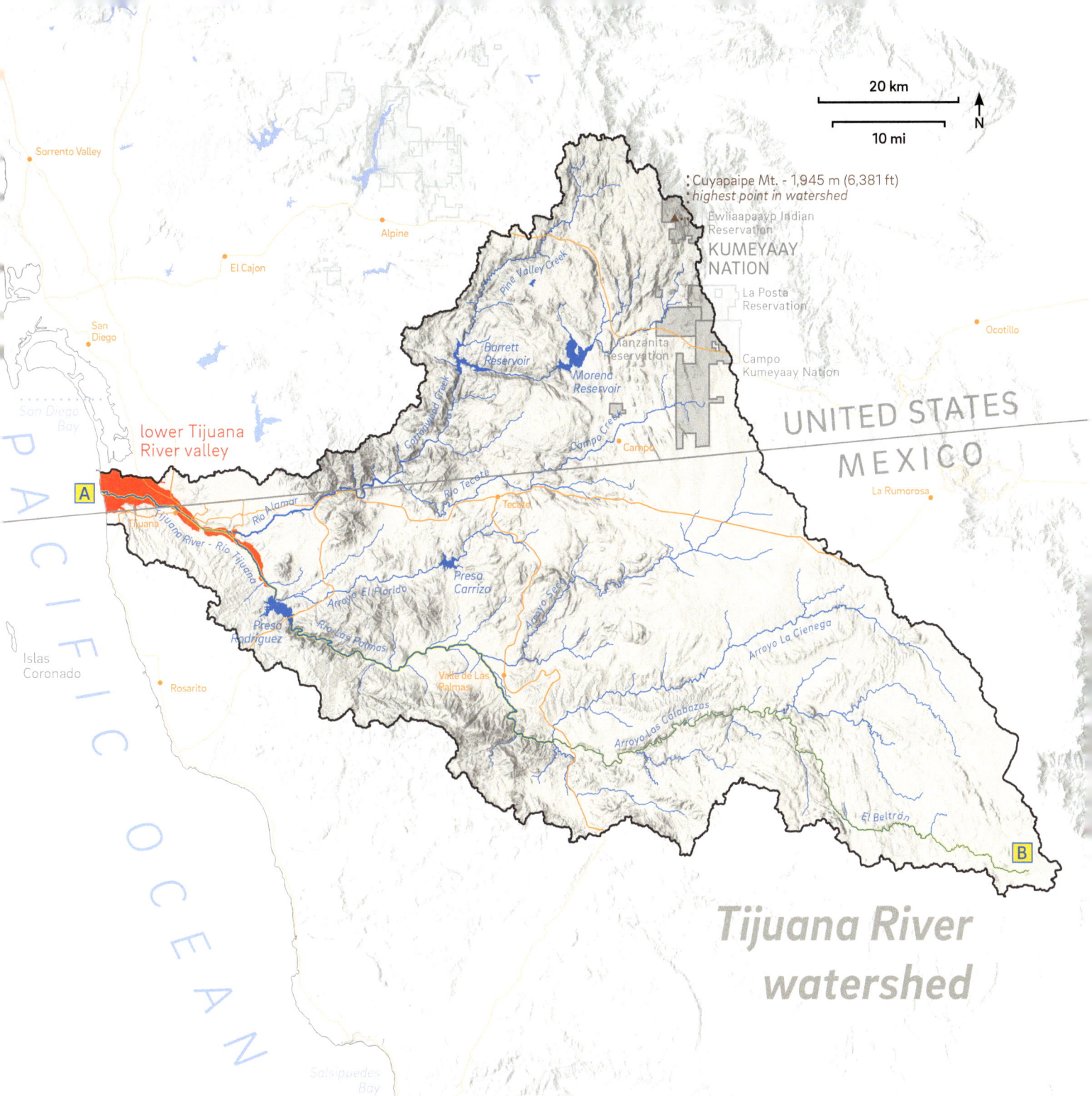

Tijuana River watershed

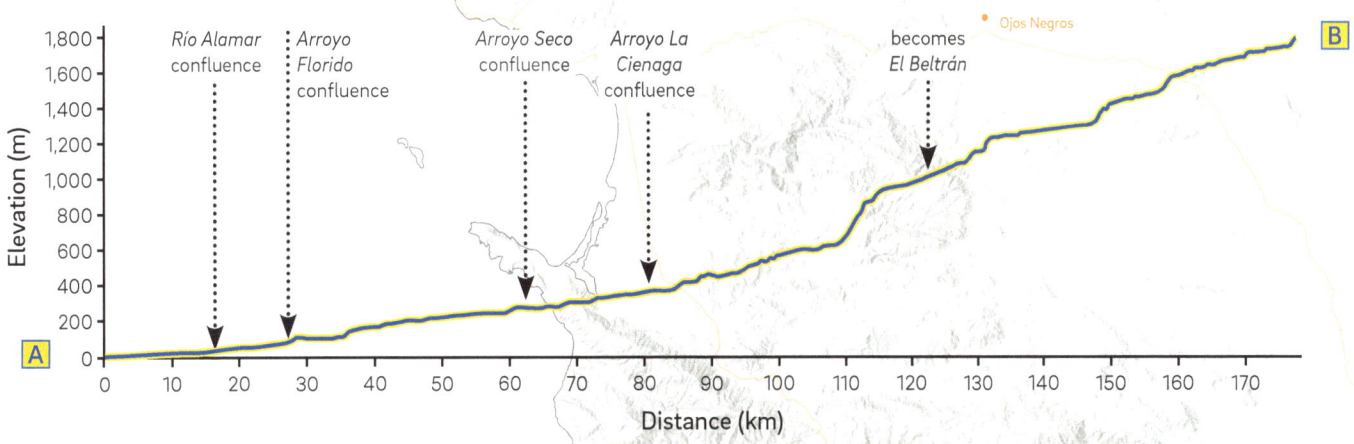

Cuyapaipe Mt. - 1,945 m (6,381 ft)
highest point in watershed

Ewiiaapaayp Indian
Reservation

KUMEYAAY
NATION

La Posta
Reservation

Manzanita
Reservation

Campo
Kumeyaay Nation

UNITED STATES
MEXICO

Ocotillo

La Rumorosa

Sorrento Valley

Alpine

El Cajon

San Diego

San Diego Bay

lower Tijuana
River valley

Tijuana

Barrett
Reservoir

Morena
Reservoir

Pine Valley Creek

Cottonwood Creek

Campo Creek

Campo

Río Tecate

Tecate

Río Alamar

Tijuana River - Río Tijuana

Presa
Carrizo

Arroyo Seco

Arroyo La Cienega

Islas
Coronado

Rosarito

Presa
Rodríguez

Arroyo El Florido

Río las Palmas

Valle de las
Palmas

Arroyo Los Calabazas

El Beltrán

PACIFIC OCEAN

Satsipuedes Bay

B

**Tijuana River
watershed**

Elevation / Distance profile

Ojos Negros

Río Alamar
confluence

Arroyo
Florido
confluence

Arroyo Seco
confluence

Arroyo La
Cienaga
confluence

becomes
El Beltrán

Elevation (m): 1,800 / 1,600 / 1,400 / 1,200 / 1,000 / 800 / 600 / 400 / 200 / 0

Distance (km): 0 10 20 30 40 50 60 70 80 90 100 110 120 130 140 150 160 170

A
B

REGIONAL CLIMATE

The study area experiences a semi-arid climate with highly seasonal precipitation.

The study area experiences a semi-arid climate characterized by warm, dry summers and cool winters (Fig. 3.2). Interior portions of the watershed experience Mediterranean or arid climate conditions, with temperature and rainfall patterns varying as result of elevation, distance from the coast, and other factors (Peel et al. 2007; see also SDSU Dept. of Geography 2005 and Pryde 2004). Annual precipitation within the watershed averages about 30 cm (12 in), ranging from as little as 20 cm (8 in) per year on average in low elevation areas near the coast to more than 101 cm (40 in) annually in mountainous areas further inland (Das et al. 2010, Aguado 2005). Precipitation is highly seasonal, with about 90% occurring between the months of October and April (Das et al. 2010). Average annual temperatures within the watershed range from about 9 °C to 19 °C (48 °F to 66 °F); near the study area, average temperatures vary seasonally from 21 °C (70 °F) in July to 14 °C (57 °F) in December (Aguado 2005).

Annual and decadal wet/dry cycles (Fig. 3.3) are driven by large-scale climate phenomena including the El Niño Southern Oscillation (ENSO) and the Pacific Decadal Oscillation (PDO; Zedler 2010). A recent analysis of southern California streamflow records shows that over the past century, large storm-induced flood flows were much more frequent during ENSO years than non-ENSO years (Andrews et al. 2004). The cycles of droughts and floods documented in the 18th and 19th centuries were influenced by these same climatic drivers (Biondi et al. 2001).

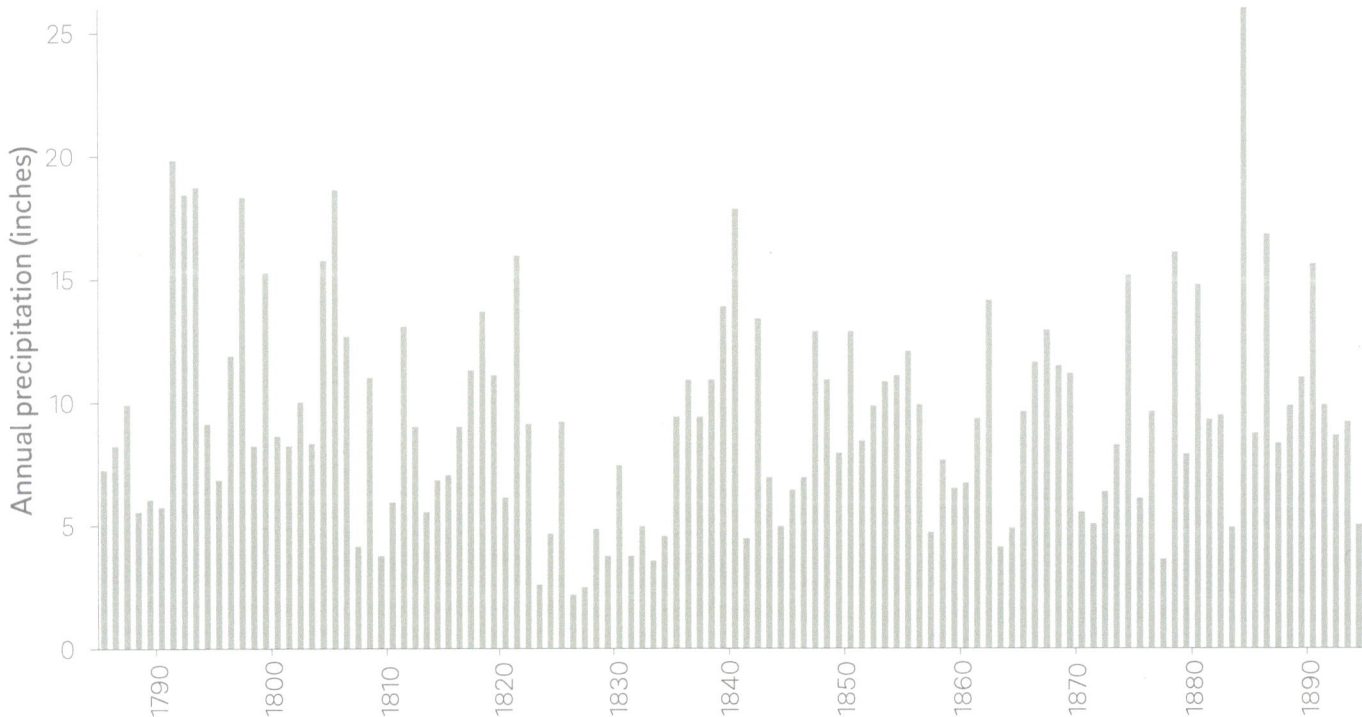

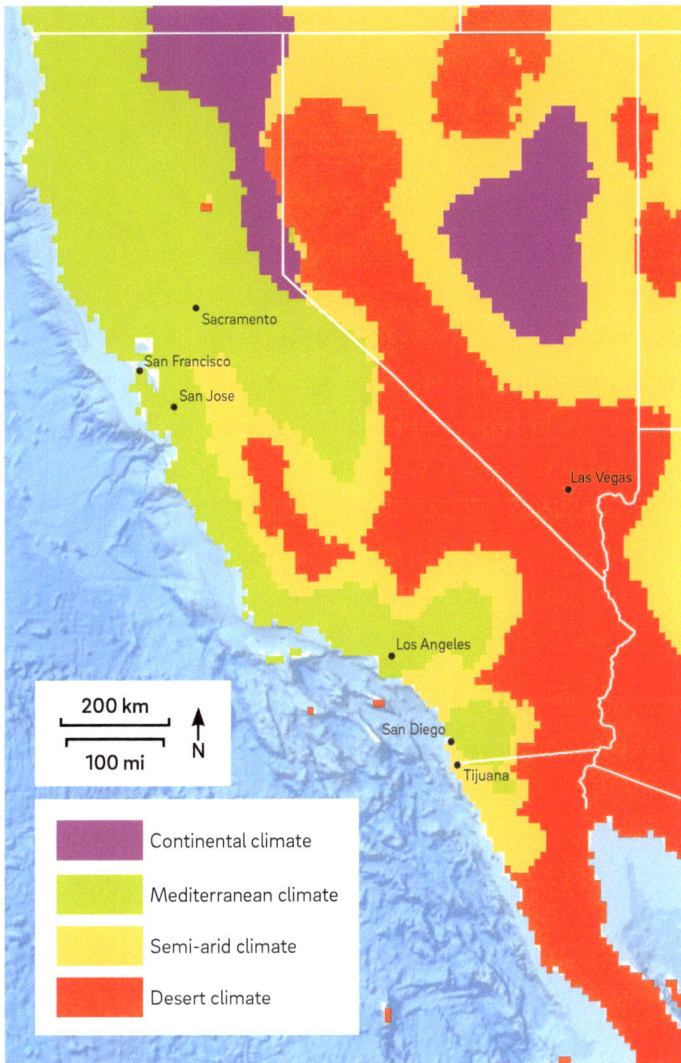

Figure 3.2. (left) The Köppen-Geiger climate zones of western North America. The Tijuana River watershed is situated at a climatic inflection point along the coast, with a Mediterranean climate to the north and a Desert climate to the south. At the continental scale, the region's coastal Semi-arid climate is relatively rare, which is one reason for the watershed's unique biological and hydrological conditions. *(Climate data: Peel et al. 2007; Base layer: courtesy ESRI, DeLorme, GEBCO, NOAA, NGDC)*

Figure 3.3. (below) Annual precipitation, 1785–2012. Values represent annual rainfall in inches for an October through September water year. Data for 1851–2012 was obtained from monthly precipitation records from San Diego Airport (WRCC 2006, NWS 2013). Additional data for 1934–2012 was obtained from daily precipitation records from Rodríguez Dam (Conagua-DGE 2014). For years with fewer than five days of missing data in the Rodríguez Dam record, missing values were linearly interpolated; years with greater than five days of missing records were omitted. Over the period of record, annual precipitation at San Diego Airport (approximately 10 miles north of the study area) averaged 1.25 inches more than at Rodríguez Dam (located at the southern end of the study area). For 1934–2012, the rainfall graph displays the higher value from the San Diego Airport and Rodríguez Dam records.

Precipitation estimates for 1785–1834 were calculated using Rowntree's (1985) rainfall index for southern California, which was constructed from crop harvest records from southern California missions. Precipitation estimates for 1835–1850 were derived from Lynch's (1931) rainfall index for the San Diego area, which for these years is based on historical diary entries describing weather conditions. Rainfall indices were translated into precipitation estimates using a mean annual rainfall value of 9.92 inches for the base period 1851–2012 (from the San Diego Airport gage data). Lower certainty is ascribed to precipitation estimates derived from the rainfall indices (particularly for the 1835–1850 period) than to the subsequent meteorological records.

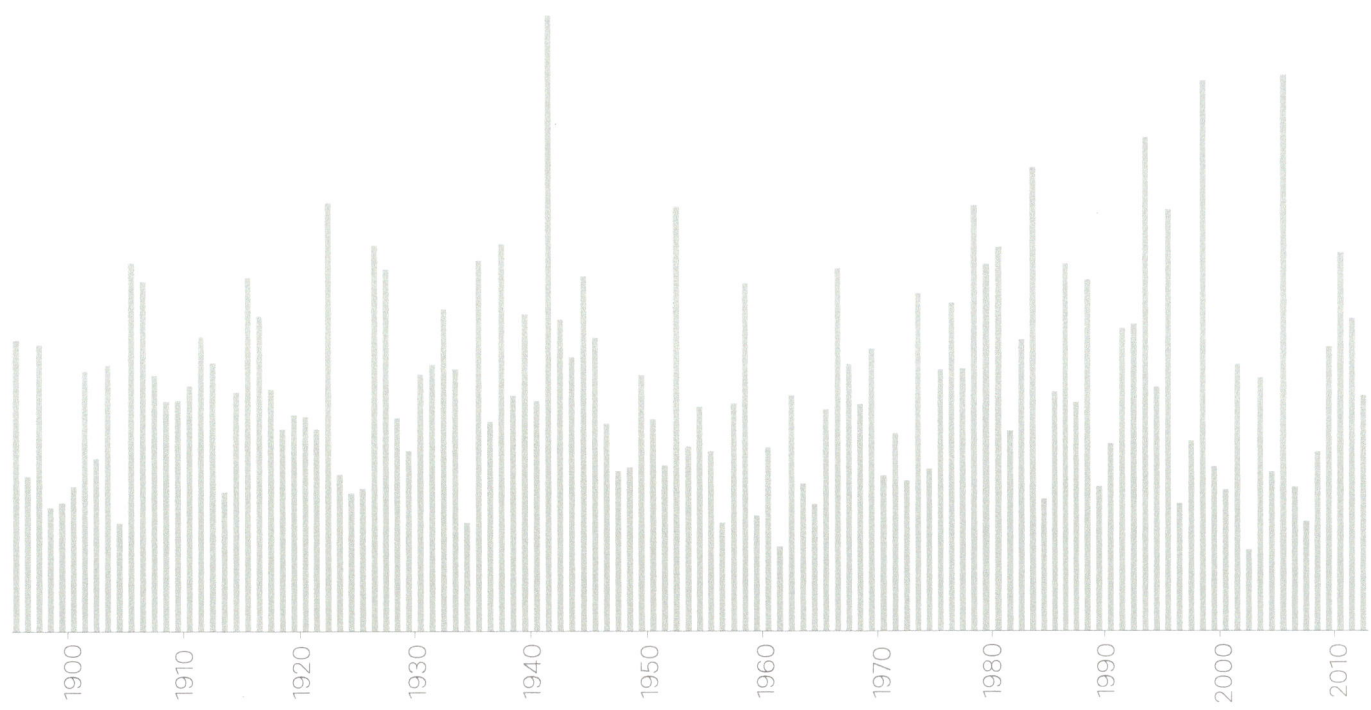

GEOLOGY & ESTUARY FORMATION

Sediment deposition and changes in sea level gave rise to the present form of the valley over the past 10,000 years.

The Tijuana River watershed lies on the western side of the Peninsular Ranges, which run north-south from southern California to the tip of the Baja Peninsula and are dominated by a Mesozoic batholith (Wetmore et al. 2003, Deméré 2005). The upper watershed consists primarily of Cretaceous Peninsular Range Granitics, with smaller formations comprised of Cretaceous Gabbro Intrusives and Metamorphic Clastic Sediments (Deméré 2005). To the west is a sequence of metavolcanic rocks known as the Santiago Peak Volcanics (Tanaka et al. 1984). Along the coast, the geology is largely comprised of Cenozoic sedimentary rocks, including those of the Otay, Rosarito Beach, San Diego, Linda Vista, and Bay Point formations. Quaternary alluvium occupies valley floors throughout the watershed (Deméré 2005).

Within the study area, surficial deposits consist primarily of late Cenozoic rocks and sediments. The valley floor is dominated by Holocene and late Pleistocene alluvium, while uplifted marine terraces fringe the valley floor to the north and south (Ku and Kern 1974, Kennedy and Tan 2008). The Nestor Terrace, to the north of the river valley, is dominated by mid- to late Pleistocene sedimentary deposits, including sandstone, siltstone, and conglomerate (Ku and Kern 1974, Kennedy and Tan 2008). The mesas to the south of the valley include outcrops of late Pliocene and early Pleistocene sedimentary rocks in the San Diego Formation, as well as early to mid-Pleistocene sedimentary deposits (Kennedy and Tan 2008).

The present form of the Tijuana River valley and estuary developed over the past 10,000 to 12,000 years, since the end of the last glacial epoch (Fig. 3.4). During the late Pleistocene, large quantities of water were trapped in glaciers, and the Pacific Ocean was 90 m (300 ft) lower than today (Swanson 1987a). The Tijuana River flowed through a steep, narrow valley and emptied into the ocean to the west of the current shoreline. Melting of the glaciers caused a rapid increase in sea level between 10,000 and 5,000 years ago, flooding the river valley and creating a coastal embayment. Over the past 5,000 years, sediment transported by the Tijuana River created a prograding delta and gradually filled in the bay, eventually giving rise to the modern river valley and estuary (Williams and Swanson 1987, Swanson 1987a, Zedler et al. 1992).

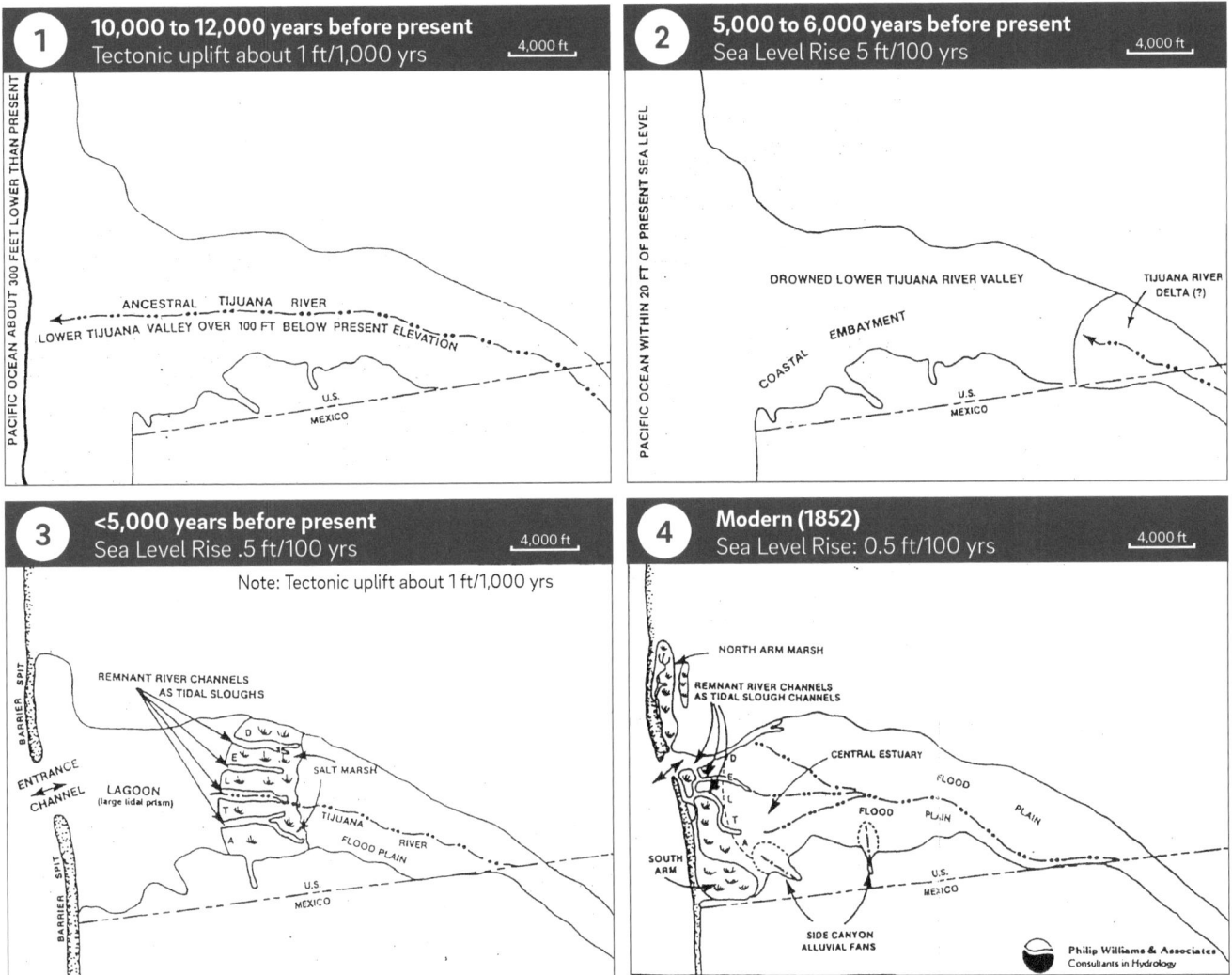

Figure 3.4. Evolution of the Tijuana River valley and estuary from Swanson 1987a.

HYDROLOGY

Changes in infrastructure and land use patterns have significantly altered hydrologic patterns for the historically intermittent Tijuana River.

Streamflow in the lower Tijuana River historically was highly variable both seasonally and interannually, with long periods of minimal or no flow punctuated by infrequent high flow events. Runoff could be heavy following large rainfall events, but surface flows rapidly percolated into the sandy substrate. Major floods periodically inundated much of the river valley, scouring away sediment and vegetation and resulting in abrupt channel movements. Sediment dynamics were closely tied to streamflow: the vast majority of sediment tranport occurred during episodic large flood events (Haltiner and Swanson 1987). Nearly all of the lower river was intermittent, with the exception of a few short reaches that supported perennial surface flows. (See pp. 96–111 for more information on historical hydrologic patterns.)

A large alluvial aquifer, in some areas exceeding 30 m [100 ft] in depth, underlies much of the lower Tijuana River Valley (Rempel 1992). Prior to widespread groundwater pumping (see pp. 52–57), groundwater levels were seasonally shallow (1–2 m [3–6 ft] below the surface) in many areas of the valley, with springs and artesian conditions found in some areas. Groundwater levels fluctuated throughout the year, in some areas decreasing by over 4 m (13 ft) during the dry season due to high rates of evapotranspiration and lack of recharge from streamflow (Ellis and Lee 1919). (See pp. 70–72 and pp. 100–104 for more information on historical groundwater patterns, and pp. 52–57 for information about groundwater extraction.)

A variety of engineering projects and land use changes have altered hydrologic patterns within the watershed. Five dams regulate streamflow, including Morena and Barrett dams on Cottonwood Creek (a major tributary of the Tijuana River) in the U.S. and Rodríguez Dam on the Tijuana River in Mexico. Below Rodríguez Dam, the Tijuana River flows through a concrete flood control channel, which terminates just beyond the international border. In addition to these direct modifications, streamflow patterns have been substantially impacted by urban and irrigation runoff, wastewater discharge, and groundwater extraction. (See pp. 52–61 for more information on these changes.) As a result of these changes, the historically intermittent Tijuana River became perennial during the 1980s (IBWC gage #11013300, Haltiner and Swanson 1987). Today, streamflow in the lower portion of the river valley is managed to be intermittent by diverting dry weather runoff and wastewater from the urban environment to a treatment plant (Wright 2005, EPA 2009). (See pp. 184–87 for some additional information on changes in streamflow over time.)

Figure 3.5 The Tijuana River flowing through Mexico towards the United States, 1943. This photograph looks downstream from just south of the international border. The city of Tijuana is visible in the image's top-left corner. (*Photo #F1391.T36 A34, 1943, courtesy Special Collections & Archives, UC San Diego Library*)

COASTAL DYNAMICS

The Tijuana Estuary lies in the Silver Strand littoral cell and is its primary source of sediment.

The California coast experiences a mixed, semi-diurnal tide regime, meaning that there are usually two high tides (a high and higher high) and two low tides (a low and lower low) per day. The tide gage just north of the Tijuana Estuary in Imperial Beach, CA (NOAA station 9410120) shows a long-term mean tidal range (difference between mean high water [MHW] and mean low water [MLW]) of 1.14 m (3.74 ft) and a mean diurnal tidal range (difference between mean higher high water [MHHW] and mean lower low water [MLLW]) of 1.64 m (5.37 ft).

The mouth of the Tijuana River is located in the Silver Strand littoral cell, which stretches more than 100 km from Point Loma to Ensenada (Scripps Institution of Oceanography 2005). To the north of the Tijuana River mouth, the direction of the longshore current is predominantly northward, while to the south the current is primarily southward (Patsch and Griggs 2007). The Silver Strand itself, a sand spit extending north from Imperial Beach and comprising the western boundary of San Diego Bay, was formed by the northward transport of sediment from the Tijuana River mouth (Fig. 3.6; Inman and Masters 1991, Scripps Institution of Oceanography 2005). The Tijuana River is the primary source of sediment to the Silver Strand littoral cell, though dam construction within the watershed has reduced sediment delivery by 30–50%, resulting in beach erosion (Haltiner and Swanson 1987, Flick 2005, Patsch and Griggs 2007).

Waves typically approach the Tijuana Estuary shoreline from the northwest during the winter and from the south during the summer (Swanson 1987b, Patsch and Griggs 2007). Wave energy is generally attenuated somewhat by the presence of the Point Loma headland to the north, the Southern Channel Islands to the northwest, and Los Coronados Islands to the southwest (Swanson 1987b, Patsch and Griggs 2007). Wave action can influence estuarine morphology, including inlet closure dynamics and inlet migration (see pp. 138–45), as well the position and stability of the barrier beach forming the western boundary of the estuary through effects on littoral sand transport (Swanson 1987b; see p. 183).

Figure 3.6. (opposite page) Looking north up the Silver Strand sand spit, which forms the western boundary of the Tijuana Estuary and San Diego Bay. *(Photo courtesy NOAA, March 2015, CCO 1.0)*

Figure 3.7. The San Diego-Arizona Eastern Railroad trestle over the Tijuana River across Matanuco Canyon, ca. 1918. *(Photo #7767, courtesy San Diego History Center)*

LAND AND WATER USE HISTORY
TIMELINE

B.C. **8000**	Approximate arrival of humans to Tijuana River Valley; initiation of native land management era.
A.D. **1542**	Cabrillo expedition passes Tijuana River valley offshore and makes landfall in San Diego Bay
1769	Portolá expedition travels overland from Baja California to San Diego
1769	Mission San Diego de Alcalá founded
1829	Rancho Tia Juana granted to Santiago Argüello
1833	Rancho Melijó awarded to Santiago E. Argüello
1835	Pueblo of San Diego founded
1851	International boundary between Mexico and the United States surveyed
1888	National City and Otay Railroad constructed
1887	Town of Imperial Beach founded
1889	Town of Tijuana incorporated
1909	Little Landers farming community founded near present-day San Ysidro
1909	Dulzura Conduit built to divert water from Cottonwood Creek to the Otay River
early 1900s	San Diego and Arizona Eastern Railway constructed (Figs. 3.7–3.8)
1912	Morena Dam completed on Cottonwood Creek
1917	Ream Field constructed on northern side of Tijuana Estuary
1921	Barrett Dam completed on Cottonwood Creek
1936	Rodríguez Dam constructed on Río de las Palmas
early 1940s	Border Field Auxiliary Landing Field built on southern side of Tijuana Estuary
1944	Treaty signed between the United States and Mexico on *Utilization of Waters of the Colorado and Tijuana Rivers and of the Rio Grande*
1970s	Portion of Tijuana River between Rodríguez Reservoir and international border channelized
1998	South Bay International Wastewater Treatment Plant begins operation

TIMELINE SOURCES: Taggart 1869, Bancroft 1888, Department of Public Works 1935, Joint Committee on Water Problems of the California Legislature 1953, Pourade 1965, USAED 1974, Brown and Pallamary 1988, Dedina 1991, Shipek 1993, Corona n.d., Michel 2000, D'Elgin et al. n.d., City of Imperial Beach n.d.

Figure 3.8. Construction of the San Diego-Arizona Eastern Railroad trestle over the Tijuana River across Matanuco Canyon, ca. 1910. *(Photo #7717, courtesy San Diego History Center)*

NATIVE LAND MANAGEMENT

Large indigenous settlements existed in the Tijuana River valley at the time of Spanish colonization.

Humans have lived in the Tijuana River watershed for at least ten thousand years (Gamble et al. 2004). At the time of European contact in 1769, the Kumeyaay were the main indigenous people living in the area. The Kumeyaay territory reached from present-day Santo Tomás, Baja California, north to Escondido, and east over the coastal mountains (Gamble et al. 2004). Estimates of Kumeyaay population size during the Spanish period range from 10,000 to 26,000, with population density averaging approximately 2–3 people/km² (5–7 people/ mi²; Shipek 1982 , Shipek 1993, Gallegos 2002). The Tijuana River valley itself was likely well-populated; upon arriving in the valley in 1769, Crespí wrote that the expedition "came near to a populous village" (Crespí and Bolton 1927), while Serra described a "gentile settlement, thickly populated" (Serra and Tibesar 1955). Both Crespí and Serra were likely referring to the village of Milejo, located on the Tijuana River's southern bank (SWCA Environmental Consultants 2004, Zaragoza 2014). Additional Kumeyaay villages were located near present-day San Ysidro and San Diego Bay (Underwood and York 2004, Zaragoza 2014).

The Kumeyaay employed a variety of hunter-gatherer, agricultural, and land management strategies. They exploited both coastal and inland food resources such as shellfish, fish, marine mammals, pine nuts, acorns, and terrestrial animals (Gamble et al. 2004). They practiced broadcast seeding of grasses, including a semi-domesticated grain in valleys across the region (Shipek 1989), and planted crops such as corn, squash, and beans on a limited scale in areas with sufficient moisture (Luomala 1978, Shipek 1993). Fire was used to clear vegetation for planting, increase yields, control plant diseases, and flush animals. Many different types of vegetation were managed with fire, including oak and pine groves, chaparral, grasslands, desert scrub, marshes, and fields of semi-domesticated grain (Luomala 1978, Shipek 1989, Shipek 1993). Large amounts of smoke, which could have come from controlled burns, were noted on land in the vicinity of the Tijuana River valley by the first Europeans to explore the region by sea (Cabrillo and Bolton 1916, Vizcaíno and Bolton 1916). Cabrillo observed "great smokes" on land in September of 1542, only three days prior

The Indians made so many columns of smoke on the mainland that at night it looked like a procession and in the daytime the sky was overcast.

— SEBASTIÁN VIZCAÍNO
NOVEMBER 9th, 1602
(IN VIZCAÍNO AND BOLTON 1916)

Figure 3.9. A Kumeyaay woman and home near Campo, California, ca. 1924. *(Photo by Edward S. Curtis, ca. 1924, Diegueño house at Campo, #LC-USZ62-98666, courtesy Library of Congress)*

to the first storm of the season. It was indeed common practice for California Indians to initiate a fire at the end of the dormant season before the first rains (C. Striplen, personal communication). For domestic fires, the Kumeyaay relied on broken branches, dead trees, or chaparral roots for firewood; they did not cut living trees (Shipek 1993).

The Kumeyaay also employed various techniques to manage the movement of water and sediment across the landscape. Rock ridges were constructed across drainages to capture sediment and runoff, boulders were placed at the heads of narrows to increase water supply in the wider areas upstream, and riparian vegetation was planted along streams to control erosion (Shipek 1993). Also to control erosion, patches of grain were burned in sequence as they dried out, which minimized the area of bare ground that existed at any one time (Shipek 1993).

GRAZING & AGRICULTURE

Ranching and farming were major land uses in the valley during the mid-19th through early 20th centuries.

Livestock grazing was the primary land use in southern San Diego County and northern Baja California in the late 18th and early 19th centuries, following Spanish colonization of the area. Cattle, sheep, goats, pigs, horses, and mules belonging to Mission San Diego de Alcalá were grazed on lands extending as far north as Valle de San José (near present-day Lake Henshaw), and at their peak in 1822 the mission's herds numbered about 30,200 head (Bowman 1947, Lightner 2013). It is possible that some grazing associated with the mission herds occurred in the Tijuana River valley during this time, though we were unable find direct evidence to quantify the intensity of mission-era grazing within the study area. An estimate of the extent of Mission San Diego's grazing lands in Bowman (1947) based on mission reports does not show the Tijuana River valley as a major grazing area.

After the collapse of the mission system and the rise of secular land grants in the late 1820s and 1830s, grazing was widespread on the ranchos in the Tijuana River valley. Rancho Melijó, which was granted to Santiago E. Argüello in 1833, occupied approximately 4,440 acres on the U.S. side of the river valley. One witness in the Rancho Melijó land grant court case testified that the Rancho supported "considerable stock of horses and cattle" (Stearns 1852), while another stated that Argüello "has had stock there ever since he occupied the ranch" (Boudini 1854). A diseño of Rancho Milejó from ca. 1840 shows several *abrevaderos* (watering holes) near the river on the U.S. side of the river valley, indicating that grazing lands extended onto the floodplain (U.S. District Court California, Southern District ca. 1840). Grazing also likely occurred on the valley's extensive alkali meadows, which were dominated by saltgrass (*Distichlis spicata*), a valuable forage plant for livestock in other systems (see "Cattle and Saltgrass" on p. 73). Argüello's holdings in November of 1854 included 2,000 head of cattle and horses, though according to Boudini (1854) he had "removed most of [the stock] to another Ranch" by that time, suggesting that grazing intensity varied seasonally or annually. Assuming that 2,000 head of stock did graze on Rancho Melijó at some point during this period, the average stocking density on the ranch would have been roughly 0.9 ha/head (2.2 ac/head; Christenson and Sweet 2008). At the much larger Rancho Tijuana further southeast, livestock holdings in 1868 included "about one thousand head of cattle and horses" (*Daily Alta California* 1868). Very little land was under cultivation during the rancho period. Boudini (1854), for instance, testified: "Except a very small garden near the house [Santiago Argüello] has not cultivated the land for the reason that it is only fit for pasture land." The adjacent Rancho Tia Juana was also not cultivated (Trujillo Muñoz 2010).

The effects of livestock grazing on habitat type, distribution, and quality in the Tijuana River valley is unknown. Potential impacts from grazing include alteration of the relative proportion of herbaceous cover and scrub on the valley floor, increased spread of invasive plant species, and changes in riparian vegetation composition, in addition to changes to the Tijuana River's channel morphology, water quality, sediment dynamics, and other hydrologic and geomorphic characteristics (Armour et al. 1991, Belsky et al. 1999). In particular, cattle grazing may have had a disproportionate impact on riparian areas given the animals' propensity for the shade, forage, and water resources riparian habitats provided (Belsky et al. 1999). However,

the relative impact of livestock compared to native herbivores such as deer and antelope historically present in the river valley (e.g., Van Wormer 2005, Schoenherr 2015) is unclear.

In the late 1860s and 1870s settlers began to arrive in the Tijuana River valley, establishing farms and displacing some of the livestock ranching. An 1868 article in the *Daily Alta California* noted, "Some twenty persons have very recently taken up claims in this lower section of the [Tia Juana] valley... [and] a conflict between the new settling tillers of the soil and the old cattle-breeding residenters is springing up" (*Daily Alta California* 1868). Various crops were grown by early settlers, including cotton, beets, yams, grapes, and vegetables (*Pacific Rural Press* 1877a, *Pacific Rural Press* 1877b, *Los Angeles Herald* 1878).

Agriculture expanded rapidly in the late 1880s and 1890s. A witness in an 1891 court case noted that "during the past five years, [the Tia Juana Valley] has become very largely occupied and inhabited" (Jones 1892). Wheat, barley, corn, and alfalfa were among the principal crops grown by settlers at this time: an article from 1889 described the valley as "beautifully fertile and highly cultivated" and referred to "great fields of corn, wheat and barley" (Ward 1889; see also *San Diego Union* 1887, Black and Smythe 1913, Unknown 1976).

The peak of agricultural production occurred in the early 20th century. By 1913 approximately 1,620 ha (4,000 ac) were under cultivation in the valley (Table 3.1; Black 1913). Though alfalfa and grains such as barley, wheat, and corn were the primary crops, a wide variety of other crops were grown including sugar beets, melons, squash, beans, lemons, potatoes, almonds, and walnuts (California State Legislature 1907, Black 1913, Böse and Wittich 1912, Unknown 1976). Much of the land to the north of the Tijuana

Figure 3.10. Cultivated land north of a levee, 1937. This image looks west along a levee on the north side of the valley, approximately 7 km (4 mi) inland, at the present day location of Interstate 5 under Dairy Mart Road. The levee, which was constructed after the 1927 flood, prevented flooding along the low-lying swale between the Tijuana River and San Diego Bay (see p. 110) and contributed to renewed agricultural expansion there between the late 1930s and 1950s. (MS 97/30 Box 1, v.5, Photo #77, courtesy Water Resources Collections and Archives, UC Riverside)

> Milijo and the land above the mouth of the Tia Juana, an extensive and wonderfully fertile valley, are now covered with the little white cabins of settlers, and some fine farms.
>
> — WILSON 1883

Year	Hectares (acres)	Source
ca. 1888	120 (300)	Black 1913
1913	1,620 (4,000)	Black 1913
1936	400 (1,000)	Joint Committee on Water Problems of the California Legislature 1953
1951	930 (2,300)	Joint Committee on Water Problems of the CA Legislature 1953
1957	960 (2,370)	Dedina 1991
1976	380 (950)	Dedina 1991
1994	200 (500)	BSI Consultants Inc. et al. 1994

Table 3.1. (right) Approximate area of cultivated land on the U.S. side of the Tijuana River valley over time. The geographic scope of the data from Black 1913 and BSI Consultants Inc. et al. 1994 is unknown, but they are presumed to apply only the U.S. side of the valley as well.

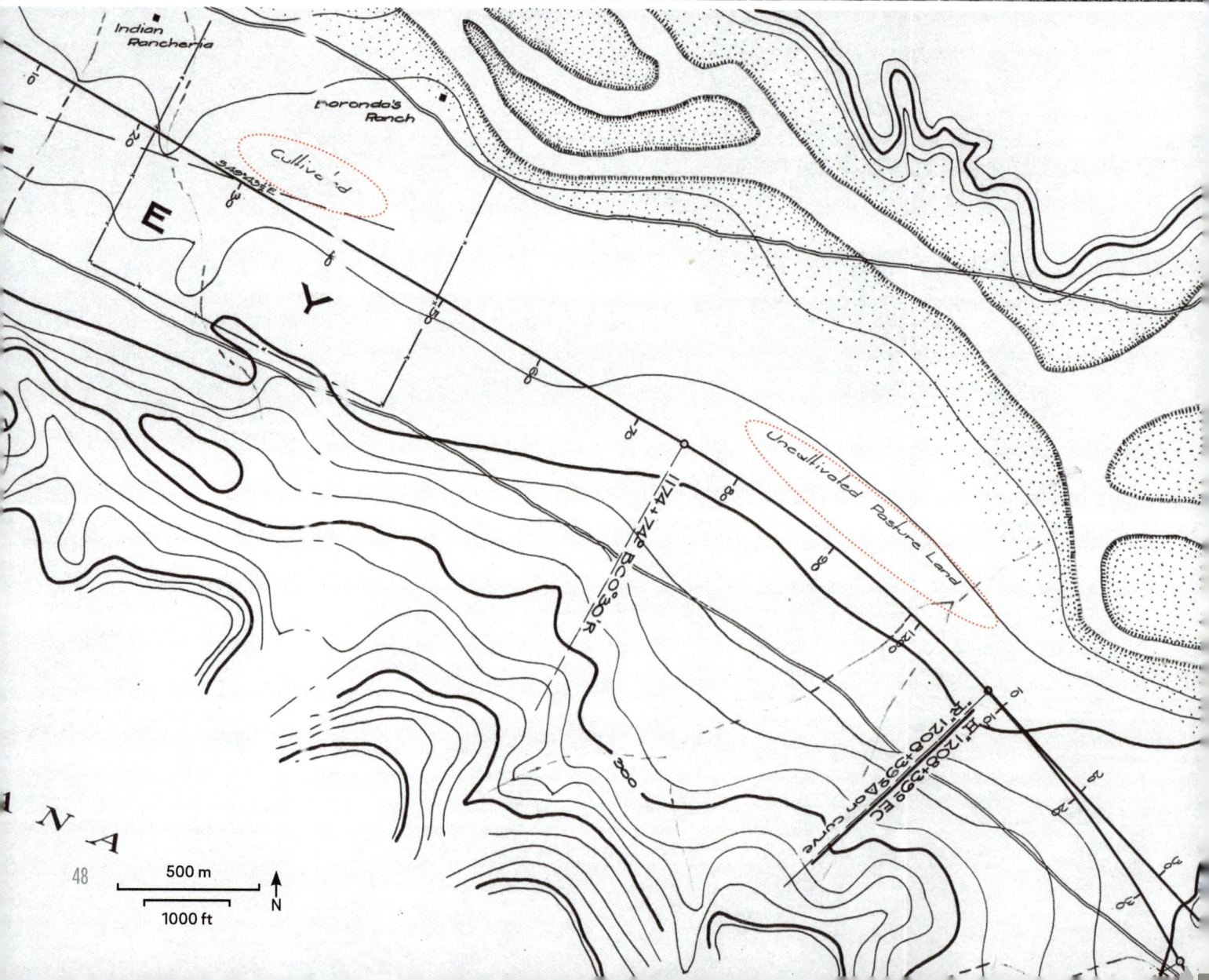

Indian Rancheria

Moronda's Ranch

Cultiva'd

E

Y

Uncultivated Pasture Land

1174+74.2 BC 0.3°R

P.I 1208+33.9 on curve

P.I 1208+33.9 EC

N A

48

500 m

1000 ft

N

River in present-day Imperial Beach was in barley production (Stephens 1912). On the eastern side of the valley, a farming cooperative called the "Little Landers" was formed in 1909 on the site that would later become San Ysidro (Pourade 1965; Fig. 3.11). Most of the cultivated land at this time was concentrated on the U.S. side of the valley: agricultural production on the Mexican side of the valley was minimal, and consisted primarily of barley, alfalfa, and beans (Bonillas and Urbina 1912).

Though ranching had significantly diminished in importance by the early 20th century, cattle and sheep grazing continued throughout the region, and in some areas ranching was still a dominant land use (*Daily Alta California* 1890a, Unknown 1890, *Los Angeles Herald* 1899, Stephens 1908, Rankin 1909, Black and Smythe 1913, Bonillas and Urbina 1912, Minnich and Vizcaíno 1998). For example, Stephens (1908) wrote that "the mesa on the south [of the Tijuana Estuary]... is used as a cattle range," and a 1909 map shows the river valley south of the international border as a combination of "cultivated" land and "uncultivated pasture land" (Rankin 1909; Fig. 3.12). The first dairy ranches in the valley also appeared during this time (Black and Smythe 1913).

The agricultural success of the late 19th and early 20th centuries came to an abrupt halt in January 1916, when massive floods wiped out much of the cropland and many of the settlements in the river valley, including farms in the Little Landers community (San Diego County Flood Control 1937, Unknown 1976). Though recovery began within a few years (Storie and Carpenter 1923, Gayman 1971, Dedina 1991), a

> We planted alfalfa on a half mile of river bed in 1912, dragging a little top soil over it. That was washed out after that in 1916; it became river bed again. Looked very much as it had before we cultivated it.
>
> — PERRY 1936

Figure 3.11. (opposite page, top) "Half acre Farm w/ Chicken pens," ca. 1912. This photo shows the Little Landers farming cooperative that formed in 1909 and farmed in the area that later became San Ysidro. *(Photo #91:18564-1440, courtesy San Diego History Center)*

Figure 3.12. (opposite page, bottom) A combination of "cultivated" land and "uncultivated pasture land" alongside the river in Mexico, 1909. *(Rankin 1909, courtesy Department of Special Collections, Stanford University Libraries)*

Figure 3.13. (right) Uncultivated lands are shown on this 1921 map. Many formerly cultivated fields in the lower Tijuana River valley remained unplowed and fallow for years following the massive 1916 floods. *(Ervast 1921, courtesy San Diego History Center)*

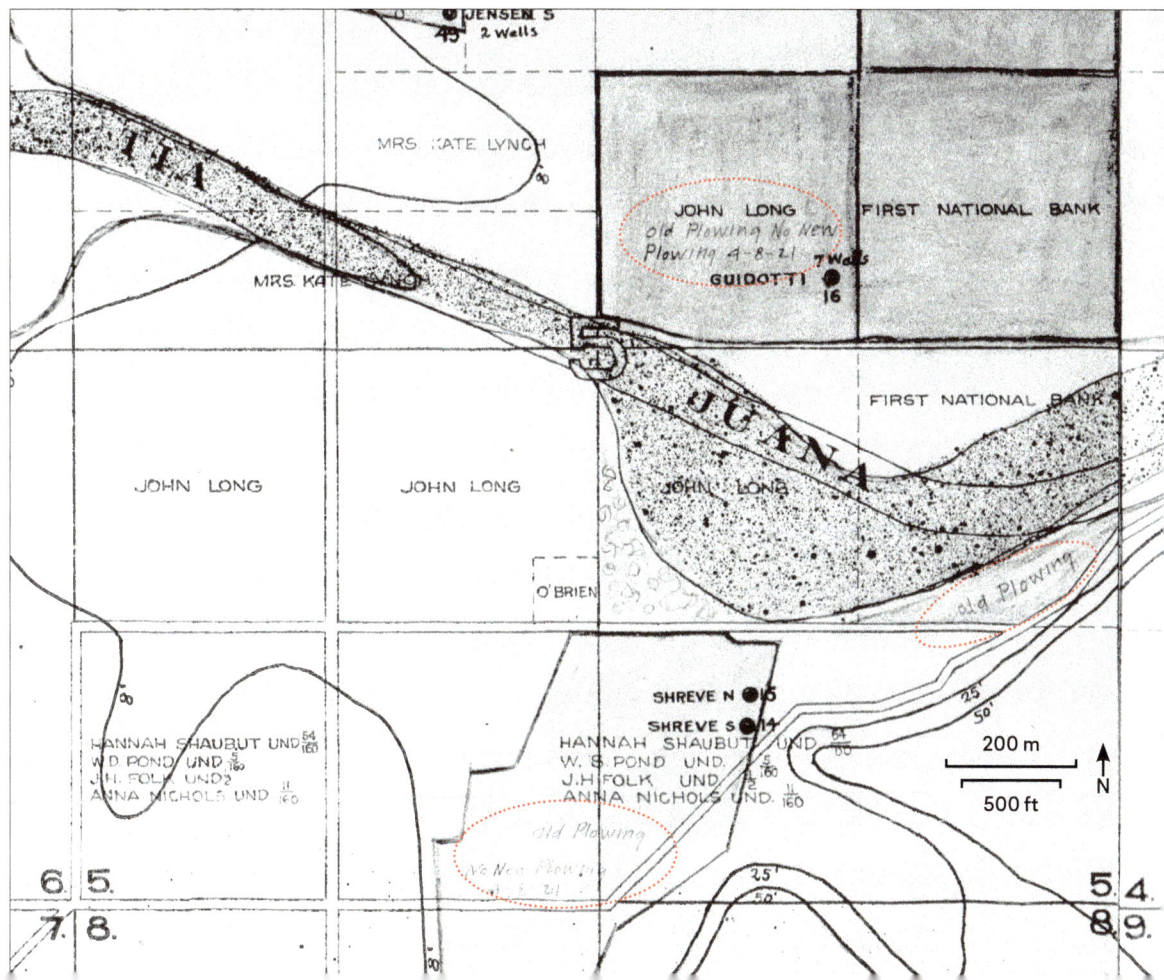

second major flood in 1927 dealt another setback to farmers, and many formerly cultivated areas remained fallow throughout the 1920s and early 1930s (Ervast 1921, Knox 1934, Van Etten 1935; Fig. 3.13). Some of the abandoned fields were used for grazing (Knox 1934). Though the flooding destroyed large areas of cropland, it also deposited sediment across the floodplain and enhanced soil fertility (Perry 1936).

The region experienced renewed agricultural expansion between the late 1930s and 1950s, during which time the acreage of cropland in the Tijuana Valley north of the international boundary increased from approximately 400 ha to 930 ha (1,000 ac to 2,300 ac; see Table 3.1; Fig. 3.14; Joint Committee on Water Problems of the California Legislature 1953, Rempel 1992, Ojeda Revah 2000). Even portions of the river bed were cultivated: one resident observed crops planted "on the first bench perhaps above the low water, being six inches or twelve above the very lowest part of the river-bed, and right above the Nestor bridge" (Perry 1936). A substantial amount of cultivated land was also found on the Mexican side of the river valley just south of the border during this period (Fig. 3.15). With more widespread cultivation and more intensive groundwater pumping for irrigation, groundwater levels began to decrease and soil salinities increased, negatively affecting agricultural productivity (Fig. 3.16; see pp. 52–59). Some farmers responded by abandoning fields or replacing grains and vegetables with alfalfa or other crops with higher salt tolerance (Gayman 1971, Dedina 1991).

By the 1960s and 1970s, agriculture in the valley had declined as a result of lowered groundwater levels, increased soil salinity, surface water diversions, and urban development (California Department of Parks and Recreation and Department of Fish and Game 1972, Cleisz et al. 1989, Rempel 1992, BSI Consultants Inc. et al. 1994; see Table 3.1). Cropland was abandoned in the western part of the valley, where soil salinities were highest, with small exceptions such as fields at the mouths of Goat and Smuggler's canyons (Gayman 1971, California Department of Parks and Recreation and Department of Fish and Game 1972, Rempel 1992). Some cropland in the eastern portion of valley was also eliminated due to the construction of flood control structures (Rempel 1992).

Figure 3.14. Cultivated land in the Tijuana River valley, 1937. Similar to the photo on p. 47, this image looks northwest from the top of a levee on the north side of the valley, approximately 7 km (4 mi) inland, at the present day location of Interstate 5 under Dairy Mart Road. The photograph shows a dairy ranch in the background, the first of which were established in the area during the early 1900s. *(MS 97/30 Box 1, v.5, Photo #2128, courtesy Water Resources Collections and Archives, UC Riverside)*

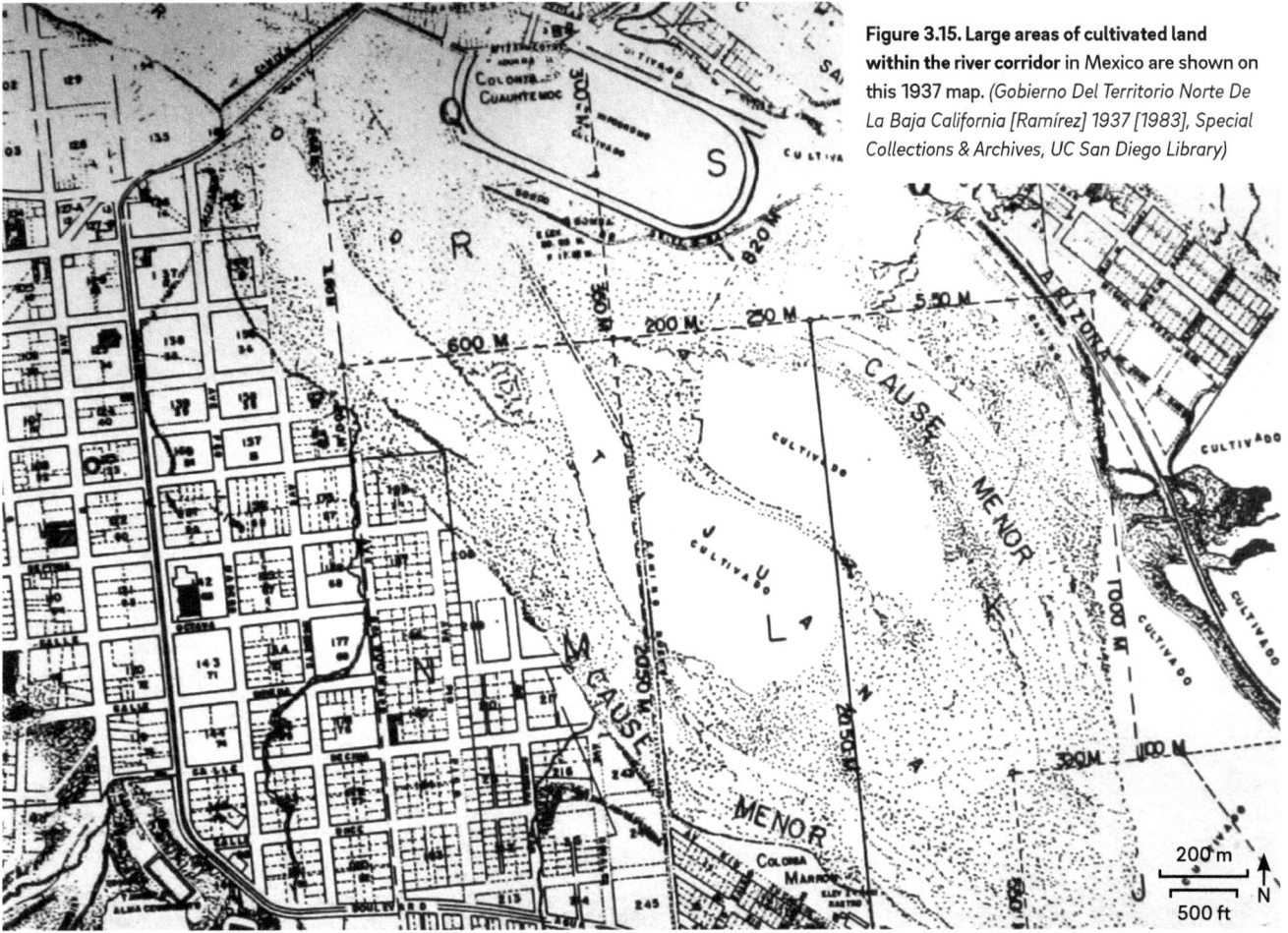

Figure 3.15. Large areas of cultivated land within the river corridor in Mexico are shown on this 1937 map. *(Gobierno Del Territorio Norte De La Baja California [Ramírez] 1937 [1983], Special Collections & Archives, UC San Diego Library)*

Figure 3.16. By the 1940s, soil salinities were beginning to increase and reduce agricultural productivity, especially in the western portion of the Tijuana Valley. Cropland still occupied a substantial amount of land near the estuary, however, as can be seen in this May 1941 aerial photograph looking east from the edge of the estuary. *(Photo #79:741-893, courtesy San Diego History Center)*

GROUNDWATER EXTRACTION

Groundwater pumping associated with agricultural production led to significant declines in groundwater levels by the mid-20th century.

The spread of farming in the lower Tijuana River valley in the 1860s and 1870s created a demand for irrigation water that prompted an increase in groundwater pumping. Records document the presence of numerous wells in the southern portion of San Diego County as early as 1869 (Pascoe 1869), and by 1872 efforts were being made to obtain "a supply of water by artesian wells for irrigating purposes in the Tia Juana valley" (*Pacific Rural Press* 1872).

The water table in many parts of the lower river valley was documented to be just a meter or two below the surface, with water "easily obtained by shallow wells and in some places standing on the surface" (Pascoe 1869, *San Diego Union* 1887, Hall 1888, Van Dyke et al. 1888, Dedina 1991). Likewise, groundwater levels upstream of the international border were reported to be within about 2 m (6 ft) of the surface near the river bed, though the water table decreased to about 12–15 m (40–50 ft) below the surface on the low terraces surrounding the river valley (Böse and Wittich 1912). Artesian conditions were found in some areas, giving rise to natural springs (Pascoe 1869, *Pacific Rural Press* 1872). For example, natural springs were found at Agua Caliente, located at a constriction in the river valley about 4.5 km (3 mi) upstream of the international border, as well as on the eastern side of the estuary (Fig. 3.17).

> The entire Tia Juana Valley is a grand and verdant belt... On inquiring I found that water was obtainable in any quantity. It can be found in abundance at some five feet below the surface, pure and adapted to any use.
>
> — SAN DIEGO UNION 1887

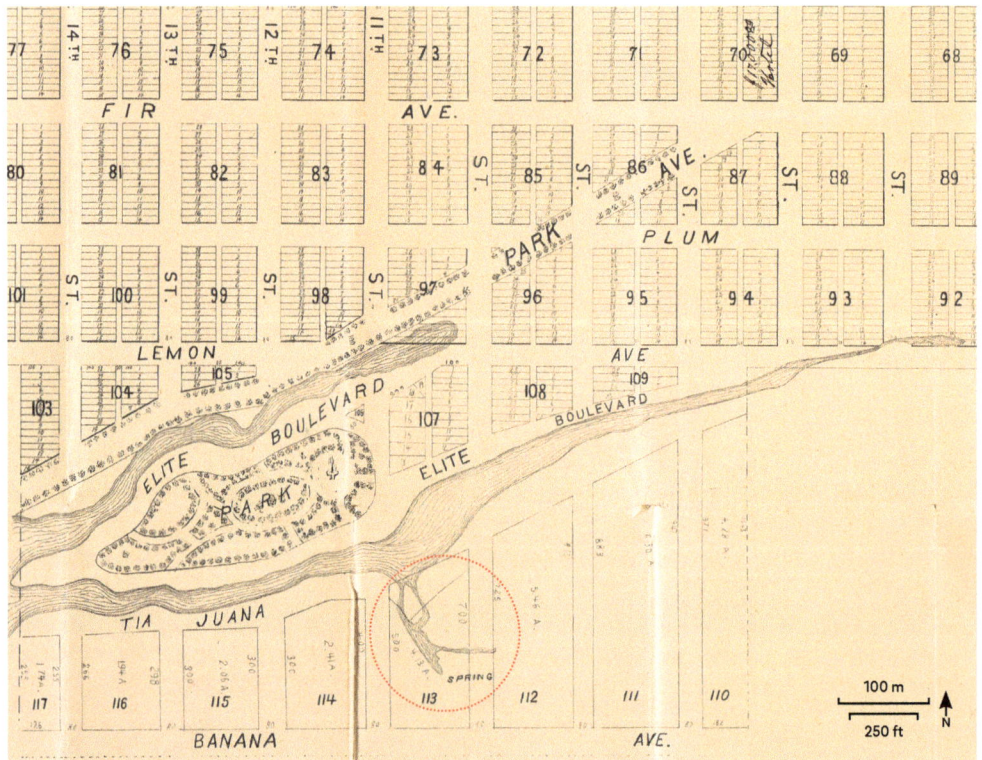

Figure 3.17. A "spring" at the end of Tijuana River Slough (right), suggesting the presence of shallow groundwater at this location. *(Los Angeles Lithographic Co. ca. 1889, courtesy Huntington Art Collections, San Marino, California)*

Figure 3.18 (facing page). An early well in Tijuana, ca. 1901. *(Photo #22533, courtesy San Diego History Center)*

Figure 3.19. A longitudinal profile of the Tijuana River valley shows seasonal fluctuations in groundwater levels between October 1914 and August 1915. The average fluctuation in water table levels during this season was approximately 2.1 m (7 ft). The accompanying map shows groundwater contours in the valley in January (solid lines) and March (dotted lines) of 1915 (in feet above mean sea level). The locations of wells shown on the longitudinal profile are indicated. (Ellis and Lee 1919)

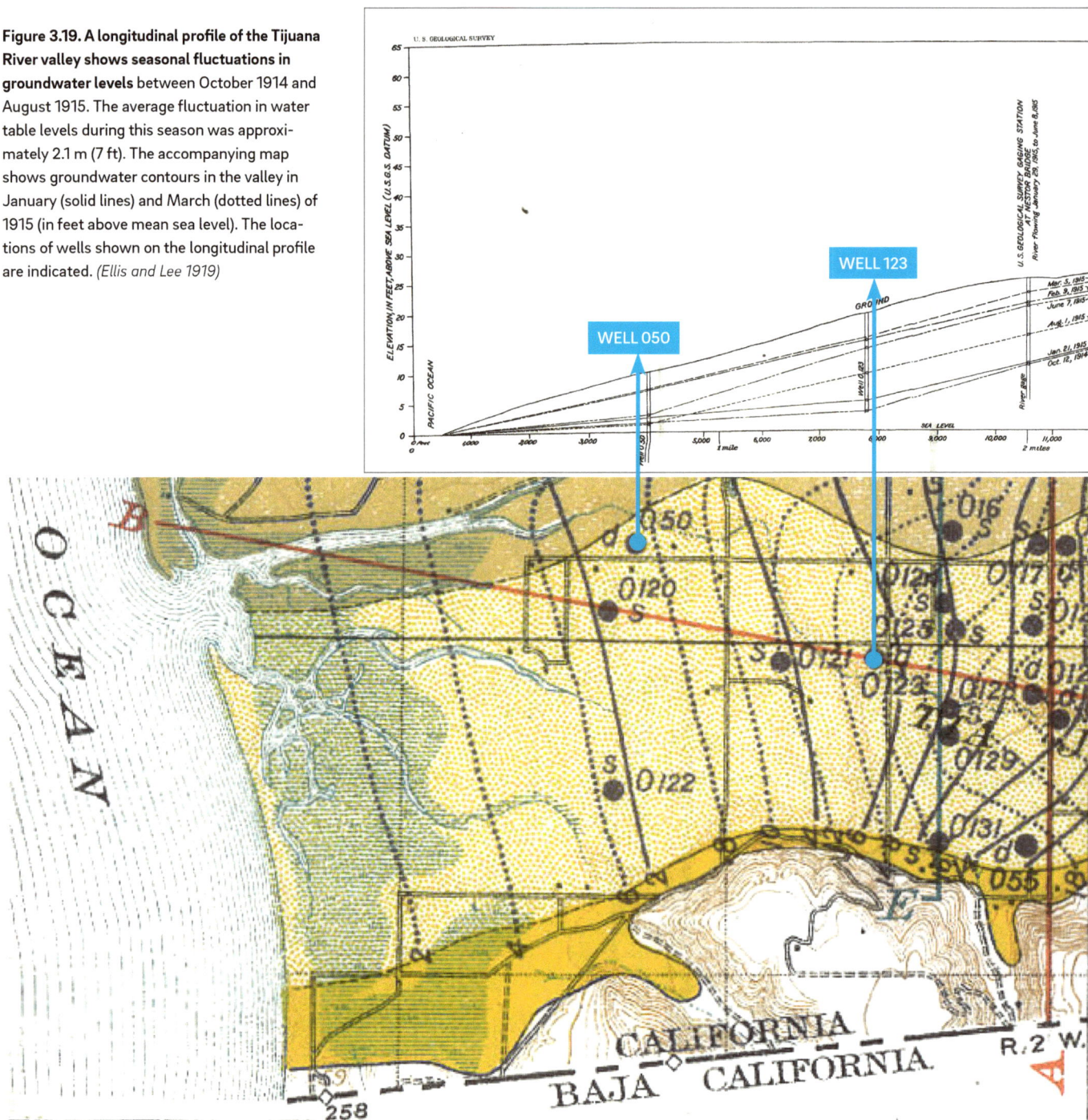

Groundwater extraction for irrigation and other uses gradually increased in the 1890s and early 1900s as agricultural development in the valley continued to expand (Fig. 3.18; Black and Smythe 1913, Black 1913, Adams 1913, Webster 1913, Ellis and Lee 1919, Hatherley 1936b, Perry 1936). N. J. Peavey, whose family moved to the region in 1893, is credited with being "the first man to develop water on a large scale in the Tia Juana valley" (Black and Smythe 1913). Ellis and Lee (1919) describe the expansion of groundwater pumping in the early 20th century: "In the period beginning with 1909 much of the Tia Juana, Otay, Sweetwater, and San Diego river valleys was put under cultivation by irrigation from individual pumping plants." Several irrigation districts were organized at this time (San Ysidro Chamber of Commerce 1915, Storie and Carpenter 1923). Groundwater extraction in the early 1900s appears to

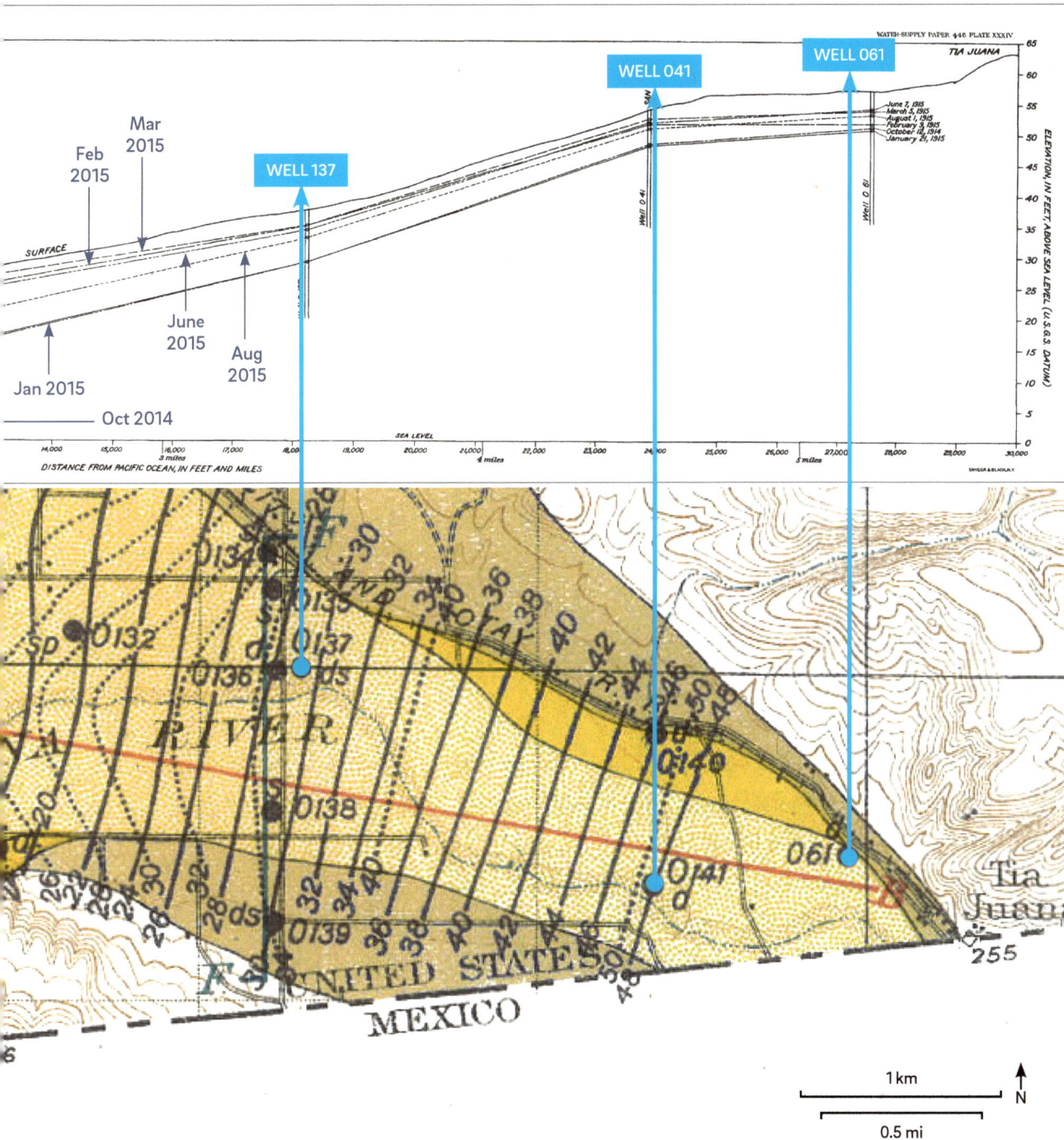

Figure: Cross-section and map showing observation wells (WELL 137, WELL 041, WELL 061) in the Tijuana River valley, with water table elevations for various months in 2014–2015. Scale bars: 1 km and 0.5 mi.

have been heaviest on the U.S. side of the border, with relatively little pumping occurring in the Mexican portion of the river valley (Webster 1913).

The extraction of groundwater from the Tijuana River valley was in part offset by recharge from rainfall and streamflow. Natural variability in precipitation and evapotranspiration, along with other factors, resulted in seasonal fluctuations in groundwater levels, with minimum levels typically occurring between October and January prior to the first major storm of the rainy season (Fig. 3.19; Ellis and Lee 1919). In 1914–15, for instance, the average fluctuation in water table levels on the U.S. side of the river valley (based on nine observation wells) was documented to be approximately 2.1 m (7 ft; Ellis and Lee 1919). Groundwater levels on the U.S. side of the border in the early 20th century ranged from less than 1 m (3 ft) to over 4.5 m (15 ft) below the surface (Ellis and Lee 1919, Waitz 1928, Haltiner and Swanson 1987).

A substantial increase in groundwater pumping occurred in the 1930s-50s, coinciding with the expansion in agricultural development during this period (Rempel 1992). Groundwater extraction on the U.S. side of the river valley increased from approximately 4,317,200 m³ (3,500 acre-feet) in 1945 to 22,202,700 m³ (18,000 acre-feet) in 1952–3 (Joint Committee on Water Problems of the California Legislature 1953, Rempel 1992). Average withdrawals from 1936–60 were an estimated 7,400,900 m³ (6,000 acre-feet) per year, with a net groundwater loss of 37,004,500 m³ (30,000 acre-feet) over this period (California Department of Parks and Recreation and Department of Fish and Game 1972). Groundwater pumping also increased on the Mexican side of the border: eight wells were drilled in the river valley in Mexico in 1950, yielding an estimated total of 4,933,900 m³ (4,000 acre-feet) during their first year of operation (Joint Committee on Water Problems of the California Legislature 1953). In 1960 the irrigated acreage in the river valley totaled approximately 1,210 ha (3,000 ac) in the United States and 300 ha (750 ac) in Mexico; throughout the basin, water for irrigation was supplied entirely from groundwater (IBWC 1960).

The increased rate of groundwater extraction resulted in a substantial lowering of the water table in the mid-1940s and 1950s. Near San Ysidro and the international border, groundwater levels dropped 9–12 m (30–40 ft), from approximately 2.5 m (8 ft) below the surface in the mid-1940s to 12–15 m (40–50 ft) below the surface by the 1960s (San Diego Regional Water Quality Control Board 1967, USAED 1974, Izbicki 1985). According to Rempel (1992), groundwater levels throughout most of the valley had dropped below sea level by the 1960s, resulting in salt water intrusion and an increase in groundwater salinity (Hatherley 1936a, California Department of Parks and Recreation and Department of Fish and Game 1972, Rempel 1992, Teng 1994). Though quantitative information on the effects of groundwater pumping on streamflow is not available, it is likely that the significant drop in water table levels that occurred during the mid-20th century reduced discharges from groundwater to surface water.

By the mid-1960s, rates of groundwater pumping had fallen, and groundwater levels began to recover (USAED 1974, IBWC 1976, BSI Consultants Inc. et al. 1994). By the late 1970s the water table in many parts of the river valley had recovered close to its historical level (Fig. 3.20; Rempel 1992).

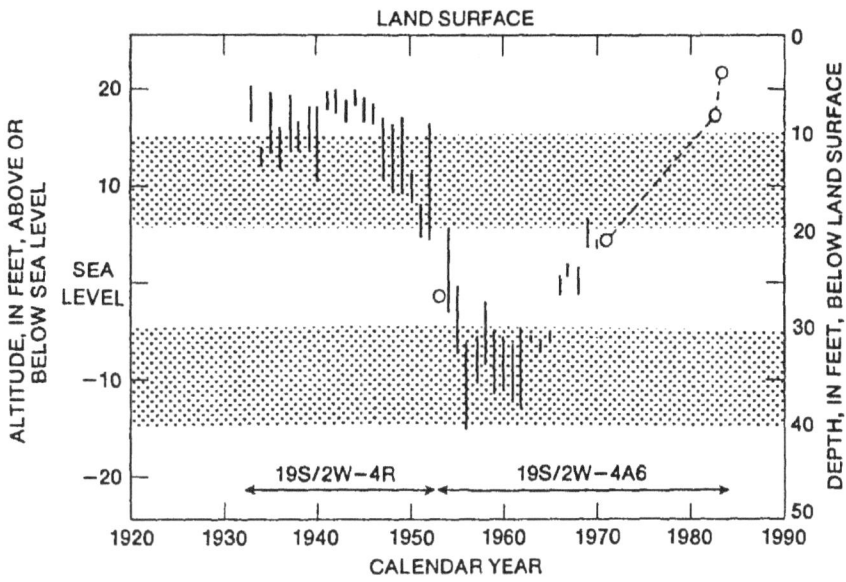

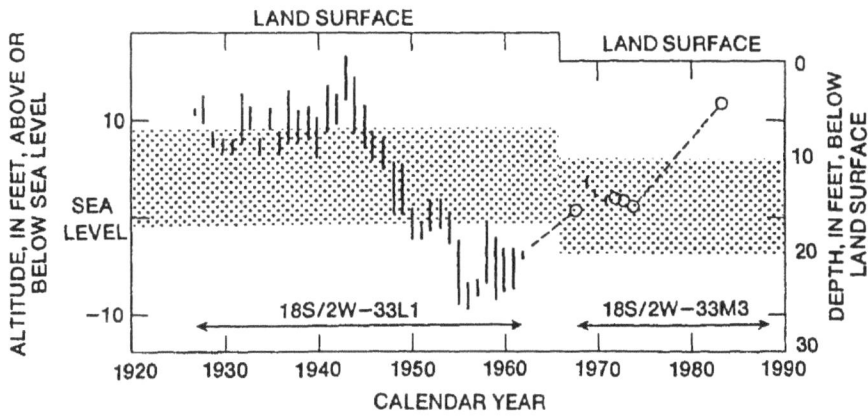

Figure 3.20. **Groundwater levels in the Tijuana valley** measured at two well locations, from Rempel 1992. Groundwater levels declined by 9–12 m (30–40 ft) from the mid-1940s to the early 1960s, but had largely recovered by the late 1970s.

DAMS & SURFACE WATER DIVERSION

Dams regulate flow from approximately 75% of the Tijuana River watershed.

Five water storage reservoirs – Morena and Barrett in the United States and Rodríguez, Carrizo, and las Auras in Mexico – capture runoff from the Tijuana River watershed. Together these reservoirs regulate about 75% of the Tijuana River watershed (Rempel 1992) and affect both the timing and volume of flows.

Morena and Barrett reservoirs were built to supply water to the City of San Diego (Adams 1913). Construction on Morena Dam, located on Cottonwood Creek in the U.S. portion of the watershed, began ca. 1896, but encountered delays and was not completed until 1912 (Fig. 3.21; Newell 1901, Department of Public Works 1935). Morena Dam captures water from a drainage area of approximately 311 km² (120 mi²; Unknown 1917). Barrett Dam, located downstream of Morena Dam on Cottonwood Creek, was completed in 1921 and captures runoff from an additional 337 km² (130 mi²; Wyman 1937a, USAED 1974).

The Dulzura Conduit was built in 1909 to divert water from Cottonwood Creek to the Otay River watershed and ultimately to the City of San Diego (Department of Public Works 1935, Wyman 1937a, IBWC 1960). After the completion of Barrett Dam, the intake for the conduit was moved to Barrett Reservoir . Average annual discharge in the Dulzura Conduit from 1937–60 was 12,449,500 m³ (10,093 acre-feet), compared with 4,150,700 m³ (3,365 acre-feet) in Cottonwood Creek below Barrett Dam, indicating that nearly 75% of total streamflow was diverted through the Dulzura Conduit (IBWC 1960).

Rodríguez Dam, completed in 1936, is located on Río de las Palmas (which becomes the Tijuana River) in Mexico about 18 km (11 mi) southeast of the international border, and regulates flows from a drainage area of 2,559 km² (988 mi²; IBWC 1960, Ojeda Revah 2000). Water diversions from Rodríguez Reservoir commenced in 1940–41; in 1949–50, 7,093,700 m³ (5,751 acre-feet) were diverted from the reservoir to the City of Tijuana (Joint Committee on Water Problems of the CA Legislature 1953). From 1937–60, average annual discharge was 24,535,200 m³ (19,891 acre-feet) above Rodríguez Reservoir and 13,050,200 m³ (10,580 acre-feet) immediately below the dam; average annual diversions from the reservoir during this period were 9,828,400 m³ (7,968 acre-feet; IBWC 1960). Rodríguez Reservoir was the sole source of water for the City of Tijuana until 1950, when the supply was supplemented by groundwater from wells drilled in the river valley downstream (see p. 56; Page 1955). A variable amount of water was also diverted from the reservoir for irrigation in the Tijuana River valley in Mexico; from the mid-1930s through mid-1950s, these diversions supplied water to approximately 2,020 ha (5,000 ac) of farmland (Williams 1933, IBWC 1960, Rovirosa 1963, Ojeda Revah 2000).

The construction of dams on Cottonwood Creek and Río de las Palmas has reduced the volume of water and sediment transported downstream and altered the timing of streamflow. Haltiner and Swanson (1987)

Figure 3.21. "Explosion at Morena dam site," December 26, 1896. Rock used to construct Morena dam was obtained from the adjacent hillsides. This photograph, taken December 26, 1896, shows an explosion at the dam site that reportedly dislodged 200,000 tons (200 million kg) of rock. *(Newell 1901, Part IV, Pl. LV, courtesy Internet Archive)*

estimate that Morena, Barrett, and Rodríguez dams reduced the total volume of streamflow at the Nestor gage (USGS gage #11013500, located at Hollister Street) by approximately 50% between 1937 and 1983. The dams also substantially attenuated peak flows, though occasional major releases from Rodríguez Dam still occurred during storm events in the 1930s-40s and again in recent decades, resulting in periodic flooding in the Tijuana River valley downstream (Zedler et al. 1986, Haltiner and Swanson 1987, Chin et al. 1991, URS Corporation 2012). In addition, dams within the watershed impound an estimated 30–50% of total upstream sediment inputs (Haltiner and Swanson 1987, Flick 2005, Patsch and Griggs 2007).

WASTEWATER DISCHARGE & URBAN RUNOFF

Sewage discharge contributed to major water quality problems in the valley.

The volume and quality of water entering the Tijuana River valley has also been affected by wastewater releases as well as agricultural and urban runoff (Fig. 3.22). A report by the Joint Committee on Water Problems of the California Legislature (1953), for instance, observed:

> As the return water from irrigation became established it was observed that the duration of stream flow both at the International Border and westerly throughout the valley was sustained for greater periods of time and in some instances was maintained at the International Border throughout the summer season.

Sewage discharge was recognized as a problem as early as the 1920s and 1930s. In 1927, sewage was observed "flowing across the International Boundary and into the dry river bed where the river crosses the Boundary" (Hull 1941), and a 1931 newspaper article noted that sewage was "polluting the Tijuana river from the Monument schoolhouse west" *(San Ysidro Border Press* 1931). The problem worsened as rapid population growth continued to outpace wastewater treatment capacity. Early efforts to address the issue were insufficient, including the construction of sewer lines to carry effluent from Tijuana and San Ysidro to the ocean, and by the mid-1950s over 15,000 m^3 (4 million gallons) of untreated wastewater were entering the estuary and river valley daily (Herzog 1990).

Discharge of sewage effluent into the Tijuana River continued to varying degrees throughout the 20th century (Spalding et al. 1999). By 1988, an estimated 18,900–37,800 m^3 (5–10 million gallons) of sewage effluent were entering the river from the City of Tijuana (Seamans 1988). Sewage was also discharged into the estuary or the river valley upstream from Ream Field and nearby wastewater treatment plants (County of San Diego 1970). Sewage discharge contributed to water quality problems, crop loss, and habitat degradation within the river valley; beaches between Imperial Beach and the international border were closed repeatedly between the late 1950s and 1980s due to public health concerns (Herzog 1990, Proffitt 1994, Spalding et al. 1999, West 2001). In recent decades, the construction of the South Bay International Wastewater Treatment Plant and other measures have improved the management of wastewater entering the river valley. However, the treatment plant can only process 95,000 m^3/day (25 million gallons/day), so flows above this rate enter the lower river valley, pass through the estuary to the ocean, and frequently necessitate local beach closures (Fig. 3.23; Minan 2002, TRVRT 2012, Lee 2012, Regional Water Management Group 2013).

Figure. 3.22. (top) Urban runoff entering the river channel in Tijuana's Zona Río, April 2015. *(Photo by Samuel Safran, April 2015)*

Figure 3.23. (bottom) Beach closure in Imperial Beach attributed to contaminated Tijuana River effluent.

URBAN DEVELOPMENT & POPULATION GROWTH

The Tijuana River valley experienced rapid population growth and urbanization during the 20th century.

As farmers and settlers moved in to San Diego County and northern Baja California in the late 1800s, a number of small towns were established in and around the Tijuana River valley. On the western side of the valley, communities established during this time included Imperial Beach, Oneonta, and Monument City (Unknown 1976, Fetzer 2005). To the east, the communities of San Ysidro (on the U.S. side of the border) and Tijuana (on the Mexican side) were organized in the late 1880s (City of San Diego and Page & Turnbull 2010, D'Elgin et al. n.d.).

Railroad and road construction accompanied the influx of settlers in the late 19th century. The National City and Otay Railroad, completed in 1888, extended south along the eastern side of the Tijuana River valley to the international border (*Los Angeles Herald* 1888, Fink 1891). The San Diego and Arizona Eastern Railway, constructed in the early 20th century, enters the Tijuana River valley near the southern end of the study area and runs northwest through the river valley into San Diego County. The railroad crosses the Tijuana River just downstream of Rodríguez Reservoir and again approximately 4 km (2.5 mi) southeast of the border; the bridge crossings were constructed between 1910 and 1917 (Rintoul et al. 1936, CDWR ca. 1942).

The population of San Diego County was less than 10,000 until the 1880s (Fig. 3.24; U.S. Census 1882, 1895). By the 1890s the population had jumped to 35,000, and by 1920 it was increased to more than 110,000. Imperial Beach began to develop rapidly in the 1930s, and by 1960 its population had reached approximately 18,000 (Unknown 1976, California Department of Finance 2012). Several military installations were established in southern Imperial Beach and the adjacent estuary in the early 1900s. The Naval Outlying Landing Field Imperial Beach (NOLF Imperial Beach, originally called Ream Field) was constructed on the northwest side of the estuary in 1917 (Dedina 1991). To the south, Border Field Auxiliary Landing Field was constructed in the early 1940s; Border Field was deactivated in 1961, and was subsequently designated Border Field State Park (D'Elgin et al. n.d., TRNERR 2015).

Tijuana remained a small town for many decades, and as late as 1919 had an estimated population of just 300 (Nelson 1922). By the 1920s Tijuana had begun to grow rapidly, and by 1950 the population had reached an estimated 60,000 (D'Elgin et al. n.d., Proffitt 1994, United Nations Department of Economic and Social Affairs/Population Division 2005). Massive development in the later part of the 20th century pushed the population to 500,000 in the 1980s and over 1,000,000 by the late 1990s (Brinkhoff 2014, INEGI 2014).

Major changes to the hydrology of the Tijuana River valley occurred as part of this rapid urbanization. In the 1970s, the Tijuana River Flood Control Project resulted in the channelization of the Tijuana River from just downstream of Rodríguez Reservoir to the international border (Ramirez 1985, Dedina 1991). Although the concrete channel ends at the international border, a variety of flood control structures, including levees and an energy dissipater, have been constructed on the U.S. side of the border; levees lining the channel extend beyond the border for approximately 760 m (2,500 ft; Chin et al. 1991, BSI Consultants Inc. et al. 1994). A portion of Río Alamar (Cottonwood Creek) upstream of the confluence with the Tijuana River has also been channelized (TRVRT 2012).

Sand and gravel extraction from the river bed and valley floor, initiated in the 1940s, also had a significant impact on streamflow patterns, sediment transport, and habitat mosaics (Haltiner and Swanson 1987). Mining activities were concentrated in the eastern portion of the valley near Dairy Mart Road, though mining also occurred in the central portion of the valley, in the Goat Canyon area, and in upland areas around Smuggler's Gulch (City of San Diego 1976, Cleisz et al. 1989, Higgins et al. 1994). Borrow pits created as a result of these mining activities still exist within the river corridor, some of which support perennial ponds (Haltiner and Swanson 1987; J. Crooks, personal communication).

Figure 3.24. Population trends in San Diego, San Diego County, Tijuana, and the state of Baja California. The population of San Diego County increased rapidly in the late 19th and early 20th centuries. Tijuana, which had a population of just several hundred in the early 20th century, began to grow rapidly in the 1920s. *(Sources: Los Angeles Daily Herald 1889, Nelson 1922, Pryde 2004, California Department of Finance 2012, INEGI 2012, Brinkhoff 2014)*

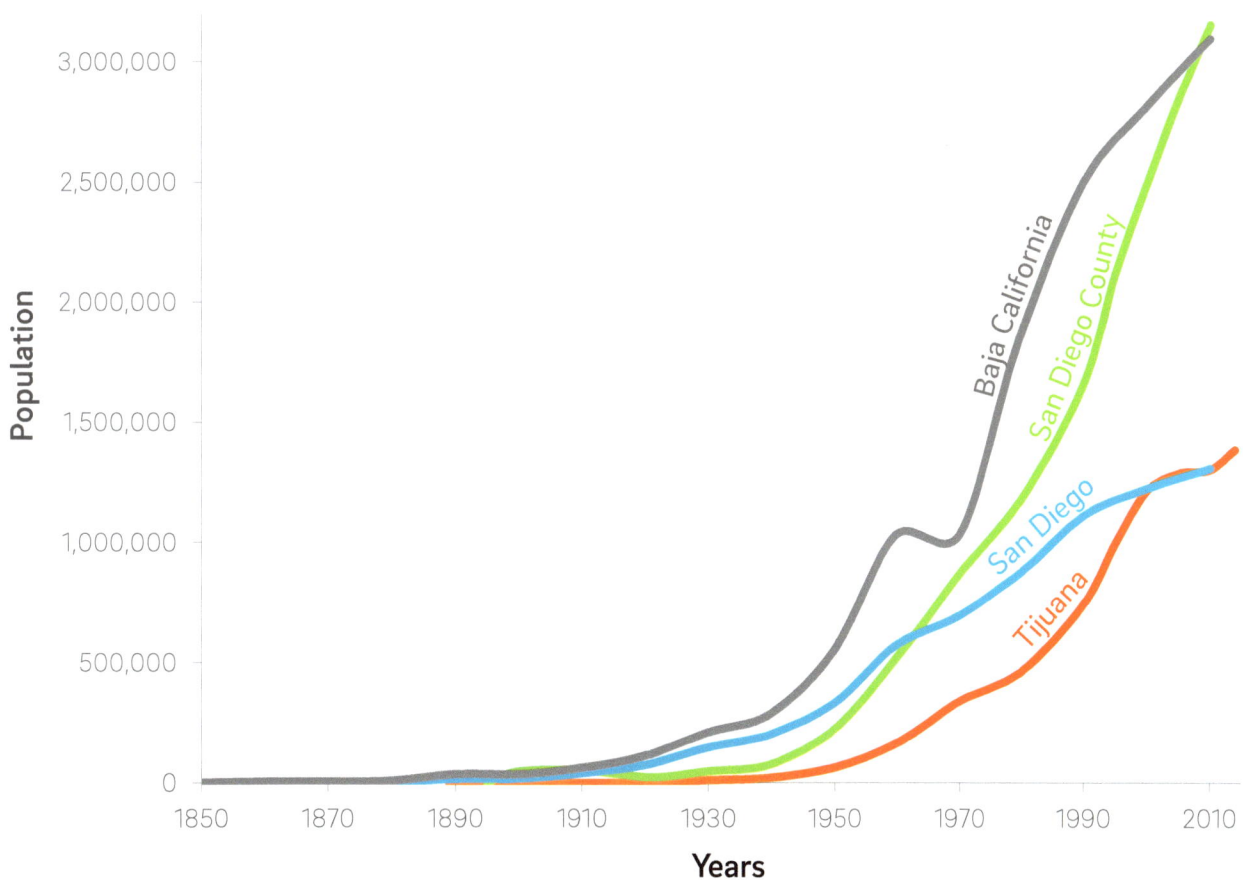

CONSERVATION MEASURES

Land preservation and ecosystem restoration have helped to protect the valley from development and degradation.

Though urban development has impacted the Tijuana River valley and estuary in many ways, various conservation measures have been implemented over the past several decades in an effort to protect and restore natural areas. The Tijuana Slough National Wildlife Refuge was established in 1980 (Figs. 3.25–3.26), and two years later the estuary was designated a National Estuarine Research Sanctuary, which later became the Tijuana River National Estuarine Research Reserve (TRNERR). The TRNERR encompasses approximately 928 ha (2,293 ac) owned by multiple federal, state, and local agencies and jointly managed by the U.S. Fish and Wildlife Service and California State Parks, with oversight and support from the National Oceanic and Atmospheric Administration (Fig. 3.25; TRNERR 2010). The TRNERR facilitates a variety of conservation activities, including scientific research, environmental education, and ecosystem restoration. Upstream of the TRNERR, the Tijuana River Valley Regional Park protects another 730 ha (1,800 ac) of riparian, terrestrial, and palustrine habitats (TRVRP 2007).

Several past and present initiatives have aimed to improve our understanding of the Tijuana Estuary and river valley and address ongoing and anticipated management challenges. For decades, the TRNERR has actively developed restoration plans, including the Tijuana Estuary Tidal Restoration Program (TETRP), which seeks increase estuary's the tidal prism and restore habitats and functions lost river channel modification, dam operation, sewage flows, unseasonal freshwater flows, and excessive erosion and sediment deposition. This adaptive restoration program has led to the multiple on the ground projects, including the Tidal Linkage project completed in 1997 and Model Marsh restoration in 2000. More recent complementary initiatives have included the Temporal Investigations of Marsh Ecosystems (TIME) project, a collaboration between the TRNERR and other partners that characterized changes in ecosystem services in the Tijuana Estuary over time to guide wetland restoration (TRNERR 2014), and the Tijuana River Valley Recovery Strategy, which brings together a variety of stakeholders to improve management around sediment and trash accumulation in the valley (TRVRT 2012).

Figure 3.25. (opposite page, top) Map of public lands in the Tijuana River valley with the boundary of the Tijuana River National Estuarine Research Reserve, established in 1980. The majority of the valley floor is publicly owned in the United States. *(TRNERR 2010)*

Figure 3.26. (opposite page, bottom) Article in the San Diego Union, January 4th, 1981, reporting on the sale of land from the Helix Imperial Harbour Development Corporation, which had sought to convert the estuary into a marina, to the U.S. Fish and Wildlife Service. The site was soon after added to the National Wildlife Refuge System. Pictured are Michael and Patricia McCoy, key figures in the conservation of the Tijuana Estuary. *(San Diego Union, January 4, 1981, Staff Photo by Cindy Lubke Romero, courtesy San Diego Union)*

Land Ownership
- California State Parks
- U.S. Fish and Wildlife Service
- U.S. Navy
- U.S. Federal
- Intl. Bndy Water Comm.
- County of San Diego
- City of San Diego

Tijuana River NERR

Tijuana River National Estuarine Research Reserve

Data Sources: TRNERR, SANGIS.org

Imperial Beach Blvd

Tijuana Estuary Visitor Center

City of Imperial Beach

Naval Outlying Landing Field

Tijuana Slough NWR

Border Field State Park

Model Marsh

Friendship Marsh

Monument Rd

Sediment Basins

Goat Canyon

Spooner's Mesa

Smuggler's Gulch

Saturn Rd

Hollister St

East Spooner's Mesa

Quarry

IBWC Treatment Plant

Dairy Mart Rd

United States

Mexico

Monument Mesa

Lichty Mesa

Yogurt Canyon

Los Sauces Canyon

Los Laureles Canyon

Matadero Canyon

City of Tijuana

1D

Pacific Ocean

N W E S

0   0.5
Kilometers

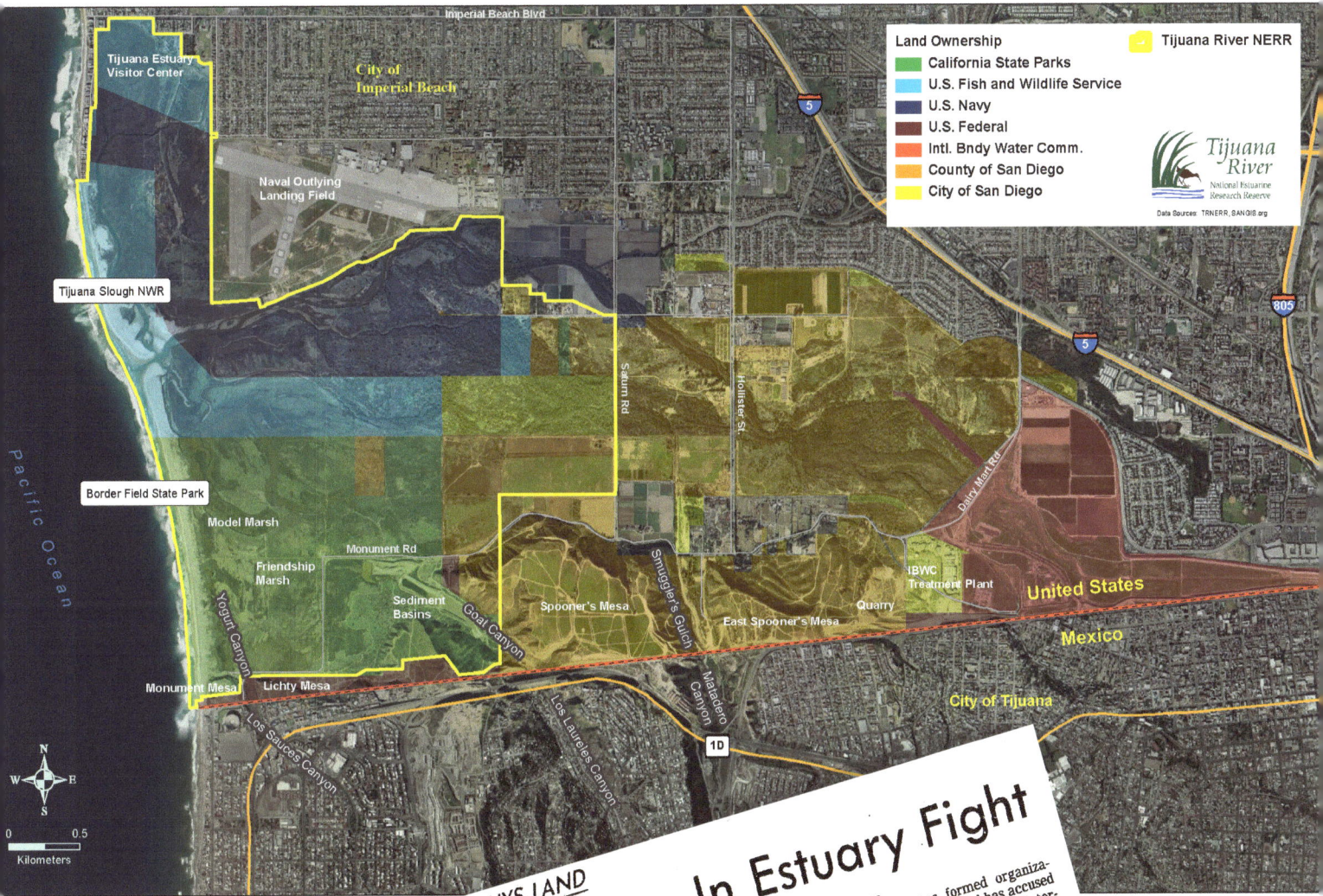

U.S. AGENCY BUYS LAND

Birds Winners In Estuary Fight

By JESUS RANGEL
Staff Writer, The San Diego Union

A misty fog set in as a flock of snowy egrets on its way south flew into the Tijuana Estuary south of Imperial Beach to feed. The quiet was broken only by the calls of the egrets and hundreds of other birds, the occasional scratchings of a squirrel and the rush of nearby ocean waves.

For almost 20 years, the tranquil, almost storybook marshland has been a noisy battleground between environmentalists who want to preserve it as a habitat for endangered wildlife and those who want to develop a marina and other commercial businesses.

The fate of the estuary has been the key issue in every municipal election since mid-1960 when the Imperial Beach City Council sold 126 acres of land to the Helix Imperial Harbour Development Corp. for the a marina.

The issue has divided the communities of Imperial Beach and South San Diego into two distinct and vocal factions.

Each side has formed organizations to raise money and has accused the other of acting in selfish interests.

But the battle appears to be over. Ted G. Lambron, president of the Helix Corp., announced on Christmas Eve that his stockholders did not have the will to fight any longer.

After a decade of fighting the federal and state governments, Lambron said he sold the company's 505 acres to the U.S. Fish and Wildlife Service.

The land now will become part of the National Wildlife Refuge system and will be managed primarily for the protection of endangered bird species inhabiting the marsh, including the California least tern and the light-footed clapper rail.

As far as the Fish and Wildlife Service and the environmentalists are concerned, the transaction killed any plans to commercially develop the estuary.

"There's no question that a marina or other type of commercial development is ruled out," said Ralph C. Pisapia, field office supervisor for the service.

It also would prevent the city of San Diego from running an outfall line from a possible proposed sewage treatment plant across the slough into the ocean, he said.

And it would prevent San Diego Gas & Electric Co. from building a power plant on about 200 acres of estuary land it owns, he added.

The environmentalists, led by Dr. Michael McCoy, a local veterinarian, and his wife Patricia, say they are elated at the sale.

They also are happy that the service will develop plans for the estuary with the help of local groups, including the Southwest Wetlands Interpretive Association, which they helped form.

The McCoys said they want to build a bio-museum within the estuary, patterned after the Arizona art Museum in Tucson

4
TIJUANA RIVER VALLEY

Spanning two nations and connecting mountains to sea and deserts to wetlands, the Tijuana River valley is a region of constant movement, exchange, and evolution. From Rodríguez Dam on the southern side of the city of Tijuana to the mouth of the Tijuana Estuary, the valley encompasses 42.5 km² (10,500 ac) and comprises a range of physical settings and habitat types.

Over millennia, the Tijuana River shaped and re-shaped its valley through the transport of water and sediment. In Mexico, the river corridor encompassed nearly the entire valley floor between alluvial terraces. In the United States, however, the valley supported a variety of habitat types outside of the river corridor, including extensive areas of alkali meadow complex flanking the river corridor, grassland and coastal sage scrub occupying slightly elevated areas to the north, vernal pools and perennial freshwater wetlands in depressions within the grassland and coastal sage scrub matrix, and the Tijuana Estuary on the western side of the valley. As one would expect given the region's semi-arid climate, xeric vegetation types such as coastal sage scrub were widespread, supporting numerous plants and animals typically associated with desert regions. However, ephemeral, seasonal, and perennial wetland complexes (riverine, estuarine, and palustrine) were also surprisingly prevalent, covering over 75% of the valley.

The earliest written accounts of the valley, from Spanish expeditions in May and June 1769, describe the valley as a "very great plain" with "everything well grass-grown with green grass" (Crespí and Brown 2001). Trees were virtually absent: Friar Juan Crespí noted that "no firewood or trees" were present in the lower valley, and Pilotín José Cañizares described how the lack of trees along the lower river forced his party to burn their tent poles for firewood to make tortillas (Cañizares et al. 1952, Crespí and Brown 2001). Other 19th century observers also remarked upon the lack of timber along the river and in the region in general (e.g., Croswell 1870, Bartlett 1963), and early General Land Office surveyors recorded extremely few bearing trees despite surveying multiple transects of the river (e.g., Pascoe 1869).

In this chapter we describe the habitat types that characterized the Tijuana River valley outside of the estuary and the river corridor (see map on pp. 68–69). Due to the somewhat less-defined nature of these habitat types relative to more prominent features such as the river and estuary, there was relatively little historical data available from which to reconstruct these portions of the valley, and thus there is greater uncertainty associated with our mapping and interpretation of these habitat types.

Homeward bound, we paused at the line long enough for a snap shot at the monument which marks the boundary between the two Republics; and to watch the sinking sun, a huge golden ball, reflecting a strange opalescent light upon the brown mesas and winding valleys of Mexico behind us and California before. And as we came up the cumbre, and saw ahead the blue line of the Pacific across the shadowy slopes, the peaks behind were rosy-purple—the lonely peaks of Mexico, whose garment's hem we had touched for one happy day.

—COLSON 1896

(Photo #FFP 836, "Four Stages – Tijuana in Background," ca. 1895, courtesy San Diego History Center)

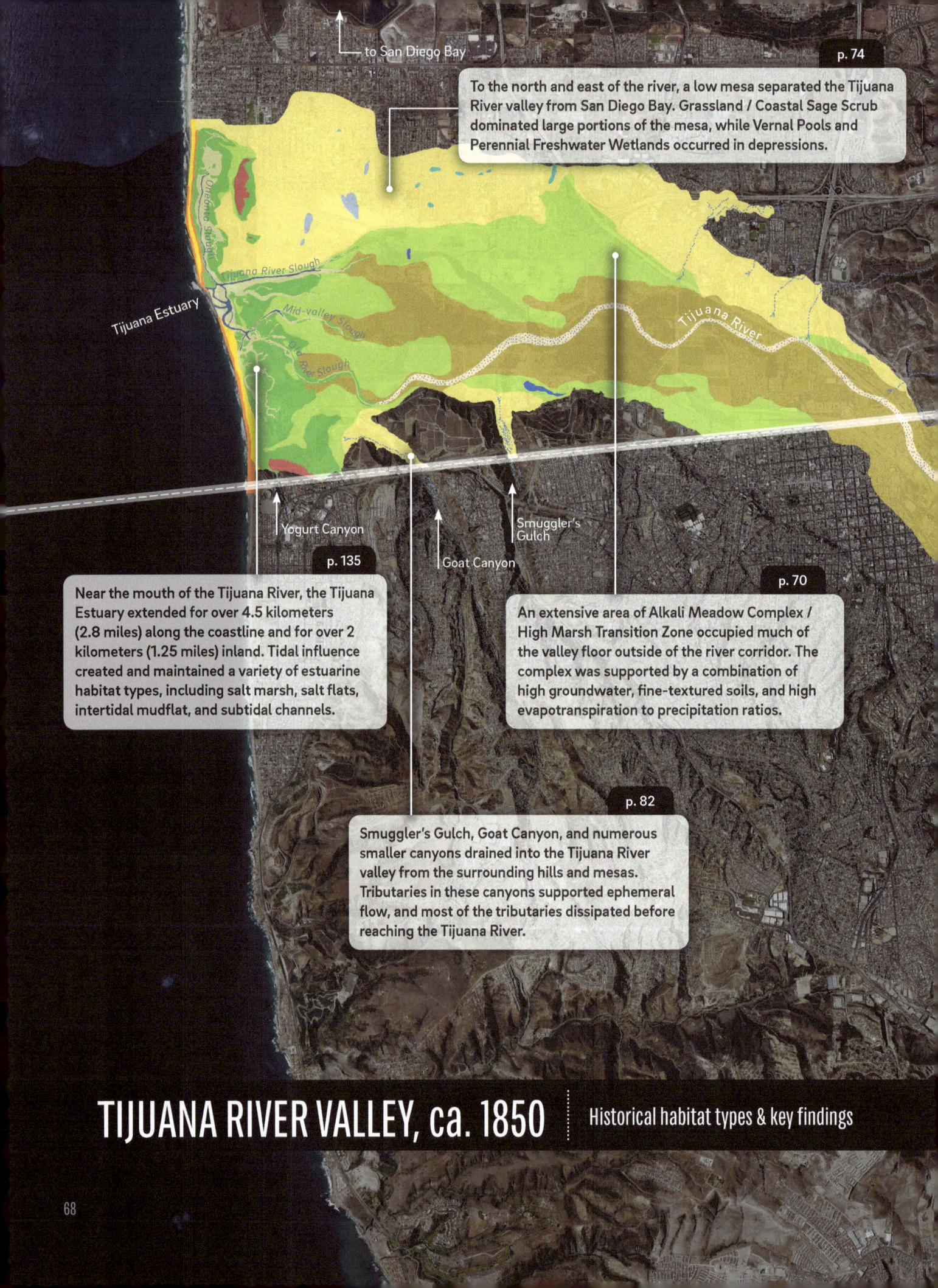

To the north and east of the river, a low mesa separated the Tijuana River valley from San Diego Bay. Grassland / Coastal Sage Scrub dominated large portions of the mesa, while Vernal Pools and Perennial Freshwater Wetlands occurred in depressions.

Oneonta Slough

Tijuana River Slough

Tijuana Estuary

Mid-valley Slough

Old River Slough

Tijuana River

Yogurt Canyon

Goat Canyon

Smuggler's Gulch

p. 135

Near the mouth of the Tijuana River, the Tijuana Estuary extended for over 4.5 kilometers (2.8 miles) along the coastline and for over 2 kilometers (1.25 miles) inland. Tidal influence created and maintained a variety of estuarine habitat types, including salt marsh, salt flats, intertidal mudflat, and subtidal channels.

p. 70

An extensive area of Alkali Meadow Complex / High Marsh Transition Zone occupied much of the valley floor outside of the river corridor. The complex was supported by a combination of high groundwater, fine-textured soils, and high evapotranspiration to precipitation ratios.

p. 82

Smuggler's Gulch, Goat Canyon, and numerous smaller canyons drained into the Tijuana River valley from the surrounding hills and mesas. Tributaries in these canyons supported ephemeral flow, and most of the tributaries dissipated before reaching the Tijuana River.

TIJUANA RIVER VALLEY, ca. 1850

Historical habitat types & key findings

p. 87

The Tijuana River occupied a wide river corridor dominated by river wash and riparian scrub vegetation. Flows were highly episodic, with little to no dry season flow in most reaches. Periodic flood events inundated much of the valley and caused major shifts in channel location.

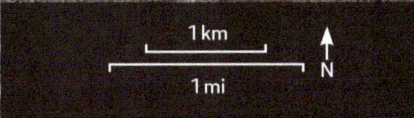

Legend

- Dune
- Beach
- Subtidal Water
- Mudflat/Sandflat
- Salt Flat / Open Water
- Salt Marsh
- Alkali Meadow Complex / High Marsh Transition Zone
- River Channel
- River Wash / Riparian Scrub
- Grassland / Coastal Sage Scrub
- Perennial Freshwater Wetland
- Pond
- Vernal Pool

1 km
1 mi
N

Rio Alamar

Cerro Colorado ▲

Matanuco Canyon →

Presa Rodríguez ↓

ALKALI MEADOW COMPLEX

An extensive alkali meadow complex flanked the river corridor in the United States.

Outside of the river corridor, historical documents suggest that the valley floor in the U.S. was dominated by an extensive alkali meadow complex. The complex was primarily comprised of saltgrass-dominated (*Distichlis spicata*) alkali meadows, though there is limited evidence for a small number of other wetland features such as perennial freshwater wetlands and ponds. Though not within the river corridor, the alkali meadow complex was subject to inundation during larger flood events. On the western side of the valley, the alkali meadow complex graded into the high marsh transition zone near the estuary (see pp. 160–65). In the U.S., the combined Alkali Meadow Complex / High Marsh Transition Zone occupied approximately 25% of the valley (7.5 km²/2.9 mi²). (Since it was not possible to distinguish Alkali Meadow Complex from High Marsh Transition Zone in the available historical data, they were mapped as a single habitat type.)

Though alkali meadows are a regionally rare habitat type today, they were a dominant wetland type historically in valley floor settings adjacent to large alluvial rivers throughout southern California, such as the Oxnard Plain (Beller et al. 2011), the Ballona Creek watershed (Dark et al. 2011), and the San Gabriel River watershed (Stein et al. 2007). An herbaceous, intermittently flooded vegetation type with a salt-tolerant plant community, alkali meadows occurred in areas characterized by fine-textured soils, high groundwater levels, and high evapotranspiration to precipitation ratios (Elmore et al. 2006).

Late 19th and early 20th century observers in the Tijuana River valley noted extensive stands of saltgrass, a dominant species in alkali meadows. One general description of the valley refers to "the salt grass meadows of Tia Juana valley" (*Pacific Rural Press* 1896), while another report notes that the coastal valleys of San Diego County "are frequently alkaline, and maintain a growth of 'salt grass' as the natural vegetation" (Hall 1888). In addition to salt grass, early records for the valley document numerous other plant species often associated with alkali meadows and other salt-tolerant plant communities, such as alkali sacaton (*Sporobolus airoides*) and spiny chloracantha (*Chloracantha spinosa*) (Table 4.1; Elmore et al. 2006). While it is possible that these observations reflect impacts from 19th century grazing in the river valley (see pp. 46–47), they are consistent with early descriptions of a "grass-grown" valley (Crespí and Brown 2001).

A few perennial wetlands and ponds also appear to have been present within the meadows of the river valley. For example, early maps of the valley depict what is apparently a large wetland feature near the mouth of Smuggler's Gulch, represented as ponds or as "low marshy land" (Gray 1849, Hardcastle and Gray 1850, Poole 1854, San Diego and Arizona Eastern Railway. ca. 1900; Fig. 4.1). Textual accounts also suggest the presence of additional wetland features within the matrix of the seasonally flooded alkali meadow: Pascoe (1869) described water "in some places standing on the surface" in the valley as late as September 1869, a year of slightly above-average rainfall. It is likely that additional wetland features were present in the valley but not captured in the historical record.

The presence of extensive alkali meadows and other wetland features is reflective of historical physical conditions on the valley floor, including fine-grained soils and high groundwater levels. In the 19th and early 20th centuries, groundwater was present within several feet of the surface in many areas of the valley (see pp. 52–55), providing a source of year-round water for facultative wetland plants such as saltgrass. This relationship was described by an early water resource report, which noted that in San Diego County coastal valleys such as Tijuana, "salt grass, yerba mansa, and swamp vegetation occupied open areas where the water table commonly stood within 5 feet of the surface" (Ellis and Lee 1919). The presence of alkaline-associated habitats also indicates the presence of at least moderate quantities of alkali in the soils, which is substantiated by early soil surveys describing the very fine sandy loams that characterized much of the valley floor outside of the river corridor as "covered with salt grass" and with alkali "in small quantities" (Storie and Carpenter 1923).

The presence of alkali meadows (as well as grasslands with assorted wetland features; see p. 78) likely explains the historical presence of wintering Sandhill Crane (*Grus canadensis*), which were observed in the valley during the winter and spring in the late 1800s (Bryant 1889). These areas—with low vegetation and shallow water—would have provided good roosting habitat with easy detection of predators. Cranes have long since been extirpated from the valley (Unitt et al. 2004). Alkali meadows also likely supported a large population of wandering skipper (*Panoquina errans*), a butterfly species of special concern that relies on saltgrass as its larval host plant. Now relatively rare, wandering skipper were once "probably more plentiful than any other butterfly" in the region during its flight period from June to October (Wright 1908). Though the species is generally considered to be limited to coastal marshes and is found in the Tijuana Estuary, it is possible that the species once also thrived further inland within the valley when alkali meadows were a major component of the valley floor. As noted by Longcore and Osborne (2015), a butterfly species' current range and habitat characteristics do not necessarily reflect its historical distribution. Recent research has found that the skippers do occupy coastal bluffs away from marshes (Greer 2014), and there is also at least one historical record of the species more than 45 miles from the coast (near Rialto; Hurd 1940, Essig Museum of Entomology record #455367) in an area that likely supported saltgrass meadows on the floor of the Santa Ana River valley.

Common name	Scientific name	Relevant excerpts	Year(s)	Notes
California sun cup	*Camissonia bistorta*	"Lower Tia Juana Valley"	1936	
Spiny chloracantha	*Chloracantha spinosa*	"San Ysidro," "near Mexican border," "San Diego near [Tia Juana]"	1902, 1918, 1937	
Spiny goldenbush	*Chloracantha spinosa var. spinosa*	"Lower Tia Juana Valley," "San Ysidro," "Tijuana River at Monument School"	1935, 1938	
California croton	*Croton californicus*	"Tia Juana," "Lower Tia Juana Valley"	1903, 1935	
Clustered tarweed	*Deinandra fasciculata*	"Tia Juana"	1903	
Saltgrass	*Distichlis spicata*	"Tia Juana"	1895	
San Diego marsh elder	*Iva hayesiana*	"Tia Juana"	1903	CNPS Rare Plant Rank 2B.2
Coulter's goldfields	*Lasthenia glabrata subsp. coulteri*	"Tia Juana Valley near ocean"	1938	CNPS Rare Plant Rank 1B.1
Alkali sacaton	*Sporobolus airoides*	"Tia Juana"	1903	

Table 4.1. Historical alkali meadow complex plants. A partial list of native plants that were historically present in the Tijuana River valley and may have been associated with the alkali meadow complex, drawn from pre-1950 herbarium records. Where available, relevant information about the locality where the specimens were collected is included. Species were selected for inclusion in the table based on a combination of the locality information provided in the herbarium records and known associations with alkali meadow or alkali sink habitat types. Species listed in the table may have occurred in other habitat types in the valley in addition to alkali meadow complex. Data were provided by the participants of the Consortium of California Herbaria.

Figure 4.1. (above and below) Wetlands on the edge of the valley floor. Mid-19th century maps document the presence of a large wetland feature, the "Well of San Antonio" on the south side of the valley, near the mouth Smuggler's Gulch. This feature is also noted on the ca. 1840 diseño as a "posa" [watering hole] (U.S. District Court California, Southern District ca. 1840; see p. 92) and falls within an area mapped as "low marshy land" by a ca. 1900 survey, approximately where the "main slough" of the marsh was denoted (San Diego and Arizona Eastern Railway. ca. 1900). *(Above: Poole 1854, courtesy The Bancroft Library; Below: Gray 1849, courtesy Coronado Public Library)*

CATTLE AND SALTGRASS

Livestock grazing was the predominant land use in the lower Tijuana River valley until well into the 19th century (see pp. 46–49). The prevalence of saltgrass in at least the U.S. portion of the river valley suggests that this plant may have been a significant component of the diet of cattle that grazed there.

Though there are no direct descriptions of saltgrass used for forage in the Tijuana River valley, saltgrass is recognized as valuable forage for cattle, sheep, and horses in many dryland areas because it remains green after many other grasses have dried during the dry season. It is also able to withstand relatively high levels of grazing and trampling (Hauser 2006, Skaradek 2010). Saltgrass was an important source of hay and pasture in many parts of the U.S. well into the 20th century, including the Atlantic Coast (Capen 1831, Sebold 1992, Skaradek 2010; Fig. 4.2), the South (Foster and Moran 1930, Chabreck 1968), and California (*Daily Alta California 1873; Pacific Rural Press* 1885; Carruthers 1912). It was recognized as particularly valuable in dryland regions or during dry periods, such as in Tulare County, where "during years of extreme drought this grass has been as a mine of wealth to the stock men of the valley" (*Pacific Rural Press 1873*).

Figure 4.2. Salt marsh hay. In this early 20th century photograph from Ipswich, Massachusetts, harvested "salt marsh hay" has been stacked on staddles to raise it above the tides while it dries. Cattle likely grazed on the saltgrass meadows of the Tijuana River valley into the 19th century. *(Photo by George Dexter, ca. 1900, Ipswich, MA, courtesy Gordon Harris)*

0.5 km
0.5 mi
N

GRASSLAND / COASTAL SAGE SCRUB ON LOW MESA

A matrix of grassland / coastal sage scrub dominated the low mesa on the north and east sides of the lower river valley.

When spring arrived and wild flowers bloomed over the fields, we were attracted more than ever to this spot. The cacti, poppies, the dainty baby blue eyes, yellow buttercups, and a pink low-growing flower that I never could find the name of, made a beautiful variegated carpet. For a background there were the taller flowers with pinkish white blooms, clumps of yellow blooming sour grass and the purple lupine with its sweet cloying perfume. The Indian paintbrush and shooting stars all combined in a riot of color.

— RICHARDS 2002

Figure 4.3. Bell's sparrow (*Artemisiospiza belli*) is a Coastal Sage Scrub specialist that is no longer found in the valley. (*Photo by Alan Schmierer, September 2009, Carrizo Plain*)

To the north and east of the Tijuana River corridor and the surrounding valley floor, a low mesa (known as the Nestor Terrace) separates the Tijuana River from the Otay River valley and San Diego Bay. The low mesa's elevation ranges from just a little above sea level near the estuary to over 18 m (60 ft) near Nestor and nearly 50 m (164 ft) on the eastern side of the mesa near San Ysidro, but has an average height of about 7.5 m (25 ft) above the valley floor (Shuirman and Slosson 1992). A low point in the mesa on the northeast side of the valley marks the passage through which the Tijuana River overflowed into San Diego Bay on several documented occasions (see pp. 108–10). The historical synthesis map includes those portions of the mesa within the alluvial plain and watershed of the Tijuana River.

The mesa was regularly described by early observers, who noted a "low table land" (Pascoe 1869) with "rolling" (Parry 1849) and "gently undulating" (Gray 1849) topography. It was dominated by a mixture of Grassland and Coastal Sage Scrub, which occupied over 9 km² (3.5 mi²), or one third of the river valley in the U.S. Early observers described the vegetation cover as "heath like" (likely a reference to coastal scrub; Gray 1849), "low brush" (Croswell 1870, Stephens 1912), or "scanty growth of desert vegetation" (Knox 1934).

As with the valley floor and the river corridor, the mesa was effectively devoid of large trees (Gray 1849, Parry 1849). Sagebrush (*Artemisia* spp.) was a dominant plant species on the mesa: botanist Charles C. Parry (1849), for instance, described the area around south San Diego Bay as an "*Artemisia* plain." The 1852 T-sheet also provides evidence for sagebrush on the mesa (Fig. 4.4). Though the map only extends two kilometers inland, a note in its margin states that "all that portion of this sheet represented as being covered with grass is also covered with low artemisia, or wild sage bushes" (Harrison 1852). A variety of other xeric-adapted plants were found on the mesa, including bladderpod (*Peritoma arborea*), clustered tarweed (*Deinandra fasciculata*), at least "5 species of the Cactus family" (Parry 1849), and various species of wildflowers (Richards 2002; Table 4.2). By the mid-19th century, the introduced iceplant (*Mesembryanthemum crystallinum*) was also common, with "extensive patches... so thickly beset with watery glands as to make the feet wringing wet in walking over it" (Parry 1849, Parry [1849]2014). (Iceplant may have been introduced to California via ships' ballast as early as the 16th century [Randall 2000].)

The Grassland / Coastal Sage Scrub on the low mesa served as habitat to numerous resident or migratory wildlife species that were recorded in the lower Tijuana River valley. These include coachwhip (*Masticophis flagellum*), common kingsnake (*Lampropeltis getula*), California glossy snake (*Arizona occidentalis*), rosy boa (*Lichanura trivirgata*), common side-blotched lizard (*Uta stansburiana*), Southern California legless lizard (*Anniella stebbinsi*; Papenfuss and Parham 2013), loggerhead shrike (*Lanius ludovicianus*), wrentit (*Chamaea fasciata*), vesper sparrow (*Pooecetes gramineus*), Bell's sparrow (*Artemisiospiza belli*; Fig. 4.3) San Diego pocket mouse (*Perognathus fallax*), and Pacific kangaroo rat (*Dipodomys agilis*; species data obtained from pre-1950 records on Vertnet and Arctos). The federally threatened coastal California gnatcatcher (*Polioptila californica californica*), which is restricted to coastal sage scrub habitats in the U.S. portion of its range, was also recorded historically within the valley as early as 1917 (VertNet specimen record, Atwood and Bontrager 2001). The valley still supports a core population of the coastal California gnatcatcher, despite the almost complete loss of Grassland / Coastal Sage Scrub on the low mesa (Mock 2004; see chapter 7). Other sage scrub specialists, including California glossy snake, red diamond rattlesnake, and Bell's sparrow have likely been locally extirpated (Fisher and Case 2000, Unitt et al. 2004).

Figure 4.4. (right) "Heath like" land on the low mesa. The 1849 Boundary Commission map describes the low mesa north of the valley floor as "Heath like; Gently undulating and destitute of Trees." Heath is not a commonly used descriptor of California vegetation, but refers to low-growing shrublands. In this instance "heath like" describes coastal sage scrub habitat, dominated by sagebrush (*Artemisia* spp.). *(Gray 1849, courtesy Coronado Public Library)*

> Yesterday I took up traps in the river bottom and set them on the divide between the Tijuana River and San Diego Bay about a mile back from the ocean, in low brush and at the edge of a barley field. Results very poor; 60 traps caught four immature *Perognathus fallax* [San Diego pocket mouse].
>
> — STEPHENS 1912

Figure 4.5. Coastal sage scrub, Los Angeles County, March 2010. *Photo by Michael O'Brien.*

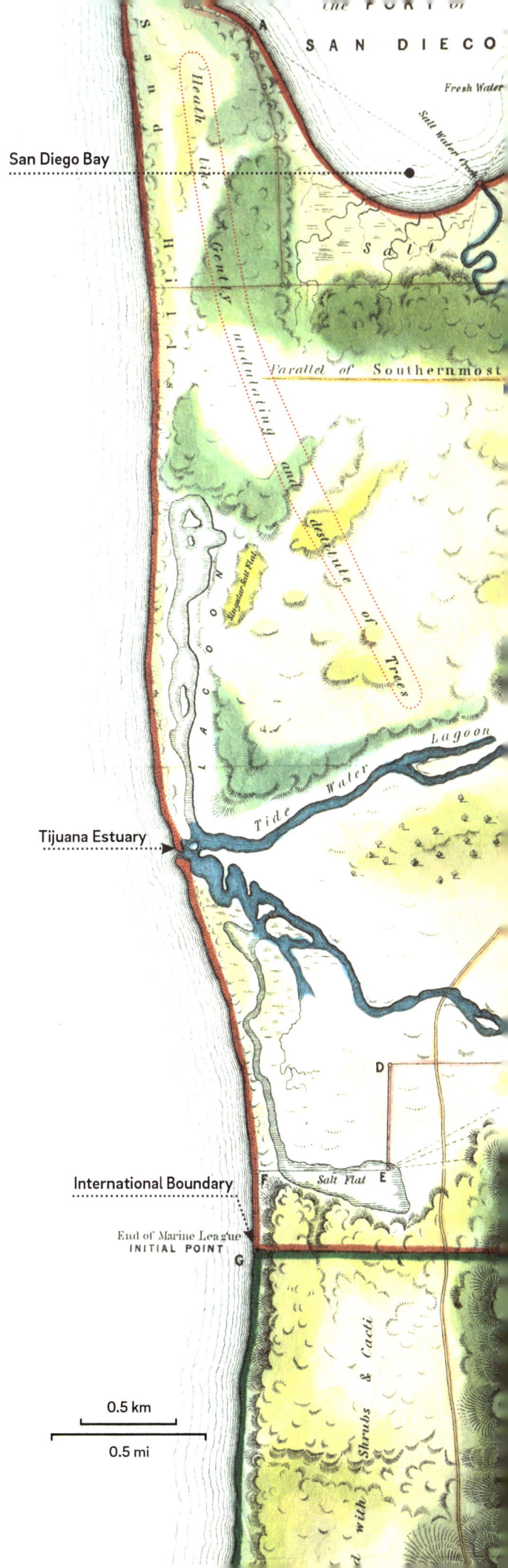

Table 4.2. Partial list of native plants that were historically present in and around the Tijuana River valley, drawn from early accounts and pre-1950 herbarium records. Many of these may have been found in the Grassland / Coastal Sage Scrub matrix on the low mesa north and east of the lower Tijuana River. Where available, relevant information about the locality where the specimens were collected is included. Species were selected for inclusion in the table based on a combination of the locality information provided in the herbarium records and known associations with Grassland or Coastal Sage Scrub habitat types. Species listed in the table may have occurred in other habitat types in the valley in addition to Grassland / Coastal Sage Scrub. All data were provided by the participants of the Consortium of California Herbaria, except for additional citations listed in the Notes field.

Common name	Scientific name	Relevant excerpts	Year(s)	Notes
Heermann's lotus	*Acmispon heermannii*	"Hills at Tia Juana," "Tia Juana river"	1903	
Strigose lotus	*Acmispon strigosus*	"San Ysidro, Near; on border"	1941	
California adolphia	*Adolphia californica*	"Hills near Tia Juana," "San Ysidro"	1903, 1931, 1936	CNPS Rare Plant Rank 1B.1
San Diego bursage	*Ambrosia chenopodiifolia*	"Tia Juana," "American side Tia Juana," "Found abundantly in the Tijuana valley, north of the U.S. boundary" (Orcutt)	1886, 1902, 1903, 1931	CNPS Rare Plant Rank 1B.1; Orcutt 1886a
Dwarf coastweed	*Amblyopappus pusillus*	"Tia Juana," "Imperial Beach"	1913, 1923, 1938	
San Diego ragweed	*Ambrosia pumila*	"Nearby [Camp Rough & Ready] in loamy waste places"	1849	CNPS Rare Plant Rank 2B.1; Parry [1849]2014
Aphanisma	*Aphanisma blitoides*	"Hills at Tia Juana, U.S. side of line"	1903	CNPS Rare Plant Rank 2B.2
Sagebrush	*Artemisia spp.*	"*Artemisia* plain," "Lower Tia Juana Valley"	1849, 1935	Parry 1849
Dwarf white milkvetch	*Astragalus didymocarpus*	"Near Tia Juana"	1903	
Santa Barbara milkvetch	*Astragalus trichopodus*	"Camp Riley," "Imperial Beach," "Tia Juana, American side"	1849, 1902, 1923, 1938	Parry [1849]2014
Fourwing saltbush	*Atriplex canescens*	"Imperial Beach," "Lower Tia Juana River"	1919, 1935	
Golden spined cereus	*Bergerocactus emoryi*	"Another beautiful columnar species was also seen growing in dense patches"	1849	CNPS Rare Plant Rank 1B.2; Parry [1849]2014
Seaside calandrinia	*Calandrinia maritima*	"Tia Juana," "Hills at Tia Juana"	1903	CNPS Rare Plant Rank 4.2
California sun cup	*Camissoniopsis bistorta*	"Camp Riley"	1849	Parry [1849]2014
Canyon clarkia	*Clarkia epilobioides*	"Tia Juana"	1913	
Ramona clarkia	*Clarkia similis*	"Tia Juana"	1913	
Clustered tarweed	*Deinandra fasciculata*	"Tia Juana," "Imperial Beach"	1903, 1919	
Parry's larkspur	*Delphinium parryi*	"Low slope about saline flat, ¼ mile from beach; Imperial Beach," "Hills at Tia Juana"	1903, 1923	
Orcutt's bird's-beak	*Dicranostegia orcuttiana*	"Tijuana, a little within the border of Lower California"	1886	CNPS Rare Plant Rank 1B.1; Gray 1886
Padre's shooting star	*Dodecatheon clevelandii*	"Imperial Beach"	1937, 1938, 1950	
Blochman's liveforever	*Dudleya blochmaniae*	"Imperial Beach"	1938	
Fingertips	*Dudleya edulis*	"Imperial Beach"	1937	
Lance-leaved liveforever	*Dudleya lanceolata*	"Imperial Beach"	1937	
Whispering bells	*Emmenanthe penduliflora*	"Hills at Tia Juana, U.S. side of line"	1903	
California ephedra	*Ephedra californica*	"Tia Juana," "Hills at Tia Juana, U.S. side of line"	1903, 1913	
California buckwheat	*Eriogonum fasciculatum var. fasciculatum*	"Tia Juana"	1903	
Golden yarrow	*Eriophyllum confertiflorum*	"Tia Juana"	1895, 1903, 1913	
California poppy	*Eschscholzia californica*	"Tia Juana"	1913	

Common name	Scientific name	Relevant excerpts	Year(s)	Notes
California primrose	*Eulobus californicus*	"Camp Riley"	1849[7]	Parry [1849]2014
Warty spurge	*Euphorbia spathulata*	"Hills at Tia Juana"	1903	
San Diego barrel cactus	*Ferocactus viridescens*	"[Found] on dry hard soil near the sea"	1849	CNPS Rare Plant Rank 1B.1; Parry [1849]2014
Narrowleaf bedstraw	*Galium angustifolium subsp. angustifolium*	"1 mi sw Otay (on road to Tijuana)"	1931	
Graceful bedstraw	*Galium porrigens*	"Tia Juana"	1903	
Shaggyfruit pepperweed	*Lepidium lasiocarpum*	"Imperial Beach"	1938	
Robinson's pepper grass	*Lepidium virginicum ssp. menziesii*	"Tia Juana"	1895	
Variable linanthus	*Leptosiphon parviflorus*	"Imperial Beach, low slope about saline flat"	1923	
Deerweed	*Lotus scoparius*	"Tia Juana," "Hills at Tia Juana, U.S. side of line"	1903	
Anderson's desert thorn	*Lycium andersonii*	"Hills at Tia Juana, U.S. side of line"	1903	
Chaparral mallow	*Malacothamnus fasciculatus*	"Lower Tia Juana Valley"	1935	
Laurel sumac	*Malosma laurina*	"Imperial Beach"	1936	
Small flowered melica	*Melica imperfecta*	"Imperial Beach"	1919	
Yellow blazing star	*Mentzelia affinis*	"Hills at Tia Juana, U.S. side of line"	1903	
California four o'clock	*Mirabilis laevis var. crassifolia*	"Hills at Tia Juana"	1903	
Lineleaf whitepuff	*Oligomeris linifolia*	"Tia Juana"	1903	
False rosinweed	*Osmadenia tenella*	"Camp Riley," "1 mi sw of Otay - Road to Tijuana"	1849, 1931	Parry [1849]2014
Bladderpod	*Peritoma arborea*	"Camp Riley"	1849	Parry [1849]2014
Common phacelia	*Phacelia distans*	"Hills at Tia Juana, U.S. side of line," "Imperial Beach"	1903, 1937	
Branching phacelia	*Phacelia ramosissima*	"Tia Juana"	1895	
Creamcups	*Platystemon californicus*	"Imperial Beach"	1939	
Redberry buckthorn	*Rhamnus crocea*	"Dry hillsides at Tijuana"	1903	
Bush senecio	*Senecio flaccidus var. douglasii*	"Tia Juana"	1903a,b	
Jojoba	*Simmondsia chinensis*	"About ¼ mile back from Imperial Beach," "1 mi SW of Otay, on the road to Tijuana," "San Ysidro," "Imperial Beach"	1923, 1931, 1936, 1938	
Blue eyed grass	*Sisyrinchium bellum*	"1 mi sw of Otay (on the road to Tijuana"	1931	
Douglas' nightshade	*Solanum douglasii*	"Hills at Tia Juana, U.S. side of line"	1903	
Desert needle grass	*Stipa speciosa*	"Tia Juana," "Hills at Tia Juana, U.S. side of line"	1903	
San Diego County viguiera	*Viguiera laciniata*	"Tia Juana," "American side Tia Juana"	1902, 1913	CNPS Rare Plant Rank 4.2

VERNAL POOLS & PERENNIAL FRESHWATER WETLANDS

Vernal pools and perennial freshwater wetlands occupied depressions within the surrounding grassland / coastal sage scrub matrix on the low mesa.

Vernal pools were common on coastal marine terraces in San Diego County and northern Baja California (Barnes 1879, Purer 1939, Cox 1984, Bauder and McMillan 1998). They occurred in areas with a relatively impermeable subsoil layer, or "hardpan," which causes water to pool on the surface, and were often associated with mima mound fields (also referred to as "hog wallows"), which are characterized by alternating low mounds and shallow depressions (Keeler-Wolf 1998, Johnson and Burnham 2012; Figs. 4.8–4.9, p. 80). The naturalist Charles Russell Orcutt, for instance, described the prevalence of vernal pools on mesas in the San Diego region based on his observations from the spring of 1884:

> Especially on our mesas were to be found thousands of miniature lagoons of large or small dimensions. The surface geology of large portions of these mesas is characterized by innumerable hillocks, or small, mound-like formations, rising from one to four feet above the intervening depressions, and ranging from ten to fifty feet in diameter. They are generally nearly circular, though often irregular; and the depressions contain in stony places, accumulations of cobblestones. These innumerable hollows were quickly filled by the persistent rains... [the water] gradually disappeared by evaporation (Orcutt 1887).

There is some indication that vernal pools, as well as perennial freshwater wetlands, occurred in depressions within the grassland / coastal sage scrub matrix on the low mesa north of the Tijuana River (Figs. 4.6–4.7). The topography and soil properties on many parts of the low mesa satisfied the conditions most commonly associated with vernal pool formation in this region: much of the area was "underlain by hardpanlike sediments," and the largest of the pools occurred in areas described as having a "hog-wallow relief" (Fig. 4.9; Storie and Carpenter 1923, 1930). Parry (1849), for instance, observed "saline depressions" just north of the valley, on the south side of San Diego Bay (vernal pools can occur on soils ranging from acidic to alkaline; Barbour et al. 2007, Bauder et al. 2011). In sum, approximately 121,000 m² (30 acres) of historical Vernal Pools and Perennial Freshwater Wetlands were documented on the low mesa, representing <1% of the historical land cover in the U.S. portion of the valley. However, small and intermittent wetlands were likely overlooked in historical surveys, so these habitat types are probably underrepresented in the historical synthesis mapping.

Vernal pools typically support a variety of widespread aquatic species as well as a number of specialized and endemic plant species not found in other habitat types (Keeley and Zedler 1996). Several early 20th century records from the lower Tijuana River valley document the presence of plant species commonly found in or associated with vernal pools (Table 4.3). Orcutt (1887) describes a number of additional plant species observed within vernal pools around San Diego, including California water starwort (*Callitriche marginata*), American pillwort (*Pilularia americana*), water pygmyweed (*Crassula aquatica*), *Elatine* sp., *Isoetes* sp., flatface calicoflower (*Downingia pulchella*), and Otay Mesa mint (*Pogogyne nudiuscula*).

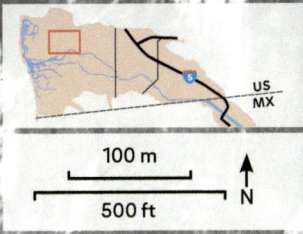

Figure 4.6. Remnant wetland features. The 1928 aerial photos show signatures of several remnant wetlands on the low mesa, though the area has clearly been disturbed by this time. *(San Diego County 1928)*

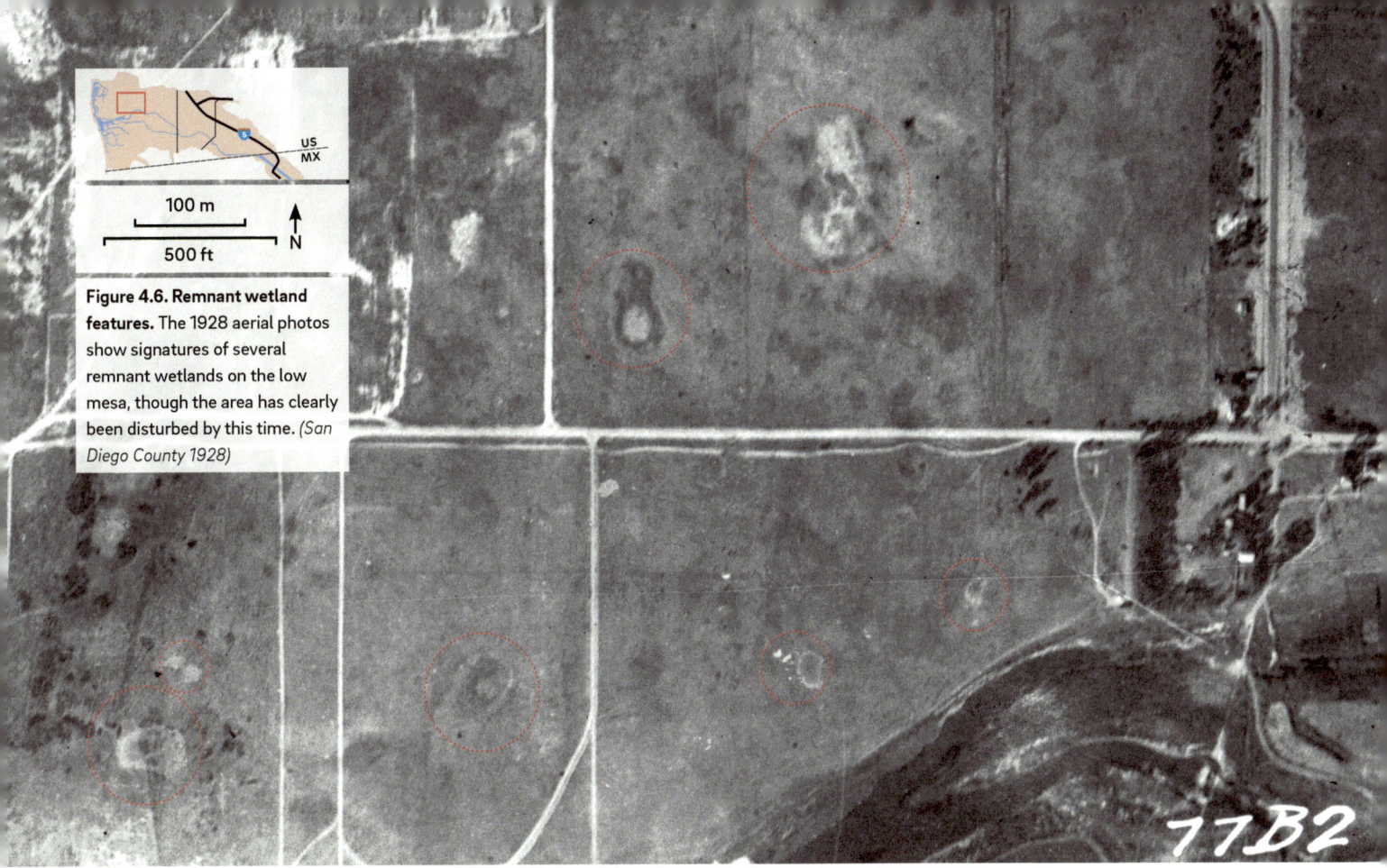

100 m

500 ft

N

77B2

Figure 4.7. Intermittent wetlands (vernal pools) and perennial freshwater wetlands on the low mesa north of the Tijuana River in 1941. *(USGS 1943)*

500 m

1,000 ft

N

Imperial Beach

PACIFIC

LANDING FIELD

LANDING FIELD

BM 19

T. 18 S.
T. 19 S.

TIA

Figure 4.8. "Hog wallows" on Otay Mesa. The mesa on the north side of the valley was characterized by mounded hog wallow topography. Vernal pools formed in the low spots between mounds. Individual mounds are easily visible in 1928 aerial photographs. *(San Diego County 1928)*

(not in mapped area)

US
MX

100 m

500 ft

N

Figure 4.9. "Hog wallows on Mesa back of San Diego," 1905. Though the exact location of the photo is uncertain, the undulating, mima mound topography is consistent with historical descriptions of portions of the low mesa on the north side of the river valley. *(Photo #25552, Mendenhall Collection, courtesy USGS Denver Library Photographic Collection)*

Table 4.3. Historical vernal pool plants. A partial list of native plants that may have been present historically in and around vernal pools in the lower Tijuana River valley, drawn from pre-1950 herbarium records. The table includes species that often occur in or near vernal pools and were recorded within the mapped area of the lower river valley. Many other vernal pool species have been documented in areas outside of the mapped study extent, but were not included in the table. Species listed in the table may have occurred in other habitat types in the valley in addition to vernal pools. Data provided by the participants of the Consortium of California Herbaria.

Common name	Scientific name	Relevant excerpts	Year(s)
Beardless wild rye	*Elymus triticoides*[2]	"Tia Juana"	1903
Toad rush	*Juncus bufonius var. occidentalis*[2]	"Imperial Beach [Saline flat back of sand dunes]"	1923
Common rock cress	*Planodes virginicum*[1,4]	"Tia Juana," "Dried pool above Tijuana, on U.S. side"	1938, 1939
Blue eyed grass	*Sisyrinchium bellum*[3]	"1 mi sw of Otay (on the road to Tijuana)"	1931

[1] *"Vascular plants that occur within pool basins and that are largely restricted to vernal pools within the study area of coastal California. These are the plants which are indicators of vernal pools"* (Zedler 1987)

[2] *"Vascular plants that are found in vernal pools, but that are more common in other aquatic, marsh, or seepage areas"* (Zedler 1987)

[3] *"Vascular plants that are common near vernal pools but which usually do not occur in the pool basin"* (Zedler 1987)

[4] *Record from mesa south of Tijuana River valley*

CANYONS & TRIBUTARIES

Ephemeral tributaries, flanked by alluvial scrub vegetation, drained into the lower river valley through side canyons.

Ephemeral tributaries drained into the lower Tijuana River valley through numerous canyons to the south and east. From the mesas to the south of the valley, streams originating in present-day Tijuana drained through Smuggler's Gulch, Goat Canyon, and the much smaller Yogurt Canyon (Fig. 4.10). Around present-day San Ysidro, at least five small tributaries drained into the lower valley from canyons to the east (Fig. 4.11). Most or all of these tributaries dissipated before reaching the river corridor. Though not shown in the historical synthesis map, there were also "numerous small arroyos" on the mesa between the Tijuana River valley and San Diego Bay (Adams 1928). In addition, numerous small tributaries drained into the river valley in Mexico, but were not included in the historical synthesis map due to the lack of an early, comprehensive mapping source (Rankin 1909, U.S. Engineer Office 1937; Fig. 4.12).

Given their small size and remoteness, relatively little early information was recovered on these tributaries from the historical record. However, available sources indicate that the tributaries to the south and east of the lower river valley would likely have been dry for much of the year, with only intermittent or ephemeral streamflow (Pascoe 1869, San Diego and Arizona Eastern Railway ca. 1900, USGS 1904). This is corroborated by textual accounts of streamflow: while surveying the area in the summer of 1869, for instance, GLO surveyor Pascoe (1869) noted "dry bed of creek" in Goat Canyon (September 3) and "dry bed of ravine" in one of the canyons east of San Ysidro (August 23). In some of the canyons, however, springs may have given rise to limited reaches with perennial flow or standing water. For example, a traveler passing through Goat Canyon in April 1904 described a "spring of pure, cold water that bubbles up from the ground," though the exact location of the spring is unclear (*Los Angeles Herald* 1904).

Early 20th century sources show braided channel segments along portions of the tributaries in Smuggler's Canyon and Goat Canyon, in contrast to the single-threaded morphology of tributaries in the smaller drainages to the east (Fig. 4.13). Sediment transported from Smuggler's Gulch and Goat Canyon during floods created alluvial fans that extended a short ways out onto the valley floor, potentially influencing channel migration patterns (Ellis and Lee 1919, Storie and Carpenter 1923; see p. 116–17).

Vegetation in the tributary canyons was likely dominated by alluvial scrub, a type of coastal sage scrub that occurs on alluvial fans and ephemeral floodplains (Fig. 4.14; see also Fig. 4.10; Smith 1980). In addition to other plant species characteristic of the Grassland / Coastal Sage Scrub habitat type, the alluvial scrub in the canyons included relatively large evergreen shrubs characteristic of chaparral communities, such as *Rhus* spp. In Goat Canyon, for instance, Stephens (1908) noted that there was "considerable brush, mostly Rhus, but there are also small patches of cactuses intermingled ('Trinas' and 'chollus' two sp. of the latter)." Parry (1849) reported that *Echinocactus* sp. was prevalent in one of the canyons draining the southern mesas. In addition, a pollen and macrofloral analysis in Goat Canyon indicated that the vegetation composition at the time of Spanish colonization likely included sagebrush (*Artemisia* sp.), other Asteraceae species, *Toxicodendron* sp., saltbush (*Atriplex* sp.), other Amaranthaceae species, cholla (*Cylindropuntia* sp.), buckwheat (*Eriogonum* sp.), grasses, chamise (*Adenostoma fasciculatum*), toyon (*Heteromeles arbutifolia*), globemallow (*Sphaeralcea* sp.), acacia (*Acacia* sp.), ash (*Fraxinus* sp.), and cattail (*Typha latifolia*; Cummings et al. 2004).

Figure 4.10. Braided channels and alluvial scrub in the valley's side canyons. This 1953 photograph, looking west over the mesas on the south side of the river valley, shows the two largest canyons and tributaries on the south side of the valley: Smuggler's Gulch (foreground) and Goat Canyon (background). The braided channel morphology in Smuggler's Gulch is clearly visible. Vegetation cover within the canyons is dominated by alluvial scrub. *(Compañía Mexicana Aerofoto, 1953, "Ciudad de Tijuana," Photo #FAO-01-10139, courtesy Fondo Aerofotográphico Acervo Histórico Fundación ICA, A.C.)*

Figure. 4.11. A tributary channel on the north side of the valley. As can be seen in this 1928 aerial photograph, an intermittent stream channel cut through the town of San Ysidro on the valley's low mesa. *(San Diego County 1928)*

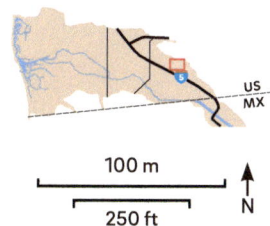

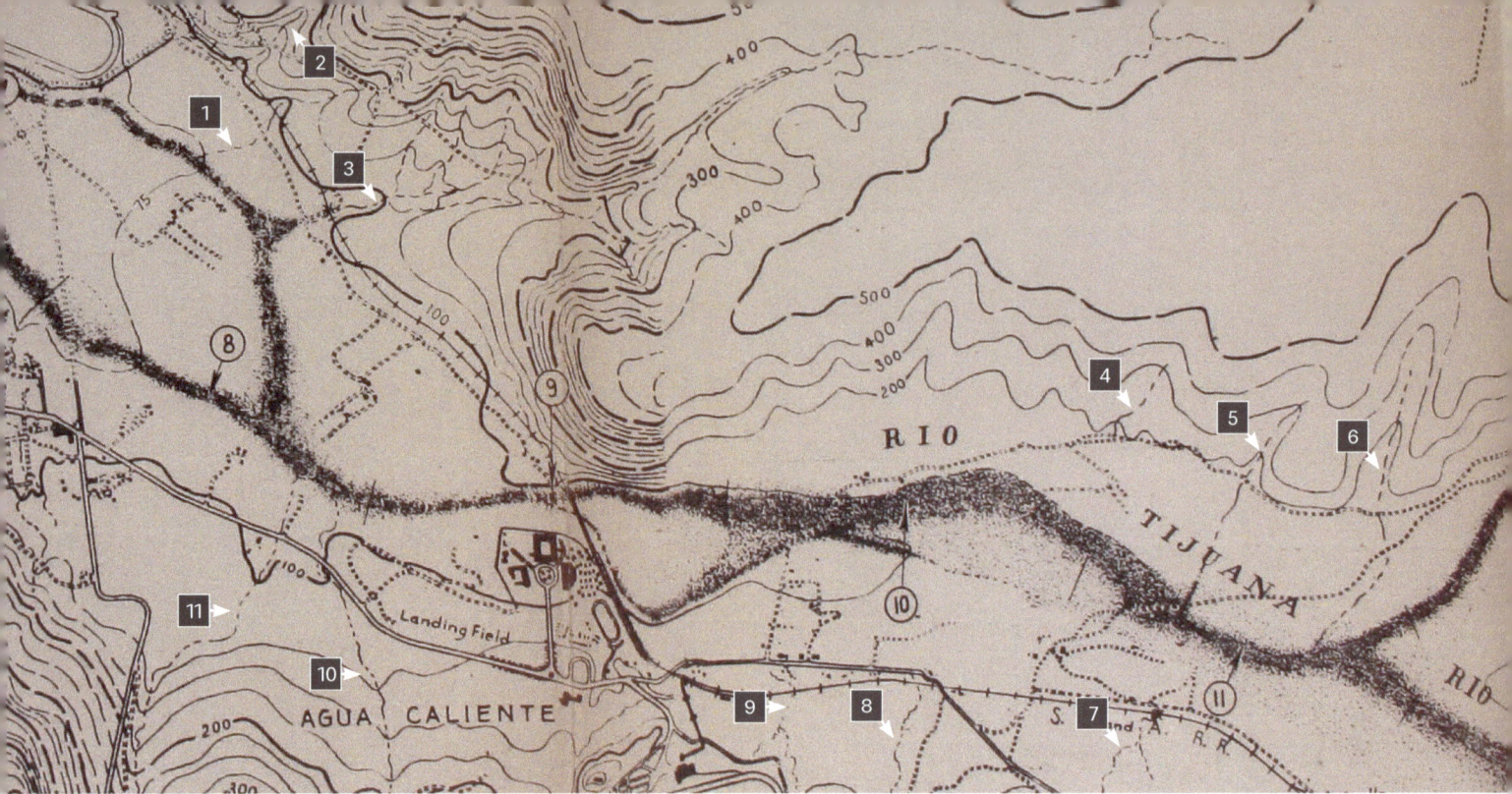

Figure 4.12. Tributaries in Mexico, 1953. At least eleven small tributaries drained into the Tijuana River valley in Mexico downstream of present-day Rodríguez Dam. Unlike tributaries in the U.S. side of the valley, early 20th century maps show many of the tributaries in Mexico extending into the river corridor and even connecting to the river channel (e.g., #3, #5, #6, #10, and #11). Note that the hand-drawn numbers original to the map label river segments, not tributaries. *(Cruse 1937, courtesy Water Resources Collections and Archives, UC Riverside)*

Figure 4.13. The braided channel morphology of the tributary in Smuggler's Gulch is clearly visible in this 1928 aerial photograph. *(San Diego County 1928)*

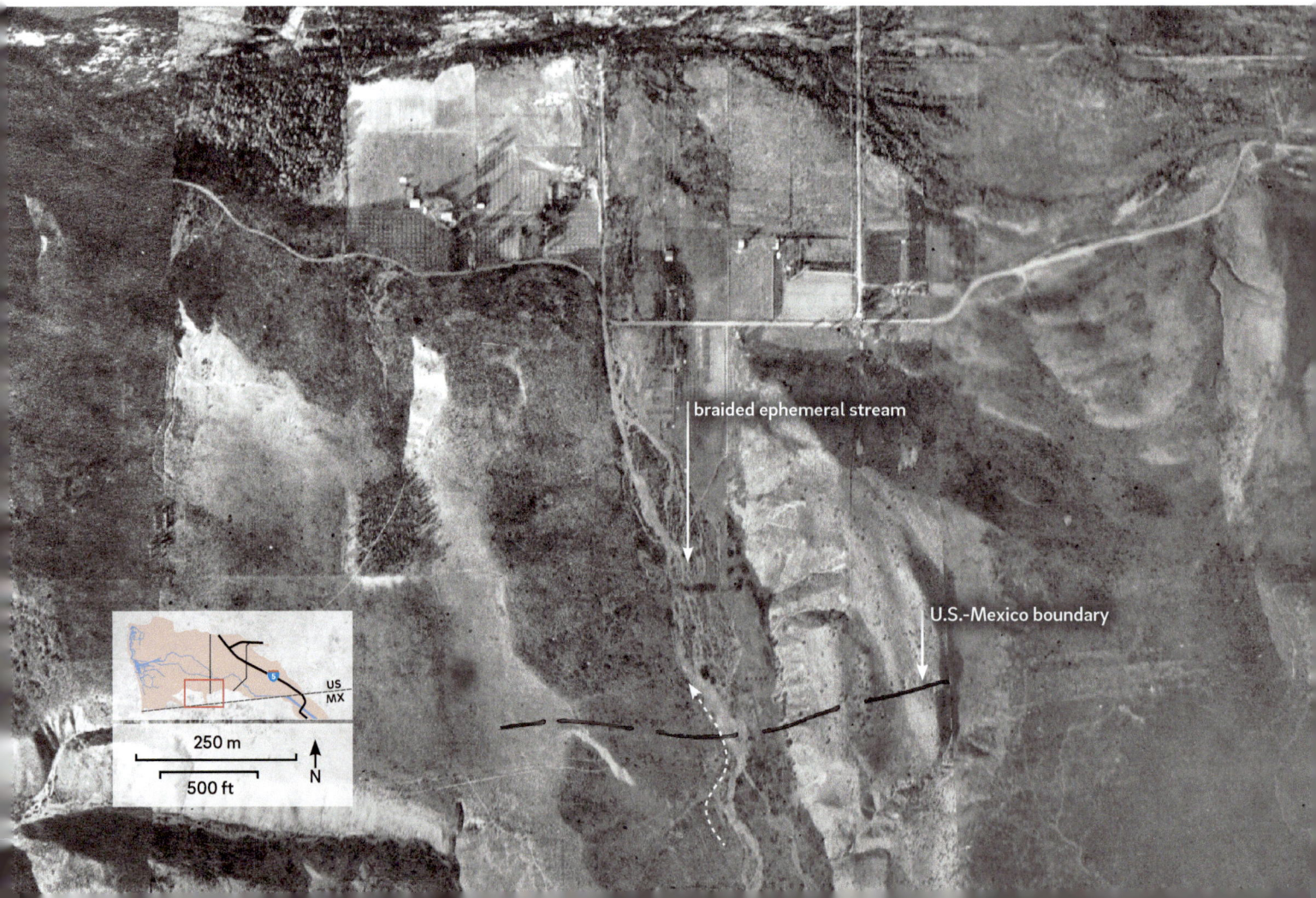

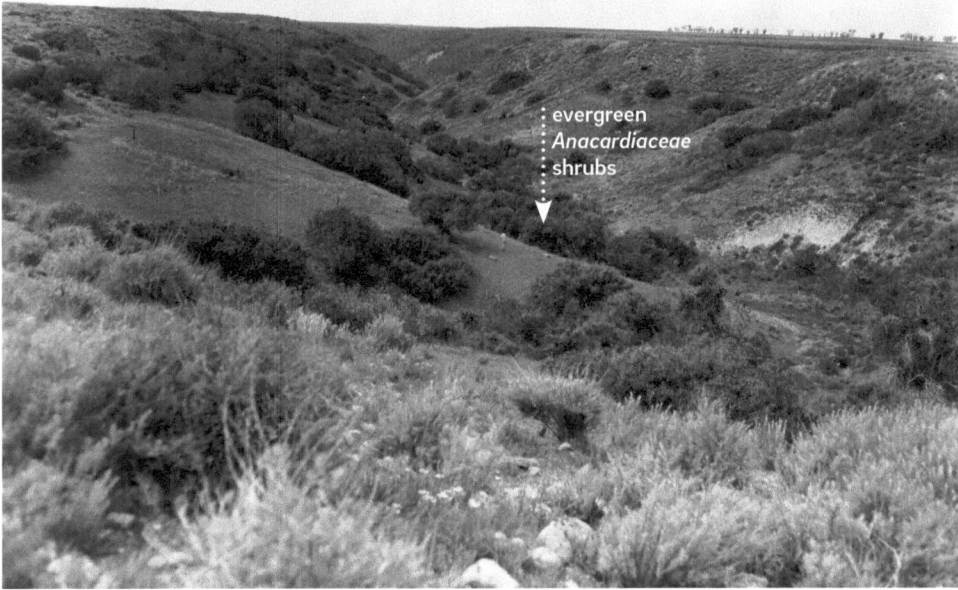

Fig. 4.14. Alluvial scrub in a small canyon on the north side of the valley. This undated photograph, labeled "San Ysidro tributary," shows one of the canyons draining towards the Tijuana River from the northeast. Sparse grassland / coastal sage scrub vegetation is visible on the mesas and canyon walls, with alluvial scrub vegetation on the canyon floor. The large shrubs appear to be evergreen species in the sumac family (*Anacardiaceae*), likely *Rhus integrifolia* or *Malosma laurina* (B. O'Brien, personal communication). *(Photo #OP 17440-162, courtesy San Diego History Center)*

Table 4.4. Historical plants of the Tijuana River valley's side canyons. A partial list of native plants that may have been present historically in Smuggler's Gulch, Goat Canyon, and other smaller canyons south of the lower Tijuana River, drawn from pre-1950 herbarium records. Vegetation communities within the canyons south of the Tijuana River likely included many of the species found within the Grassland / Coastal Sage Scrub habitat type on the low mesa north and east of the river (see Table 4.2), but species were only listed here if the locality information provided in the herbarium records supports their presence in this portion of the valley. (Species documented in the canyons east of the valley were not differentiated from grassland and coastal sage scrub species on the low mesa, and are included in Table 4.2.) Species listed in the table may have occurred in other habitat types in the valley in addition to Grassland / Coastal Sage Scrub. Data were provided by the participants of the Consortium of California Herbaria.

Common name	Scientific name	Relevant excerpts	Year(s)	Notes
Southern California milkvetch	*Astragalus trichopodus* var. *lonchus*	"Coones Ranch Mexican Boundary about 2 miles from the ocean"	1949	
Golden spined cereus	*Bergerocactus emoryi*	"American side of the boundry [sic]," "South side of Tia Juana Valley, bluff about 1 miles from ocean"	1903, 1940	CNPS Rare Plant Rank 2B.2
Spiny goldenbush	*Chloracantha spinosa* var. *spinosa*	"Tijuana River at Monument School"	1938	
Canyon clarkia	*Clarkia epilobioides*	"Hillside on so. Side of Tia Juana Valley, about one mile from ocean"	1937	
Chinese houses	*Collinsia heterophylla*	"Near Monument School"	1936	
Parry's larkspur	*Delphinium parryi*	"Near Monument School"	1936	
Lance-leaved liveforever	*Dudleya lanceolata*	"Bluff ca. 1 mile from ocean on south side of Tia Juana Valley"	1940	
Blunt leaved lupine	*Lupinus truncatus*	"Near Monument School"	1936	
Baby blue eyes	*Nemophila menziesii* var. *integrifolia*	"Near Monument School"	1936	
Rillita pellitory	*Parietaria hespera* var. *hespera*	"Hillside on south side of Tia Juana Valley, about one mile from the ocean"	1937	
Creamcups	*Platystemon californicus*	"Hillside above Tia Juana, U.S. side," "Near Monument School"	1936, 1938	

If one should fly above the beaches about a mile and a half north of the Mexican border, there could be seen a dry river bed winding and twisting down from the brown cactus covered hills of old Mexico as it approached the green salt marshes through which it meandered until it finally reached the sea.

—HARWOOD 1931

5
TIJUANA RIVER

It is tempting to compare the appearance of the Tijuana River as it flows through the city of Tijuana (embedded in a concrete channel and surrounded by urban development) to its appearance as it flows through the U.S. (embedded in a lush riparian forest surrounded by wide open spaces) and to conclude that one is "modified" and the other "natural" or "unmodified." But this conclusion would not be correct. A careful look into the river's ecological history reveals many ways in which the river's form and function have been altered over time, even downstream of the U.S.-Mexico border.

This chapter describes the river's historical morphology, hydrology, and habitat mosaics. We also discuss some of the key physical processes, like flooding and channel avulsions, that structured the river and its valley. Our findings highlight the dynamic nature of the river, which exhibited significant spatio-temporal variability in streamflow, channel position, and the structure and composition of riparian habitats. The historical perspective offers a range of new insights about the river and in the process raises important questions about the best way to manage this critical resource moving forward.

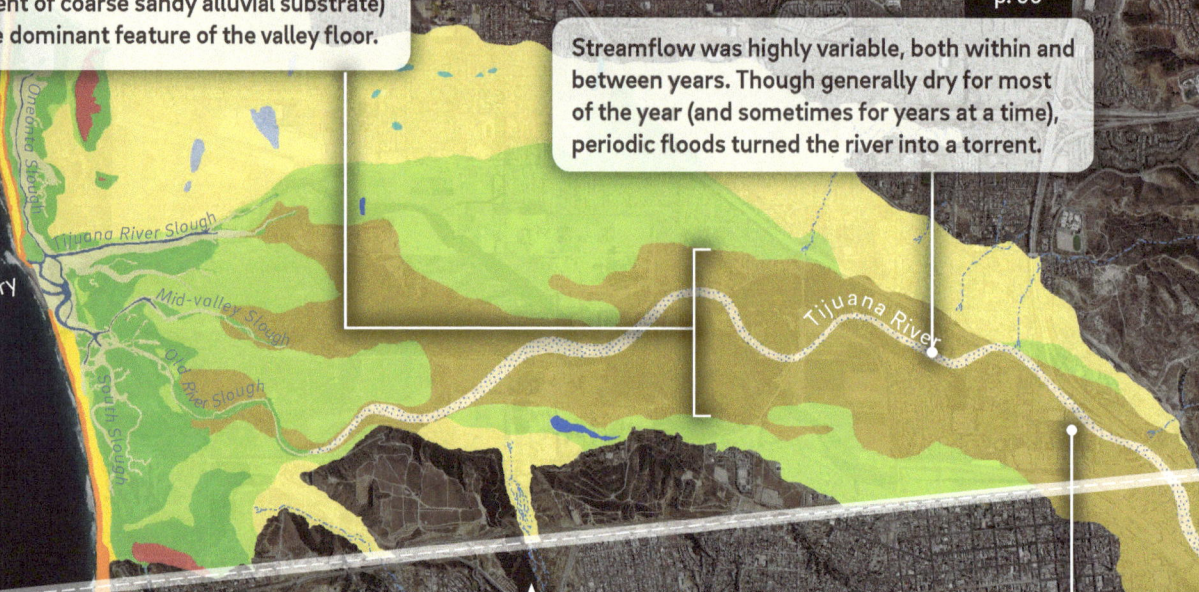

to San Diego Bay

p. 90

Nearly a kilometer wide along most of its length, the Tijuana River corridor (defined by the extent of coarse sandy alluvial substrate) was the dominant feature of the valley floor.

p. 96

Streamflow was highly variable, both within and between years. Though generally dry for most of the year (and sometimes for years at a time), periodic floods turned the river into a torrent.

Oneonta Slough

Tijuana River Slough

Tijuana Estuary

Mid-valley Slough

Old River Slough

North Slough

Tijuana River

Yogurt Canyon

Goat Canyon

Smuggler's Gulch

p. 106

Large floods drove channel avulsions, which altered the course of the river with some frequency, but some reaches were more stable than others.

TIJUANA RIVER, ca. 1850

Historical habitat types & key findings

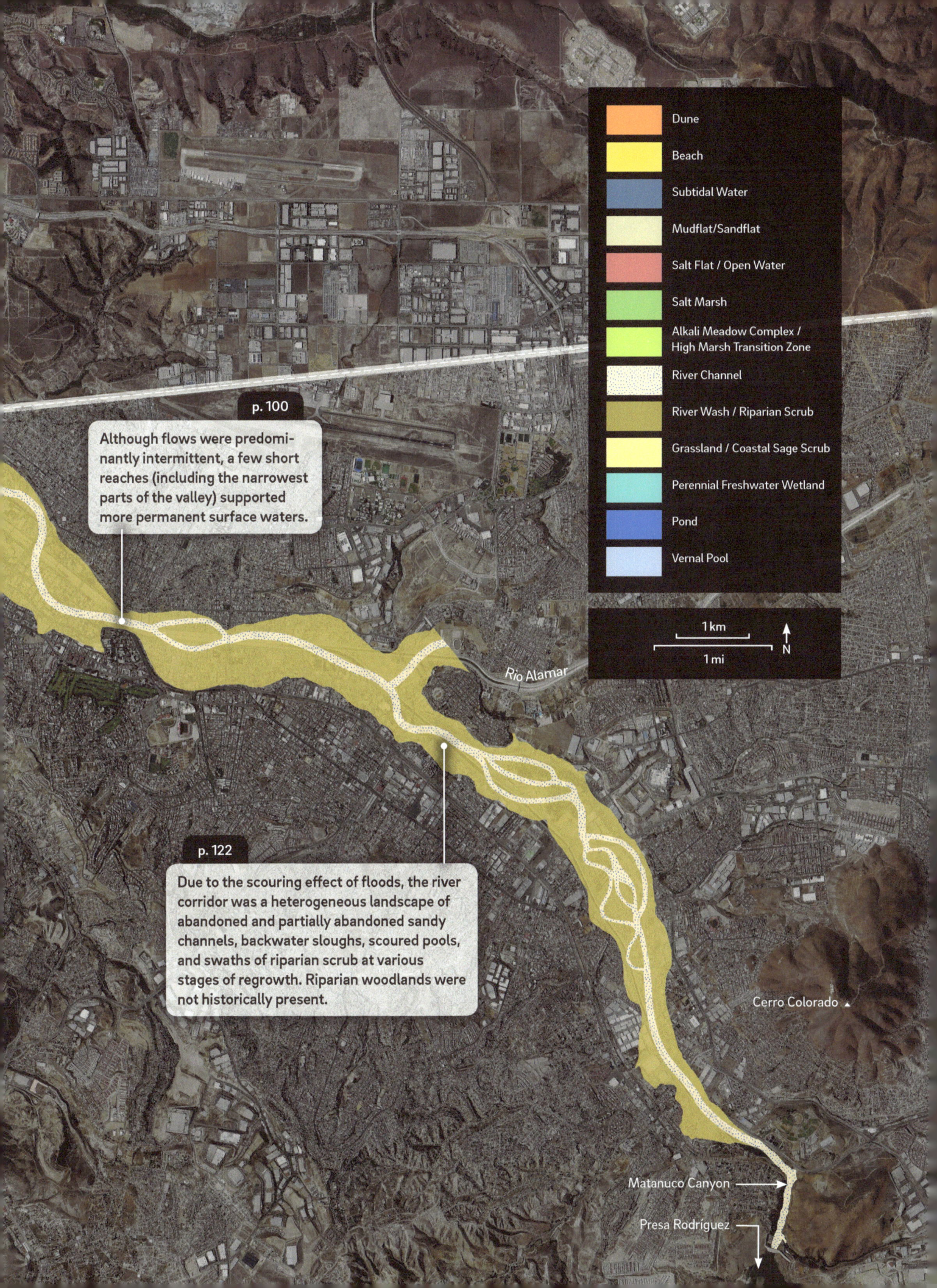

Legend

- Dune
- Beach
- Subtidal Water
- Mudflat/Sandflat
- Salt Flat / Open Water
- Salt Marsh
- Alkali Meadow Complex / High Marsh Transition Zone
- River Channel
- River Wash / Riparian Scrub
- Grassland / Coastal Sage Scrub
- Perennial Freshwater Wetland
- Pond
- Vernal Pool

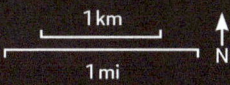

1 km
1 mi
N

Rio Alamar

p. 100

Although flows were predominantly intermittent, a few short reaches (including the narrowest parts of the valley) supported more permanent surface waters.

p. 122

Due to the scouring effect of floods, the river corridor was a heterogeneous landscape of abandoned and partially abandoned sandy channels, backwater sloughs, scoured pools, and swaths of riparian scrub at various stages of regrowth. Riparian woodlands were not historically present.

Cerro Colorado ▲

Matanuco Canyon

Presa Rodríguez

Highly variable flows produced a complex and dynamic assortment of geomorphic features in the river corridor.

Although many early maps depict the Tijuana River as a single narrow meandering line (e.g. Gray 1849, Ilarregui and de Chavero 1850, Poole 1854, International Boundary Commission 1901), the historical morphology of the river was much more complex. Consistent with other dryland rivers, where form is largely shaped by high-magnitude infrequent flood events (Tooth 2000), the Tijuana River was a dynamic, shifting mosaic of nested landforms (Figs. 5.1–5.2). It included one or more sparsely vegetated, meandering low-flow channels inset within a wider, flood-washed, and often-braided high-flow channel that, in turn, was situated within an extensive vegetated floodplain. We define this entire area as the river corridor, a term that includes all active channel surfaces (which were generally sandy, recently scoured by high flows, and sparsely vegetated) as well as the floodplain (which was characterized by flood-deposited sandy alluvium covered with varying densities of riparian scrub). The term excludes some areas that were prone to flooding, but were characterized by finer soils and herbaceous vegetation (see p. 70). Within the Tijuana River valley, the river corridor was quite broad, generally exceeding one kilometer (0.6 mi) in width, but ranging from 70–1,400 m (230–4,590 ft). As mapped, the historical river corridor occupied an area of 21.0 km² (5,200 acres), approximately 60% of which was situated within Mexico, making it the dominant feature of the valley floor. This section describes geomorphic features of the river corridor in more detail.

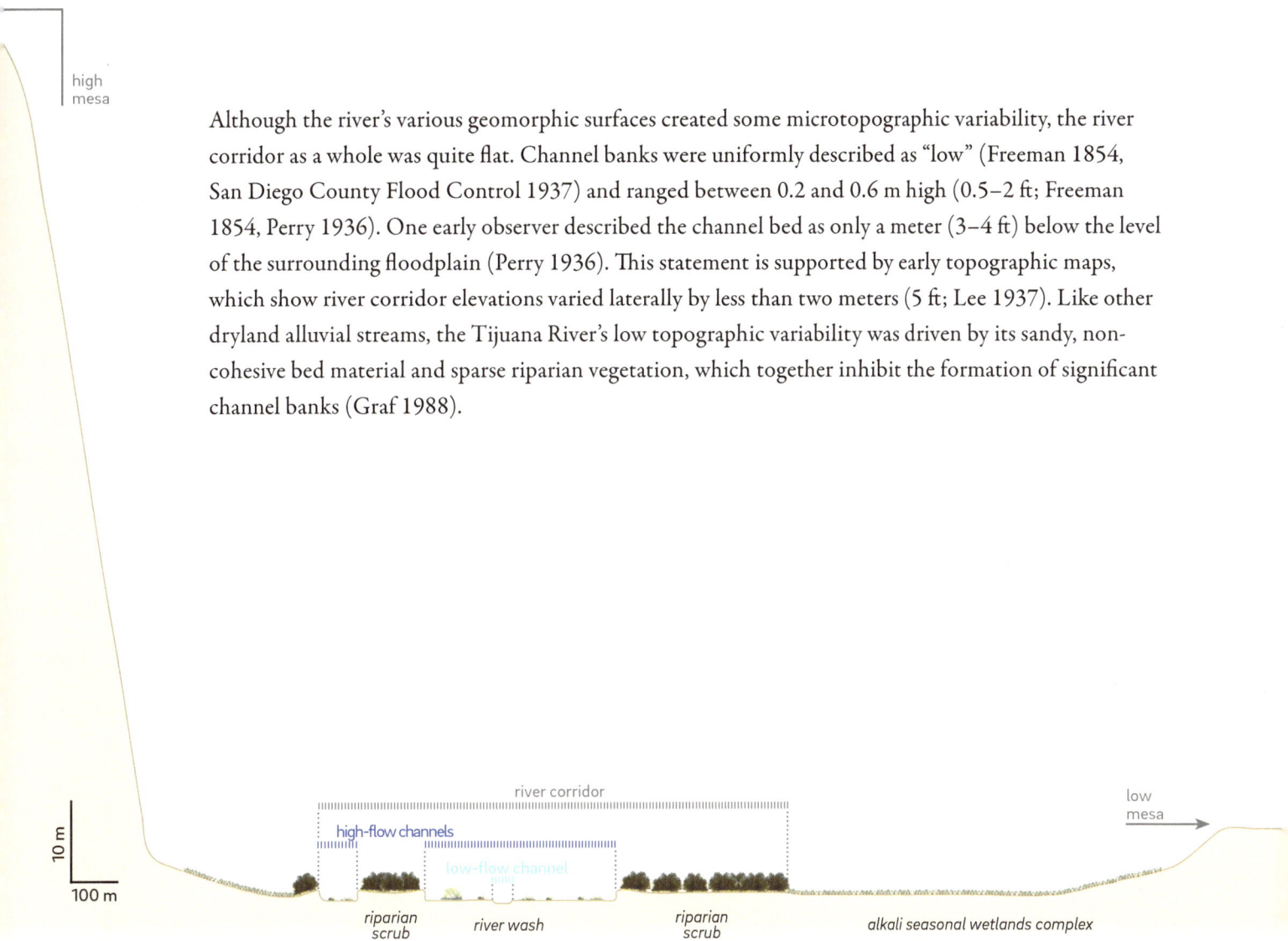

Although the river's various geomorphic surfaces created some microtopographic variability, the river corridor as a whole was quite flat. Channel banks were uniformly described as "low" (Freeman 1854, San Diego County Flood Control 1937) and ranged between 0.2 and 0.6 m high (0.5–2 ft; Freeman 1854, Perry 1936). One early observer described the channel bed as only a meter (3–4 ft) below the level of the surrounding floodplain (Perry 1936). This statement is supported by early topographic maps, which show river corridor elevations varied laterally by less than two meters (5 ft; Lee 1937). Like other dryland alluvial streams, the Tijuana River's low topographic variability was driven by its sandy, non-cohesive bed material and sparse riparian vegetation, which together inhibit the formation of significant channel banks (Graf 1988).

high mesa

river corridor

high-flow channels

low-flow channel

low mesa

10 m

100 m

riparian scrub

river wash

riparian scrub

alkali seasonal wetlands complex

Figure 5.2. Different geomorphic surfaces of the river corridor. A variety of surfaces, including a densely vegetated floodplain and a variety of more active channels, are visible in these adjacent photographs from the late 1920s. (Photos #90:18138-428, #90:18138-429, & #90:18138-430, all courtesy San Diego History Center)

Tijuana

In most years and locations, the majority of the river corridor was occupied by the river's extensive floodplain. The floodplain was broad and relatively flat, with a fine sandy soil that supported extensive areas of riparian scrub (Storie and Carpenter 1923). The floodplain is explicitly represented on the on the earliest known map of the Tijuana River valley, which shows *"vega del arroyo"* [floodplain] covered in *"monte de saus"* [willow scrub] extending for kilometers along both sides of the river (Fig. 5.3; U.S. District Court California, Southern District ca. 1840). Though rarely shown on other early maps (but see Ruhlen 1900, Knox 1933, CDWR ca. 1942), this vegetated surface is obvious in many early photographs (Figs. 5.2 and 5.4; also see sections starting on p. 122 for a full discussion of the river's riparian vegetation). During large floods (see p. 106), high flows often cut across the floodplain, forcefully uprooting riparian scrub and carving new channels. As a result, the floodplain was a heterogeneous landscape of abandoned and partially abandoned channels, backwater sloughs, scoured pools, and swaths of riparian vegetation at various densities and stages of regrowth (p. 132).

Nested within the floodplain was a series of sparsely vegetated channels. Channels existed at a variety of scales and reflected a range of flow conditions. Large floods removed wide swaths of riparian scrub, creating a large sandy channel that early residents referred to as the "high water channel" (Rice in Wyman 1937b) or area of "river wash" (Storie and Carpenter 1923, Abbey in Wyman 1937b); low flows were carried in smaller channels inset within the high water channel (Fig. 5.5). Often then, the historical form of the Tijuana River was analogous to what is today termed a "compound channel," considered to be the most common channel type in large alluvial dryland systems (Vyverberg 2010). Compound channels are defined by a relatively narrow, meandering "low-flow" channel inset within a wider, braided "high-flow" channel (Graf 1988, Tooth 2000).

Figure 5.3. The earliest known map of the Tijuana River valley. The Tijuana River's floodplain, or *"vega del arroyo,"* is shown along both sides of the river in the ca. 1840 diseño. The map also notes willow scrub ("*monte de saus*") and shows two watering holes ("*posa*") within the floodplain. Note that the map was drawn with north down. *(U.S. District Court California, Southern District ca. 1840, courtesy The Bancroft Library)*

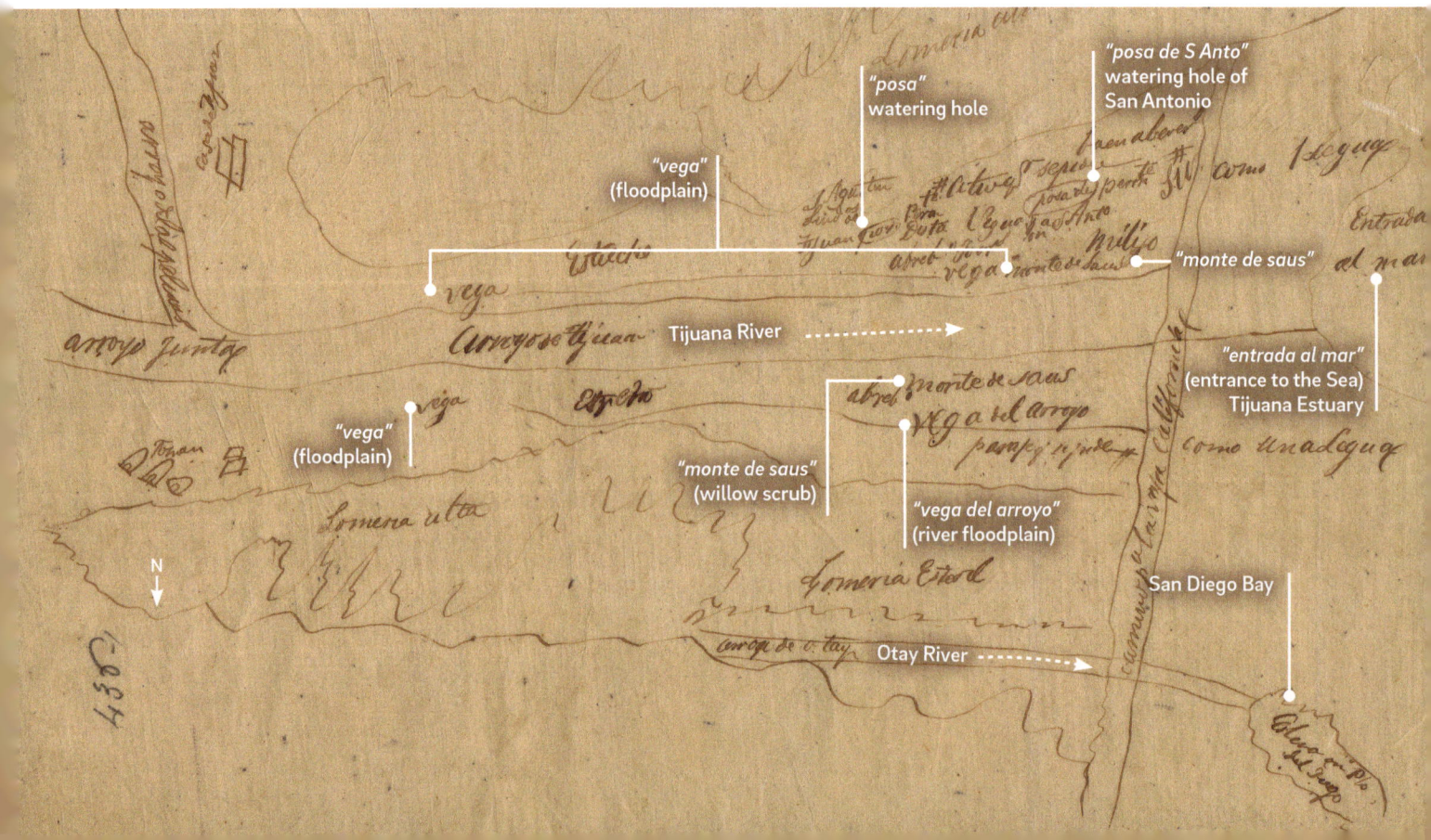

To some observers on the ground, the high-flow channel of the Tijuana River was an obvious geomorphic feature (Rice in Wyman 1937b), though, over time, its precise boundaries could blur as portions of the channel re-vegetated between floods, or as the channel shifted altogether (Harwood 1931). The high-flow channel of the Tijuana River took a variety of forms, sometimes appearing as a series of large braids and at other times as a single wide meandering channel. Although the high-flow channel ranges from 100–600 m (300–2,000 ft) wide in the earliest aerial photographs (see Figs. 5.5 & 5.30), the width of the largest channel likely varied over time, depending on both the size of and time since the last major high flow event (Tooth 2000). Other early spatially accurate sources record channels with widths on the order of 50–150 m (e.g., Freeman 1854, Pascoe 1869, Orozco 1889, USGS 1904, Rankin 1909, Ervast 1921)—it is unclear if these narrower measurements apply to smaller low-flow channels inset within the high-flow channel, or simply reflect natural temporal variability in the high-flow channel's width.

It is ultimately impossible to neatly classify and distinguish the various features depicted by each and every one of the historical sources. This difficulty reflects the challenges inherent to interpreting historical data, as well as the natural complexity and dynamism of the Tijuana River itself. GLO surveyors crossed the Tijuana River valley in multiple locations in 1854 and 1869 but never mentioned separate components of the channel. This could suggest that at the time of the surveys the various geomorphic surfaces described above were not distinct from one another, or simply that the surveyors only interpreted one of these surfaces as the true "channel." Mapping the river was apparently a challenging task: USCGS surveyor Robert G. Knox (1934) acknowledged that "there was some question about the charting of the bed of the Tia Juana River, but it was best to show it as a wash."

In our mapping, we depict a representative channel from the year ca. 1850. Although we have high confidence in the location of the channel within the U.S. in the late 1840s and early 1850s, we do not know the precise location of the channel at that time in Mexico (it was mapped from slightly later sources). Additionally, in both the U.S. and Mexico we have a relatively low confidence in the precise interpretation of the mapped channel (i.e., whether it represents a low-flow or a high-flow channel); the maximum width of the sparsely vegetated channel almost certainly exceeded the width of the mapped channel at certain places and times. Similarly, the channel would have often been narrower than what is mapped. We refer readers to the GIS metadata for notes on the sources (and the interpretations of these sources) we used to map the historical river channel.

Figure 5.4. **Vegetated floodplain surfaces** dominate the river corridor in this photograph taken on May 1st, 1944. Areas scoured during the flood of 1927 have largely re-vegetated, but are still distinct from surrounding parts of the floodplain, with different density of riparian vegetation. *(Photo #79:741-895, courtesy San Diego History Center)*

scoured during flood of 1927

Hollister Street bridge

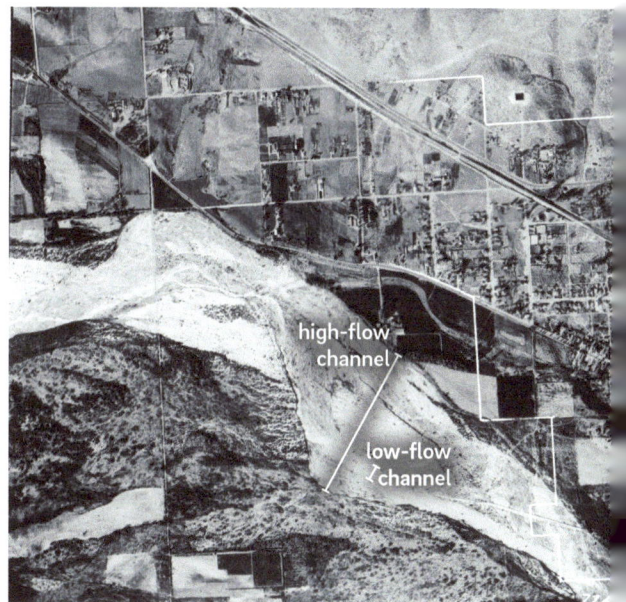

high-flow channel

low-flow channel

Figure 5.5. **Inset channels, 1928.** A channel sized to carry low flows is apparent within the high-flow channel. *(San Diego County 1928)*

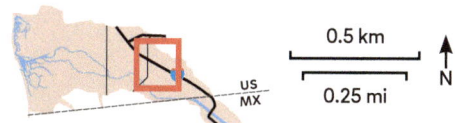

0.5 km

0.25 mi

US MX

N

COTTONWOOD CREEK

Cottonwood Creek ("Rio Alamar") is one of the two primary branches of the Tijuana River. The stream originates in the U.S. and flows southwest before entering Mexico south of Dulzura. From there, the creek flows another 20 km (12 mi), at which point it joins the southern branch of the Tijuana River ("Rio Las Palmas"). The combined branches then flow back into the U.S. and to the Pacific.

Although our study extent only captures the lowest portions of Cottonwood Creek, we do not wish to downplay the importance of this stream for driving physical and ecological processes in the lower valley. Though it was generally described as the "principal tributary" of the Tijuana River (e.g., Wyman 1937a), Cottonwood Creek likely delivered a disproportionate amount of water to the lower valley (the creek's drainage includes only about 25% of the total Tijuana River watershed area, but precipitation in this portion is generally much greater than in the remainder; c.f. SDSU Dept. of Geography 2005).

Although they were not analyzed in detail, archival data suggest that Cottonwood Creek—like the southern branch of the Tijuana River (see p. 96)—was a spatially intermittent stream, with alternating reaches of perennial and intermittent stream flow. C.S. Alverson (1914) wrote that the stream flowed year-round to a point approximately 10 miles downstream of Barrett Dam, "where it ceases to be classed as a living stream, the water sinking below the surface for several months in the year"; perennial flow was said to resume one mile below the junction with Rio Tecate, at Marron Canyon. This is supported by the 1903 USGS Quadrangle (USGS 1903; Fig. 5.6) and the account of Chase and McLean (1928), who found 50 inches of surface flow in the Cottonwood just above its junction with the Tijuana River on May 10th, 1928. From this point west, however, "the water disappeared, percolating through the gravel as underflow." The presence of perennial flows immediately upstream of our study extent on Cottonwood Creek adds to the evidence of spatial variability in surface flow on the Tijuana River (see p. 100). This hydrogeological variability, once common in coastal California, is known to influence the distribution and diversity of riparian vegetation in other systems (Lite and Stromberg 2005, Stromberg et al. 2005, Orr et al. 2011, Beller et al. 2011, Beller et al. 2016).

Lower Cottonwood Creek has been dramatically altered in recent years. Between 2008 and in 2016, the creek was channelized from its junction with the Tijuana River to a point 7.5 km (4.7 mi) upstream (Fig. 5.7). This process converted the alluvial streambed and remnant patches of riparian vegetation into a concrete flood control channel, much like was done along the Tijuana River in the 1970s.

Figure 5.6. (left) Flow permanence along Cottonwood Creek. According to this 1903 map, streamflow on Cottonwood Creek alternated between perennial (solid blue symbol) and intermittent (brown stipple symbol) as it flowed south towards the border, making this branch of the Tijuana River a spatially intermittent stream. Our study extent begins about 20 km (12 mi) downstream of this map. *(USGS 1903)*

Figure 5.7. (below) Cottonwood Creek (Rio Alamar) just above its confluence with the Tijuana River in 1955 and in 2015. The older photograph shows that Cottonwood Creek had a broad, compound channel, with substantial riparian vegetation. The contemporary photograph shows the result of the recent channelization work. *(1955: Photo #4-55, 1178-30 & #4-55, 1178-32, courtesy Fondo Aerofotográfico Acervo Histórico Fundación ICA, A.C.; 2012: NAIP 2012)*

The Tijuana River exhibited pronounced interannual and seasonal variability in flow volume. Although the river carried sustained flows and large floods during portions of many years, the lower reaches of the river (i.e., in our study extent) were most often dry, frequently for months or even years at a time (Figs. 5.8–5.10). Numerous early observers commented on the intermittent or ephemeral nature of the river, some even characterizing the system as "a river only part of the year" (Ecarg 1912; Figs. 5.8 and 5.10). Only a few short reaches supported perennial surface water (see p. 100 and map on p. 103). The river's extremely limited perennial flows were typical for the region, where streams were said to flow "bottom upwards — meaning that the water sinks quickly into the ground, leaving a few feet sand on top" (Ward 1889; also see Emory 1857, Warner 1891, Nelson 1922).

The volume of runoff in the lower Tijuana River varied widely between years. By some accounts, annual variation in streamflow is greater in the San Diego region than anywhere else in the United States (Pyrde 1976). Although wet years could bring extensive flooding and sustained flows (see the Flooding section starting on p. 106), surface flow was insignificant during many years and sometimes for multiple years at a time (Fig. 5.11; Williams 1933, Knox 1934). Based on gage data from 1937–2010, the Tijuana River carries approximately 630 times more water during very wet years than it does during very dry years (90th and 10th percentile of annual discharge, respectively). This extreme annual variability creates large discrepancies in flow: for example, in the early 20th century, the river carried more flow during a single two hour period than during a separate period of seven consecutive years (International Water Commission 1930). Annual variability was

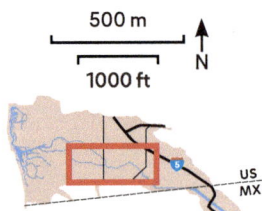

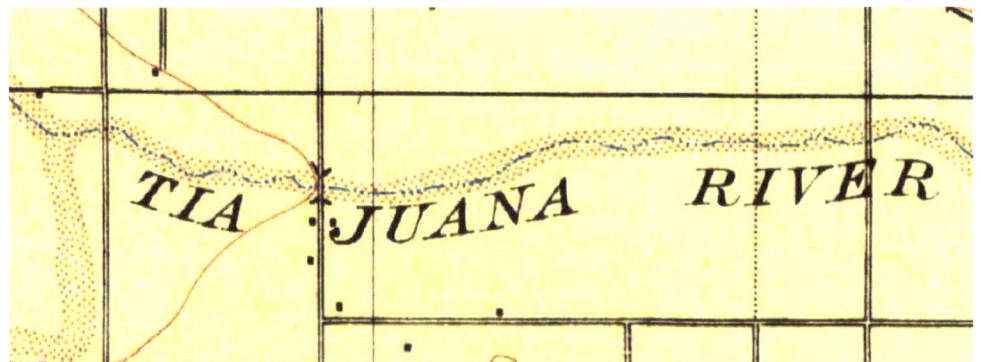

> The rivers of this area, including the...Tia Juana, are normally dry washes. There is an underground flow...but water is visible only at rare intervals, usually several years apart.
>
> — KNOX 1934

Figure 5.8. Intermittent/ephemeral flow is indicated by the dashed blue line used to draw the Tijuana River on this USGS map, surveyed in 1902. Though the river flowed year-round in parts of the upper watershed, the water was said to sink below ground "some distance back from the coast" (Nelson 1922). The stream flowed only intermittently along most of its length in the lower valley. *(USGS 1904)*

Figure 5.9. Temporal flow variability is well-illustrated by these two views across the international boundary, both ca. 1918. Extended periods during which streamflow was minimal or absent (top) were punctuated by large floods (bottom). *(Top: Everrett Photo #9578, courtesy Sociedad de Historia de Tijuana; Bottom: Photo #89:17358-6, courtesy San Diego History Center)*

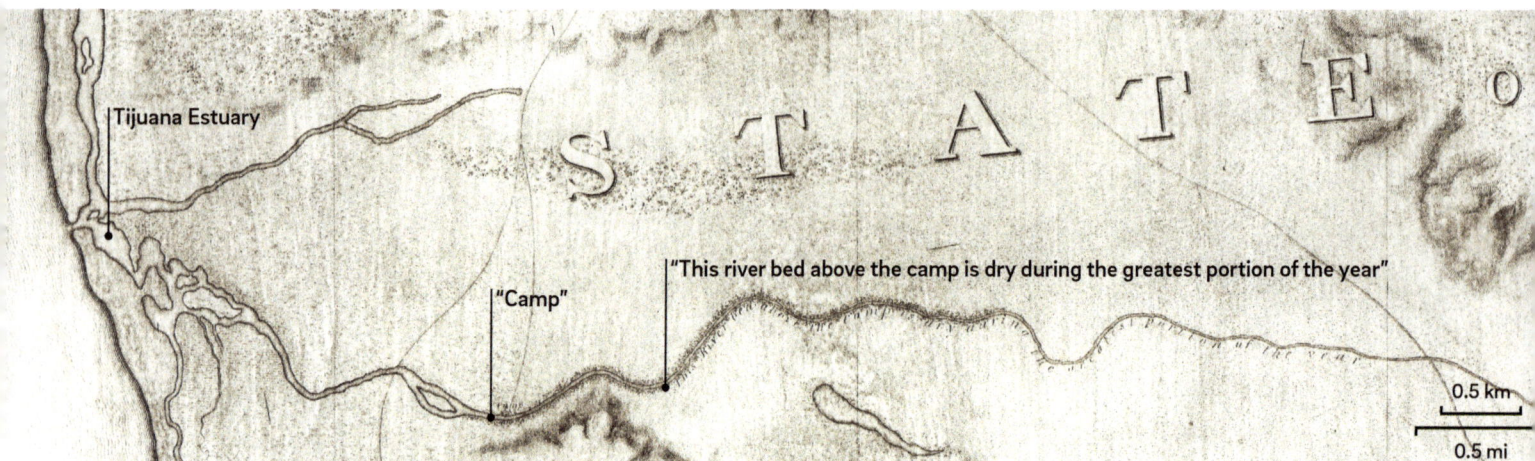

Figure 5.10. The river was "dry for the greatest portion of the year" upstream of the estuary, according to a note on this U.S. Boundary Commission map. *(Hardcastle and Gray 1850, courtesy National Archives and Records Administration)*

driven at least in part by large-scale climate cycles such as the El Nino-Southern Oscillation (ENSO) and the Pacific Decadal Oscillation (PDO), which have a strong influence on discharge in southern California streams (Andrews et al. 2004, Milliman and Farnsworth 2011). For instance, between 1937 and 2009, there was an approximately 23% probability of annual peak flows exceeding 10,000 cfs in ENSO years, but only a 7% probability of exceeding this value in non-ENSO years (Fig. 5.13).

The presence of streamflow was also highly seasonal. Temporary flows occurred during the winter and spring but were largely absent the rest of the year (Irelan 1888, *Los Angeles Herald* 1897, USGS 1904, Cicourel 1921, Nelson 1922, International Water Commission 1930). Between 1936 and 1979, for example, 98% of the Tijuana River's average annual discharge occurred between the months of December and May (Fig. 5.12), as is typical in semi-arid climates. Seasonal variability in streamflow ultimately reflects the highly seasonal nature of precipitation in the watershed, 90% of which falls between the months of October and April (Das et al. 2010; see p. 34).

Even during the rainy season, much of the total runoff was concentrated in relatively infrequent high flow events, with large fluctuations occurring over very short periods of time. A *San Francisco Call* newspaper article from 1895 captures the flashy nature of streamflow:

> Tuesday morning the Tia Juana River...was crossed with ease, there being little but sand to indicate its course. That evening horses were compelled to swim in crossing at the same place.

Although flows sometimes continued into the dry season (see p. 105), they also sometimes ceased just a few days after rain stopped falling (Cicourel 1921). Flashy conditions were at least partially attributable to the sandy, porous nature of the river bed, which promoted rapid infiltration of surface water and often precluded substantial surface flow even following moderate rainfall events (Irelan 1888, Wyman 1937a, URS Corporation 2012).

> The soils are sandy and the absorption of rainfall is rapid. Consequently, surface run-off occurs only when the rate of precipitation is in excess of a high percolation rate or when the large storage capacity of both surface and subsoil has been exceeded.
>
> — WYMAN 1937A

Figure 5.11. Tijuana River average daily discharge (dark blue), October 1936 to December 2009. The hydrograph and cumulative discharge plot (light blue) shows the lack of base flow and the flashy nature of flows. Through 1961, streamflow measurements were taken at the USGS gage (#11013500) at the Nestor Bridge; later measurements were taken at the IBWC gage (#11013300) at the international boundary. Although all measurements post-date the construction of upstream water supply dams, which together regulate approximately 70% of the watershed, the data are still useful for understanding general patterns of flow variability.

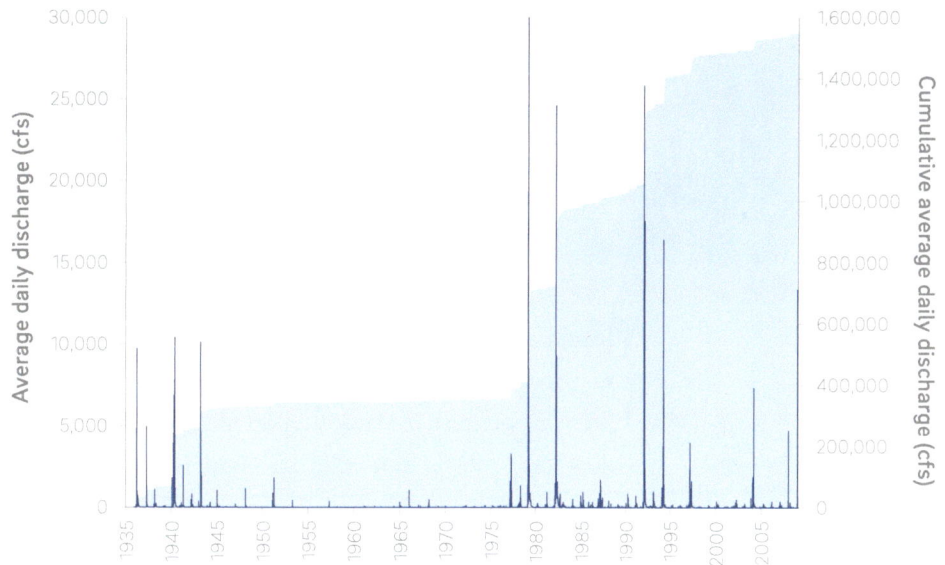

Figure 5.12. Average monthly discharge, October 1936 to December 1979. This chart illustrates the Tijuana River's pronounced seasonal flow variability, with more than 90% of the average annual discharge occurring between the months of December and April. Data from after 1979 are excluded since the discharge of treated wastewater created perennial flow conditions after that date.

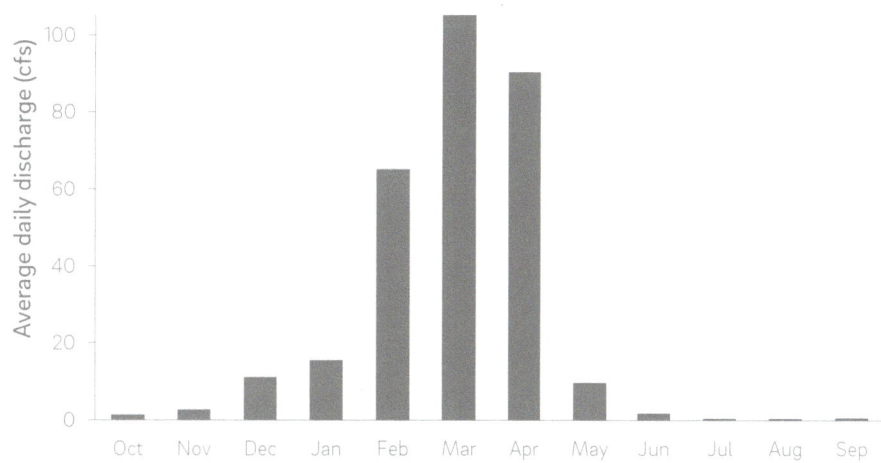

Figure 5.13. Flow exceedance probability for ENSO vs. non-ENSO years for the Tijuana River (water years 1937–2009). The largest annual peak flows were more likely to be exceeded in ENSO years compared with non-ENSO years. Annual instantaneous peak discharge data for 1937–1981 was obtained from USGS gage #11013500. Instantaneous peak discharge for 1982–2009 was estimated from mean daily discharge data from IBWC gage #11013303 (based on a second-order polynomial regression between USGS and IBWC gage data for 1962–1981, years with overlapping data).

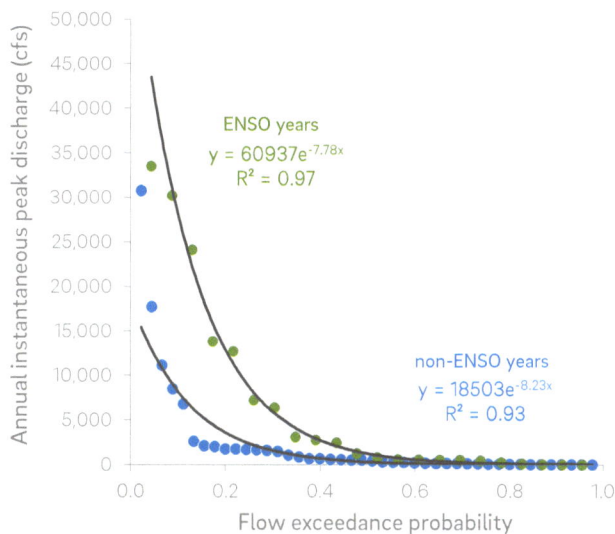

ENSO years
$y = 60937e^{-7.78x}$
$R^2 = 0.97$

non-ENSO years
$y = 18503e^{-8.23x}$
$R^2 = 0.93$

SPATIAL FLOW VARIABILITY

El lecho del río no lleva agua la mayor parte del año; sólo en lugares angostos del valle aparecen pequeñas corrientes superficiales constantes.

[The bed of the river does not carry water most of the year; only in narrow parts of the valley do small constant surface flows appear.]

— BONILLAS AND URBINA 1912

In addition to dramatic temporal variability, the historical Tijuana River experienced spatial variability in flow conditions. Although flows were predominantly intermittent (see previous section), the river supported a few short reaches with more permanent surface flow (see Table 5.1 and Fig. 5.16 on pp. 102–03). Perennial flow was found where the river flowed through narrow portions of the valley (Bonillas and Urbina 1912) as well as in locations where the high groundwater table intersected the river bed, such as near the estuary. This section describes these sources of year-round surface water in greater detail.

From upstream to downstream, the first constriction in the valley that supported perennial flow during many years was in Matanuco Canyon, at the current site of Rodríguez Dam (Fig. 5.14). Here, the valley was confined within steep bedrock walls and limited to a width of approximately 50 m (160 ft). During the dry season, water reliably accumulated within this narrow reach before flowing into the broader valley and sinking subsurface into the sandy bed of the unconfined river (Böse and Wittich 1912; see also Ecarg 1912). Although the canyon appears to have had a relatively high degree of flow permanence, it is important to note that flows were not perennial there in every year (Chase and McLean 1928).

A second geological constriction in the valley with more persistent surface water was located 12 km (7.5 mi) downstream from Matanuco Canyon, at what was termed in at least one source the Agua Caliente Narrows (Fig. 5.15). Here, the river corridor was only 200 m (650 ft) wide and featured a small group of eponymous hot springs that emerged from the sands of the river bed (Orcutt 1886b; Rodriguez Galeana 1920). Water from the springs mixed with surface water from upstream (Bonillas and Urbina 1912), and during the summer—when the river otherwise ceased to flow—the springs continued to feed channels with a perennial flow (Rodriguez Galeana 1920; Unknown 1921). In addition to a significant tourism industry, the unique conditions at Agua Caliente supported in-channel herbaceous wetlands and a nearby cottonwood grove (see p. 131).

Not all areas that supported more perennial flow conditions were associated with narrow parts of the valley. Early observers also noted the presence of flowing water at the upper edge of the estuary, where the groundwater plane intersected the river bed, even when the river was dry upstream. The earliest account of the Tijuana River, from mid-May 1769, describes how it "broke forth" near the estuary as a "handsome stream running with a good sized flow of water that with great force issues up out of the ground," suggesting streamflow generated from rising groundwater and hyporheic exchange (Crespí and Brown 2001; see also Cañizares et al. 1952). A century later, subsequent observers also recorded running water at the upper edge of the estuary during late summer, months after the river ceased flowing elsewhere in the valley

There is no running water except at the mouth of the... creek.

— JAMES PASCOE, SEPTEMBER, 1869, DESCRIBING THE TIJUANA RIVER WITHIN THE U.S.

Figure 5.14. Dry-season flow in Matanuco Canyon. Shallow surface flow is visible in photographs of the river taken within and immediately upstream of Matanuco Canyon during the summer of 1920. This confined reach of the river was one of the few locations where perennial surface flow could be found during many years. Once it exited the canyon, the water seen in this photo would likely have sunk back into the sands of the riverbed. *(Photo by Hiram Savage, July 1920, MS 76/16, Box 2, Folder 44, #385, courtesy Water Resources Collections and Archives, UC Riverside)*

Figure 5.15. Dry-season flow at Agua Caliente. Permanent water and a marsh ("*cienega*") are noted along a channel adjacent to the Agua Caliente hot springs, August 1921. These springs were found within another narrow portion of the river valley and were also a source of year-round surface water. *(Photo #AAS-723-10513-1-83 (3), courtesy Archivo Historico de Agua)*

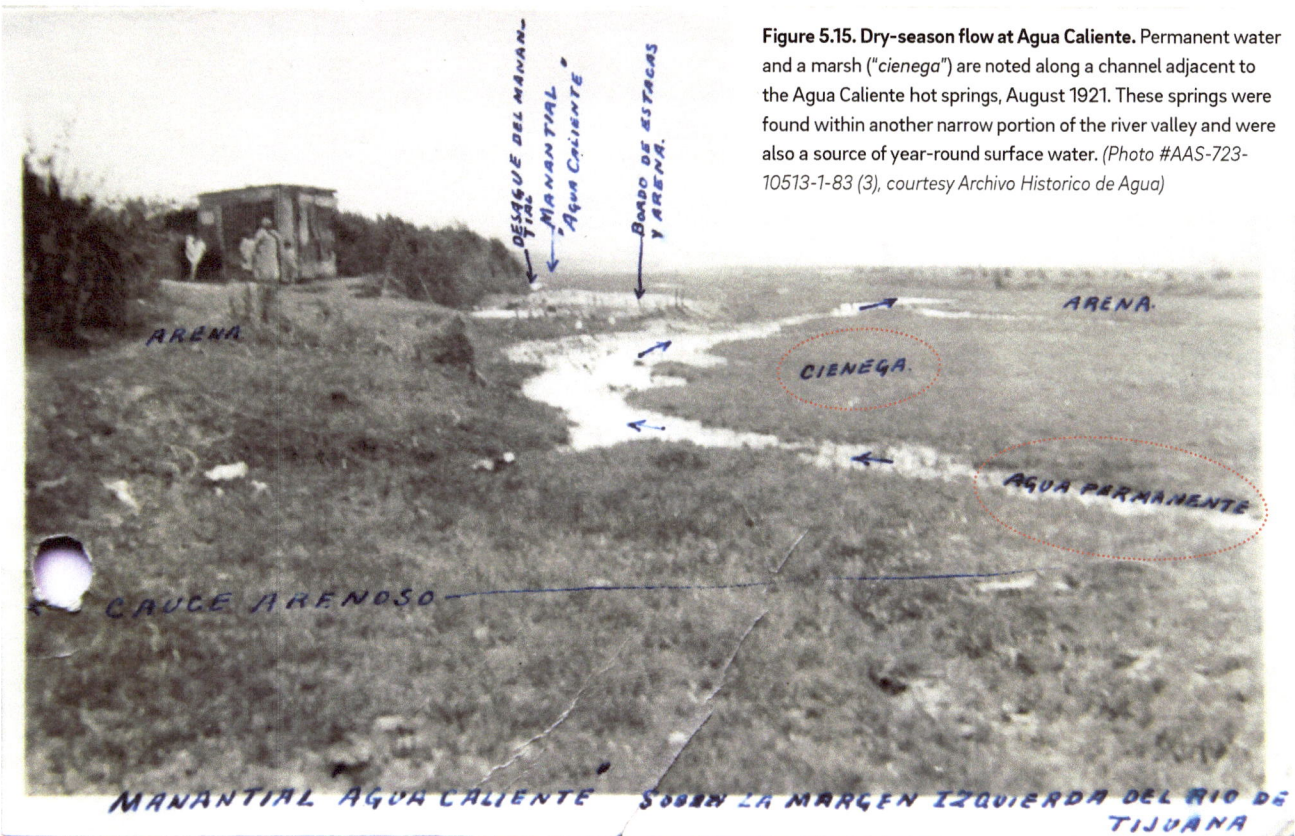

ID	Contemporary location	Condition	Date/Season	Evidence	Source
1	Tijuana Estuary	Wet	September 3rd, 1869	"Tia Juana river with running water 1 chain wide"	Pascoe 1869
2	Tijuana Estuary	Wet	August 22nd, 1937	"south fork near the ocean was still flowing about .3 second foot"	Hull 1937
3	Head of Tijuana River Slough	Wet	1889	"spring"	Los Angeles Lithographic Co. ca. 1889
4	Near mouth of Goat Canyon	Wet	May 13th, 1769	"a handsome stream running with a good sized flow of water that... issues up out of the ground"	Crespí and Brown, 2001
5	Head of Old River Slough	Wet	July 12th, 1894	"at the head of one of the tide creeks... water that could be drank was then to be had"	Stephens 1908, Mearns 1898
6	0.5 km west of 15th St. along Sunset Ave.	Dry	June 17th, 1937	"ceased flowing"	Hull 1937
7	Goat Canyon at International Boundary	Dry	September 3rd, 1869	"dry bed of creek"	Pascoe 1869
8	East of Goat Canyon	Dry	1850	"this River bed above the camp is dry during the greatest portion of the year"	Hardcastle and Gray 1850
9	Along river bed within 4 km of ocean	Wet	Summer, 1931	"pot holes" with water	Harwood 1931
10	1 km west of Saturn Blvd.	Dry	August 22nd, 1937	"flow does not extend easterly to the point where the south fork leaves the main stream"	Hull 1937
11	Hollister St. Bridge	Dry	July 4th, 1916	"water flowing... until... almost the Fourth of July"	Perry 1936
12	Hollister St. Bridge	Dry	September 3rd, 1869	"dry bed of Tia Juana river"	Pascoe 1869
13	Hollister St. Bridge	Dry	June 7th, 1915	"flow ceased"	Ellis and Lee 1919
14	1 km east of Hollister St.	Dry	June 9th, 1854	"in dry time sink in sand"	Freeman 1854
15	0.4 km west of Dairy Mart Rd.	Dry	September 3rd, 1869	"bed of river dry"	Pascoe 1869
16	Via De la Bandola at Via De La Melodia	Dry	August 23rd, 1869	"dry bed of ravine"	Pascoe 1869
17	Sipes Ln. at Anella Rd.	Dry	September 3rd, 1869	"bed of river dry"	Pascoe 1869
18	International Boundary	Dry	July 11th, 1937	"main river near the boundary ceased flowing"	Hull 1937
19	International Boundary	Dry	August 2nd, 1903	"waterless Tia Juana River"	Webster 1903
20	Agua Caliente	Wet	July 29th, 1910	surface flow visible in photograph	Unknown 1910
21	Agua Caliente	Wet	August 13th, 1921	"agua permanente" [permanent water]	Unknown 1921
22	Agua Caliente	Wet	Summer, 1920	"en el Verano [el rio] permanece completamente seco y es por lo que el manantial... se mantiene a medio río" [in the summer the river remains completely dry, such that the spring... is sustained midstream]"	Rodriguez Galeana 1920
23	Cottonwood Creek confluence	Dry	May 10th, 1928	"from this point west the water disappeared, percolating through the gravel as an underflow."	Chase and McLean 1928
24	Cottonwood Creek confluence	Wet	May 10th, 1928	"surface flow in the Cottonwood just above its junction with the Tia Juana"	Chase and McLean 1928
25	Blvrd. Bernardo O'Higgins	Wet	Dry season, 1913	"agua corriente... salir en tramos de nuevo cerca del rancho de Boronda [running water... appears again in sections near the Boronda ranch]"	Böse and Wittich 1912
26	Downstream of Matanuco Canyon	Dry	Dry season, 1913	"el río muestra agua corriente que se pierde más abajo... en el ancho lecho de arena [the river has running water that is lost further downstream... in its wide sandy bed]"	Böse and Wittich 1912
27	Mouth of Matanuco Canyon	Wet	Dry season, 1913	"se acumula el agua en el citado cañón, pues en su salida el río muestra agua corriente [water accumulates in this canyon, while at its mouth the river has running water]"	Böse and Wittich 1912
28	Matanuco Canyon	Wet	January 12th, 1910	"there will be water in it [the river] for some months to come"	Garretson 1910, in Ecarg 1912
29	Matanuco Canyon	Dry	May, 1928	"Tijuana River was dry at the Dam site"	Chase and McLean 1928
30	Matanuco Canyon	Wet	July 16th, 1920	surface flow visible in photographs	Savage 1920

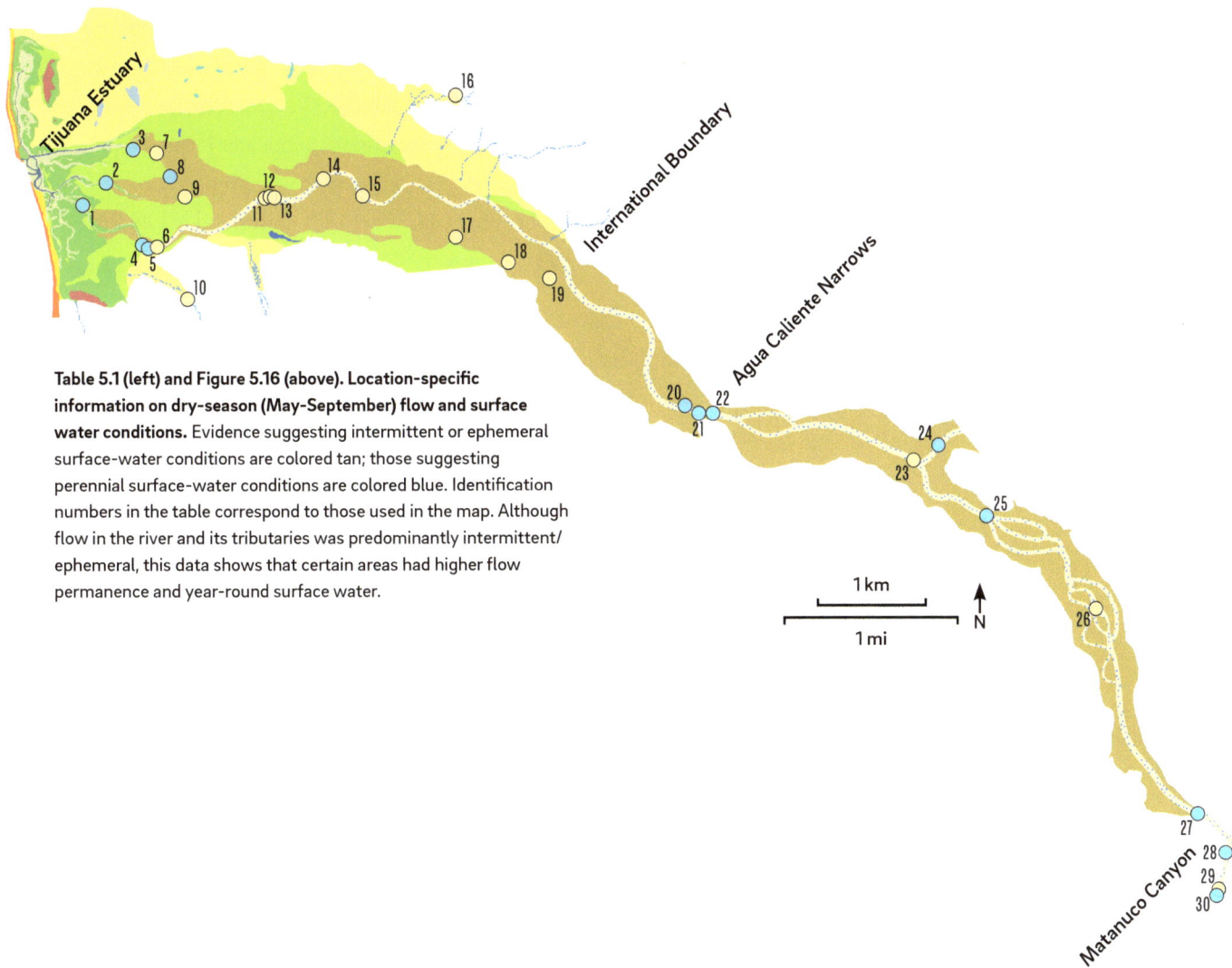

Table 5.1 (left) and Figure 5.16 (above). Location-specific information on dry-season (May-September) flow and surface water conditions. Evidence suggesting intermittent or ephemeral surface-water conditions are colored tan; those suggesting perennial surface-water conditions are colored blue. Identification numbers in the table correspond to those used in the map. Although flow in the river and its tributaries was predominantly intermittent/ephemeral, this data shows that certain areas had higher flow permanence and year-round surface water.

STREAMFLOW SHIFTING WITH THE SUN

In spatially intermittent streams, the longitudinal extent of surface flow can fluctuate over the course of a single day, creating small-scale spatiotemporal variability in streamflow. This phenomenon was observed by Major William Emory in southern California, who noted in the mid-1800s how the time of day and cloud cover affected the point in space where streams ceased flowing: "[it] is by no means a fixed one; thus, during the night it extends further downwards than in the daytime; in cloudy weather, for the same reason, its course is more prolonged than under a clear sky" (Emory 1857). This daily variation was likely tied to short-term fluctuations in groundwater table elevations, which rise and fall with the evapotranspiration rate of groundwater-dependent vegetation. The evapotranspiration rate, in turn, is dependent on meteorological factors—such as humidity, incident solar radiation, and temperature—that fluctuate over the course of a day (see Lundquist and Cayan 2002, Lautz 2008, Gribovszki et al. 2010). This phenomena highlights the complex interactions between physical and biological processes in dryland alluvial rivers.

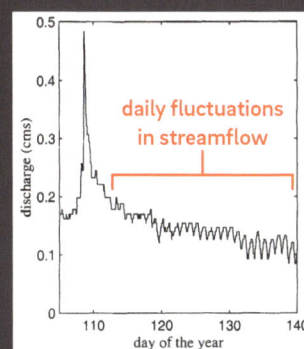

Figure 5.17. Daily fluctuations in stream discharge along Temecula Creek in Riverside County, driven by cycles of night and day (from Lundquist and Cayan 2002). This pattern, which can create daily spatial variability in the presence of streamflow, is common in southern California (Lundquist and Cayan 2002).

(Hardcastle and Gray 1850, Pascoe 1869, Hull 1937). Other indications of gaining stream reaches and perennial surface water in this area include a spring at the head of Tijuana River Slough (Los Angeles Lithographic Co. ca. 1889; see Fig. 3.17) and the presence of potable water 1894 at the head of Old River Slough during July of 1894 (Stephens 1908; Mearns 1898).

In addition to these reaches with a high degree of flow permanence, standing water also persisted into the summer in pools within the river corridor that were scoured and filled by floodwaters. Near the estuary, landowner Herbert Perry recalled swimming holes 5–10 feet deep within the river channel during the late 19th century (Perry 1936). Similarly, the biologist Robert Harwood noted five pools during the summer of 1931 "at the side of the regular stream bed" in a report on the estuary's plant and animal life. Harwood determined that the features were formed five years prior (during the floods of 1926) and had maintained a continuous supply of water since, despite a lack of any surface connection to the river during the intervening period. The pools, which he termed "pot holes," ranged from 15–25 feet wide, 30–50 feet long, and 5–10 feet deep, and were surrounded by tules (*Schoenoplectus acutus*). Since the water level in the pool fluctuated with the groundwater table (Perry 1936), it is probable that they were sustained into the dry season by inflowing groundwater.

These pools were important habitat for a range of species. Harwood described them as "headquarters for…animals that require fresh water" and noted an abundance of macroinvertebrates, fish, birds, and reptiles, including the Western pond turtle (*Emys pallida*; a Species of Special Concern in California). Although Western pond turtles have since been extirpated from the lower river valley (Fisher and Case 2000), Harwood netted a total of 20 in just a single day's work.

Even in locations that did not consistently support perennial flows, surface water could in some years persist into the spring and early summer. At the U.S.-Mexico border, for instance, a party traveling to Tijuana in mid-May 1892 described crossing a "brook which went by the name of the Tia Juana river" on a plank footbridge so narrow that "several of the party came near toppling over into the water" (Parkinson 1894). Though annual precipitation was average in 1892, that particular May was one of the wettest Mays on record (96th percentile, based on monthly precipitation data for San Diego from NOAA). In 1915, a year with high precipitation, the river flowed at the Hollister Street bridge from January 29th through June 7th (Ellis and Lee 1919). This broadly agrees with the recollections of Perry (1936), who reported that in years with significant flooding the Tijuana River could support surface flow into late spring (about June 1), while in other years there would be only a few days of flow. He recalled that, after the landmark floods of January 1916, there was flow under the Hollister Street bridge until nearly the 4th of July.

FLOODING

Large floods periodically inundated the majority of the valley and were a major driver of landscape form and function.

While surface flow over most of the Tijuana River was limited or nonexistent for much of the year, periodic floods transformed the river into a torrent that inundated large portions of the valley (Figs. 5.18–5.20). Floods were one of the primary drivers of landscape change in the Tijuana River valley, reworking the configuration of the river channel and floodplain (see p. 112), clearing vegetation (p. 132), and scouring and depositing sediment (p. 111).

In the U.S., the area subject to flooding encompassed roughly 2,450 ha (6,050 acres), including most of the low-lying portions of the river valley west of present-day I-5, north of Spooners Mesa, and south of Imperial Beach (Cruse 1937, San Diego County Flood Control 1937, USAED 1963; Fig. 5.18). This flood-prone area included wide swaths of river wash, riparian scrub, and alkali meadow habitats, as well as vernal pools, grassland, and coastal sage scrub on lower portions of the Nestor terrace.

Because the valley floor in Mexico is narrower and bounded by steeper topography, flooding there was more confined than in the U.S. That said, the area subject to overflow still covered the vast majority of the valley

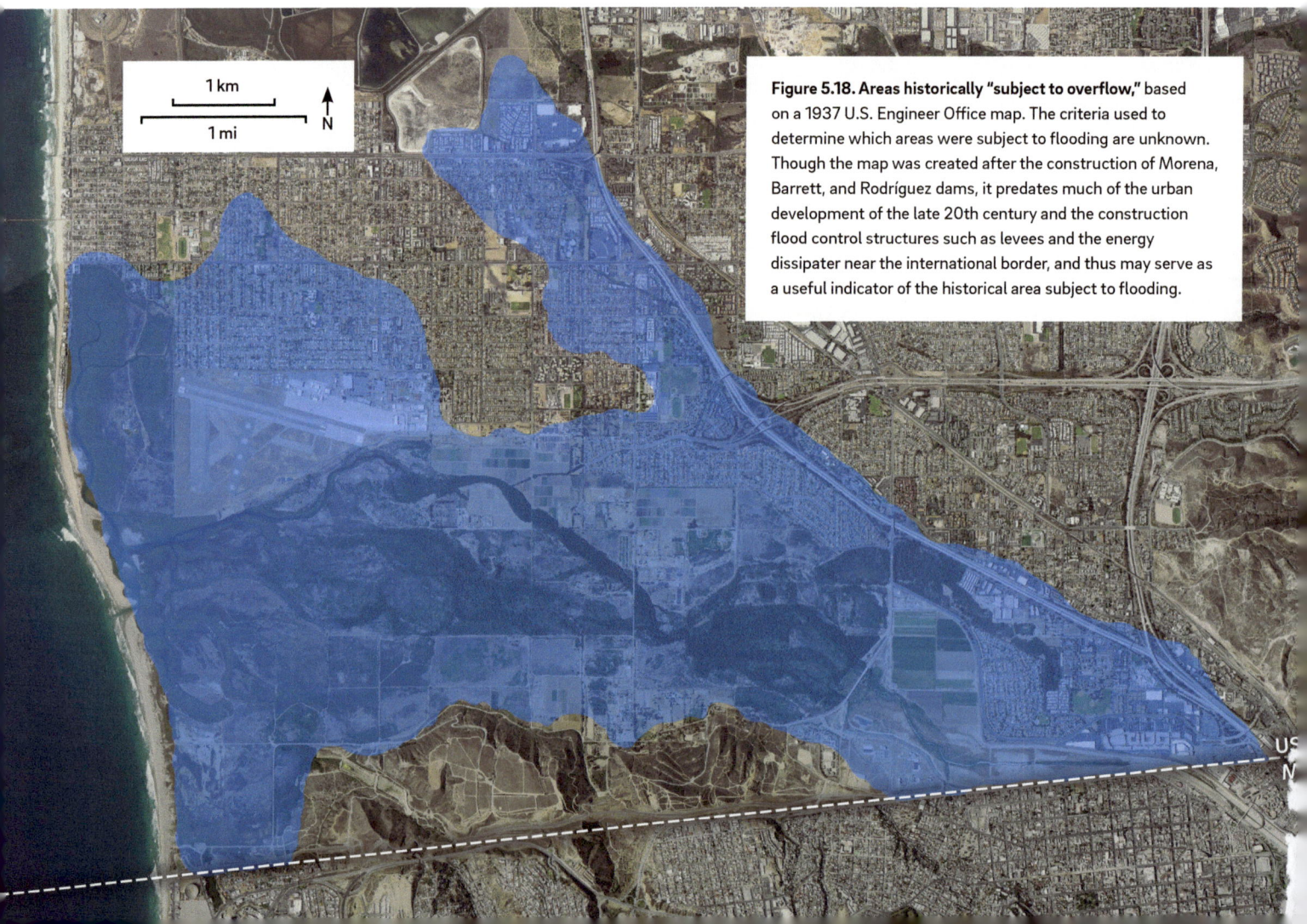

Figure 5.18. Areas historically "subject to overflow," based on a 1937 U.S. Engineer Office map. The criteria used to determine which areas were subject to flooding are unknown. Though the map was created after the construction of Morena, Barrett, and Rodríguez dams, it predates much of the urban development of the late 20th century and the construction flood control structures such as levees and the energy dissipater near the international border, and thus may serve as a useful indicator of the historical area subject to flooding.

Figure 5.19. People cross between Mexico and the U.S. during a flood, ca. 1890. Based on the height of the horses, the water was probably at least two feet deep. The photo was probably taken during the flood of 1886, 1891, or 1895. *(Photo #80:1910, courtesy San Diego History Center)*

Figure 5.20. A person crosses from Mexico to the U.S. using a makeshift cable bridge during the flood of 1916. Remnants of the automobile bridge are visible on the far bank. The river was presumably too deep and strong to ford (note the large river rapids beneath the cable). *(Fotos del Puente Mexico Rio Tijuana y Lluvias, Sobre No. 15 de 22, courtesy Sociedad de Historia de Tijuana)*

floor (1,300 ha [3,250 acres]), which was up to kilometer (0.6 mi) wide (Unknown 1965). Early records describe major floods inundating the customs house, the Tijuana racetrack (Fig. 5.21), large sections of the town of Tijuana, portions of the railroad, and the Agua Caliente hotel and hot springs (*Daily Alta California* 1886, *Daily Alta California* 1891, *San Diego Sun* 1891, *San Francisco Call* 1895, *Sausalito News* 1916, Plasencia Navarro 2011).

The extent of flooding varied depending on the timing and volume of streamflow, the location along the river corridor, and other factors (such as soil moisture). The valley's relatively flat topography and the river's relatively low banks meant that a rise of just a few feet could create extensive flooding (*San Diego Sun* 1891). During major floods, such as the 2,100 m³/s (75,000 cfs) event in 1916, the river could swell to well over 1 km (0.6 mi) wide, filling the valley floor (*The Arizona Sentinel* 1874, Durán 1989, Morin 1916, Plasencia Navarro 2011). Inundation was less extensive during more moderate flood events: the floods of February and March of 1905 were approximately 100 m (300 ft) wide (*Los Angeles Herald* 1905; *San Francisco Call* 1905). Flood depths as high as 5.5 m (18 ft) were recorded by historical observers, although they also varied significantly over time and space (Table 5.2).

Portions of the floodwaters spilled out of the Tijuana River valley and flowed north into San Diego Bay on several occasions, following a low-lying swale across the Nestor terrace (Hertlein 1944). Overflows occurred during the flood of January 1916 and are visible in the earliest known aerial photograph of the Tijuana River valley (Fig. 5.22). Similar overflows occurred in 1891 and 1927, and may have also occurred in 1825 and 1862 (McGlashan and Ebert 1918, San Diego County Flood Control 1937). By 1933 a levee disconnecting the river from this part of its floodplain was in place (Knox

Figure 5.21. Flooding at the Tijuana racetrack in January, 1916. The racetrack was located just south of the international border and east of the low-flow channel of the Tijuana River. *(Photo #95:19385-21, courtesy San Diego History Center)*

1933) and by the 1980s the swale had been filled and urbanized (Haltiner and Swanson 1987).

Annual peak discharge provides one measure of the intensity and frequency of flooding in the Tijuana River valley (Figure 5.23). Since 1937 (the period with continual gage data) peak annual discharge has ranged from 0 to 852 m³/s (30,088 cfs), with a mean of 68 m³/s (2,407 cfs). Although the mean peak annual discharge exceeds the flows necessary to transport sediment to the estuary (42 m³/s; Moffatt and Nichol Engineers 1987), floods of this magnitude only occurred during 21% of years, a reflection of high interannual flow variability. Floods capable of transporting sediment to the ocean (>283 m³/s) occurred in only 8% of years. The flood of record, January 17th, 1916, had an estimated peak discharge of >2,000 m³/s (75,000 cfs; City of San Diego 1973), more than twice the peak of any flood since. When considering these trends, it is important to remember that major dams (completed in 1912, 1921, and 1936) have likely attenuated peak flows and influenced discharge during the post-1937 period with reliable gage data.

Table 5.2. Data describing historical flooding depths (1874–1980). Recorded flood heights range from less than one to more than five meters. Peak discharge numbers are from the City of San Diego (1973), with the exception of the 1980 event, which was recorded by the IBWC gage at the international border (#11013300).

Year	Recorded depth	Approx. depth (m)	Relevant excerpts and notes	Source	Peak discharge of relevant flood
1874	18 ft	5.5	"by January 29 the Tia Juana River was eighteen feet deep in some places"	Pourade 1964	unknown
1886	12 ft	3.7	Agua Caliente hot springs	*Daily Alta California* 1886	unknown
1891	3 ft	0.9	"some distance below" town of Tijuana	*Sacramento Daily Union* 1891	20,000 cfs
	shoulder height	~1.4	"he stood in the water up to his shoulders for twelve hours "	*Daily Alta California* 1891	
	bridge height	~5.0	"the railroad bridge was soon covered"	*San Diego Sun* 1891	
1895	horse height	~2.0	Tuesday morning [Jan. 15] E. Waideman crossed the bed of sand; in the evening...his horses were compelled to swim"	*San Francisco Call* 1895	38,000 cfs
1905	5 ft	1.5	river 400 ft (121 m) wide	*San Francisco Call* 1905	unknown
	2–6 ft	0.6–1.8	river 250 ft (76 m) wide	*Los Angeles Herald* 1905	
1916	27 in	0.4	2 km [1.2 mi] inland; "twenty-seven inches above the floor of my pump house"	Perry 1936	75,000 cfs
	4 ft	1.2	near mouth	*Morning Oregonian* 1916; *Sausalito News* 1916	
	9 ft	2.7	"High Water Jan 28, 1916" marked on schematic of first railroad crossing	San Diego and Arizona Railway 1910	
	house height	~3.0	~0.5 mi [0.8 km] downstream of Tijuana; "the depth of water may be judged by the fact that not a tree, house or fence is to be seen"	Morin 1916	
1927	17 in	0.4	2 km inland; "seventeen inches...above the floor of my pump house"	Perry 1936	25,000 cfs
	several meters	~3.0	river 1 km [0.6 mi] wide	Plasencia Navarro 2011	
	9 ft	2.7	area on Palm Avenue between 13th Street and 15th Street	Unknown 1976	
1941	5.4 m	5.4	river >1 km [0.6 mi] wide	Plasencia Navarro 2011	10,400 cfs
1944	2 ft	0.6	"barracks [at Border Field] were two feet under water"	Tipton 2008	13,800 cfs
1980	4 ft	1.2	"this new channel...averaged about 500 ft [152 m] wide and 4 ft deep"	Chin et al. 1991	30,088 cfs

Figure 5.22 (above). Overflow from the river to San Diego Bay. An aerial photograph taken in the aftermath of the January 1916 deluge shows floodwaters flowing from the Tijuana River into south San Diego Bay. The photograph, taken by Raymund Morris from his so-called "flying boat" during relief efforts, is the earliest known aerial image showing the Tijuana River valley (Morin 1916). *(MS 97/30, Box 1, v.5, courtesy Water Resources Collections and Archives, UC Riverside)*

Tia Juana River higher than it has been for 40 years. The river ran over into the Otay basin with such force as to tear all the railroad track out....So much water run over that where the falls of six feet were there was no perceptible evidence of them, the water running smooth over.

— TRUSSELL 1891

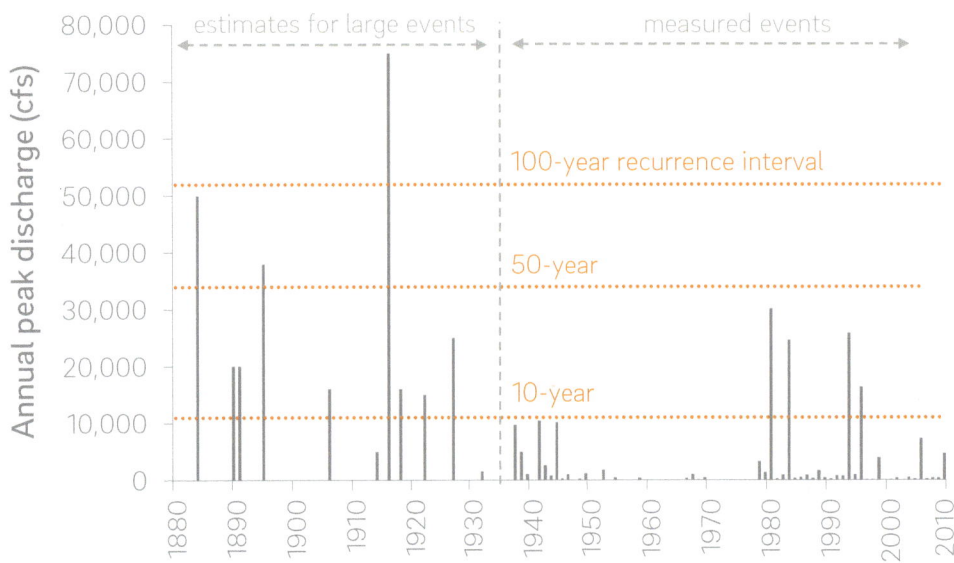

Figure 5.23 (left). Peak annual discharge (1880–2005), highlighting the years and magnitudes of major flood events. The estimated magnitude of floods with return intervals of 10 years (312 m³/s [11,000 cfs]), 50 years (963 m³/s [34,000 cfs]), and 100 years (312 m³/s [52,000 cfs]) are provided for reference (URS Corporation 2012). Data prior to 1937 are estimates from the City of San Diego (1973) based on historical information and data from nearby streams. Data from after 1937 are derived from the gauge data shown and described on p. 99.

In the lower valley, flood events were said to move and deposit large amounts of sand, silt, and large woody debris, which likely originated from higher up in the watershed (Rintoul et al. 1936; Perry, Rice, and Bruhlmeier in Wyman 1937b). An early 20th century flood reportedly deposited more than a meter (4 ft) of sediment in parts of the valley (Hull 1937). In some events, sediment was transported through the valley and out to the Pacific; during the floods of 1916, the river's sediment plume reached at least five miles into the open ocean (Morin 1916). This is notable since prior to the construction of dams in the watershed, material delivered to the ocean by the Tijuana River was the principal source of sediment to the Silver Strand littoral cell, which formed and nourished the barrier systems of the Tijuana River Estuary and San Diego Bay (Inman and Masters 1991, Dingler and Clifton 1994). These areas are now experiencing significant erosion, a fact often attributed to the lack of sediment from the Tijuana River (Moffatt and Nichol in Dingler and Clifton 1994). Large flood events accounted for the vast majority of sediment transported by the river. In the 46 years between 1937 and 1983, two single flood events accounted for approximately 60% of the total sediment (15.3 million m³ [20 million cubic yards]) moved past the Nestor gage (Haltiner and Swanson 1987). Conversely, during low-flow years the river transports little to no sediment.

Finally, flooding also had a direct influence on groundwater levels. Large floods that activated the river's floodplain were said to be required for complete replenishment of the groundwater supply (Lee 1940).

> In the flood of 1916 there were two peaks – on the 17th and again on the 27th of January. The first flood brought down from six inches to a foot of very, very valuable fertilizer, and the second flood, on the 27th of January, carried out into the ocean all of that valuable fertilizer and left virtually quicksand which was of little value.
>
> — HERBERT PERRY 1937
> IN WYMAN 1937B

> The most complete replenishment of ground water supply occurs during periods of maximum flood flow, during which the stream temporarily leaves its banks and extends out over the flood plane with a long subsequent period of strong flow in the main channel. Annual replenishment during a season of short flow period, when the stream does not break over its banks, is often incomplete.
>
> — LEE 1940

Figure 5.24. Crossing the flooded valley in Mexico. This photo (date unknown) was taken near Agua Caliente. *(Photo #OP 10834-50, courtesy San Diego History Center)*

CHANNEL MOVEMENT

The Tijuana River often shifted courses during floods, but some reaches were more stable than others.

> The characteristics of that river are in the nature of a gopher. When is grows tired of one channel, it makes another.
>
> — FOSTER 1937
> IN WYMAN 1937B

The historical course of the Tijuana River was never static. Large floods overtopped the river's low banks and, to the dismay of early American settlers, frequently cut new paths across the sandy floodplain (e.g. Rice in Wyman 1937b). The frequent channel movement was a key driver of landscape form and function, helping to maintain habitat heterogeneity on the valley floor. Newly scoured channels, abandoned reaches, and relatively undisturbed portions of the floodplain created a mosaic of riparian environments with a great deal of diversity in vegetation structure, density, and species composition, creating a varied landscape that supported a diverse community of native plants and animals (pp. 122–27).

Numerous early observers provided detailed accounts of channel movement. The rancher Emil Bruhlmeier, for example, described how "this river has never been confined to any one bed" and that "the right-of-way is wherever the river happens to go" (Bruhlmeier in Wyman 1937b). The president of a local irrigation district, describing the Tijuana River's propensity to move, compared the river to a gopher: "when it grows tired of one channel, it makes another" (Foster in Wyman 1937b; also see Rice in Wyman 1937b). The instability described by these residents is a common characteristic of dryland alluvial rivers. The limited resistance offered by sandy bank materials and a relative paucity of restraining vegetation mean that these rivers are highly susceptible to the erosive effect of high flow events (Tooth 2000). As a result, large floods frequently led to changes in channel form, including rapid lateral channel movements, or avulsions.

The Tijuana River has experienced at least 10 major channel avulsions since 1848 (see p. 114–15). These events were often quite dramatic, shifting the channel laterally by more than a kilometer (0.75 mi; Perry 1936; also see *San Diego Sun* 1891 and Bruhlmeier in Wyman 1937b, which describe channel movements of approximately half a kilometer). Avulsions occurred with a relatively high frequency, sometimes recurring more than once per decade (Chanecellor in Wyman 1937b, Harwood 1931). A report by C.S. Alverson (1914) describes a series of three channel avulsions near the border over a period of just 15 years:

> Originally the channel of the River commencing at the US and Mexico Boundary Line followed near the north line of the present valley....After the flood of 1890–91 Mr. Teavan put in a Brush and Rock wing wall...and turned the channel in a southwesterly direction. Later floods [likely those of 1895] made a second channel farther to the south and the flood waters of 1905–06 cut a channel still farther to the south....Very high water may cut the barrier of sand and silt near the Boundary Line and divert the river into one of the old channels.

The frequency of channel avulsions was at least partially driven by the frequency of high flow events; extended periods without floods sometimes led to periods of relative stability. During the mid-20th century, for example, the Tijuana River maintained a single course in the U.S.

for more than 35 years (1941 to at least 1977). This period coincided with a "cool" phase of the Pacific Decadal Oscillation (PDO) regime (1947–1976; Mantua and Hare 2002), during which time average daily discharge at the Nestor gage never exceeded 51 m³/s (1,800 cfs). Since discharge in southern California streams is strongly related to large-scale climate patterns like PDO (Milliman and Farnsworth 2011), the Tijuana River's avulsion frequency is likely at least in part also driven by these cycles.

Avulsion events were sometimes only "partial" (i.e., only a portion of the river's flow was transferred to a new channel; see Slingerland and Smith 2004). Two such events were described by Herbert Perry (1936), who owned land just upstream of the estuary. Perry said the flood of 1895 cut a new channel near his farm, but "for several years that channel was not the entire channel" and "water still flowed along the [old] channel." Similarly, after the flood of 1916, there was water in both the new and the old channels, which effectively made Perry's farm a large island. After partial avulsions the original channel generally filled on over time, a process that is seen in sequential USGS quads (USGS 1930, 1953, 1967). Residents of the valley, wanting the river to be located in one place or another, often affected this process by constructing levees (see p. 119).

(continued on p. 118)

Figure 5.25. **Multiple courses of the Tijuana River** are evident in this photograph taken during the floods of 1944. The river often created new channels during flood events. Since flow has not recently scoured the bed of the historical courses, riparian vegetation has encroached into what used to be sparsely vegetated channels. *(Photo #79:744-890, courtesy San Diego History Center)*

Tijuana Estuary

Pacific Ocean

HISTORICAL CHANNEL COURSES

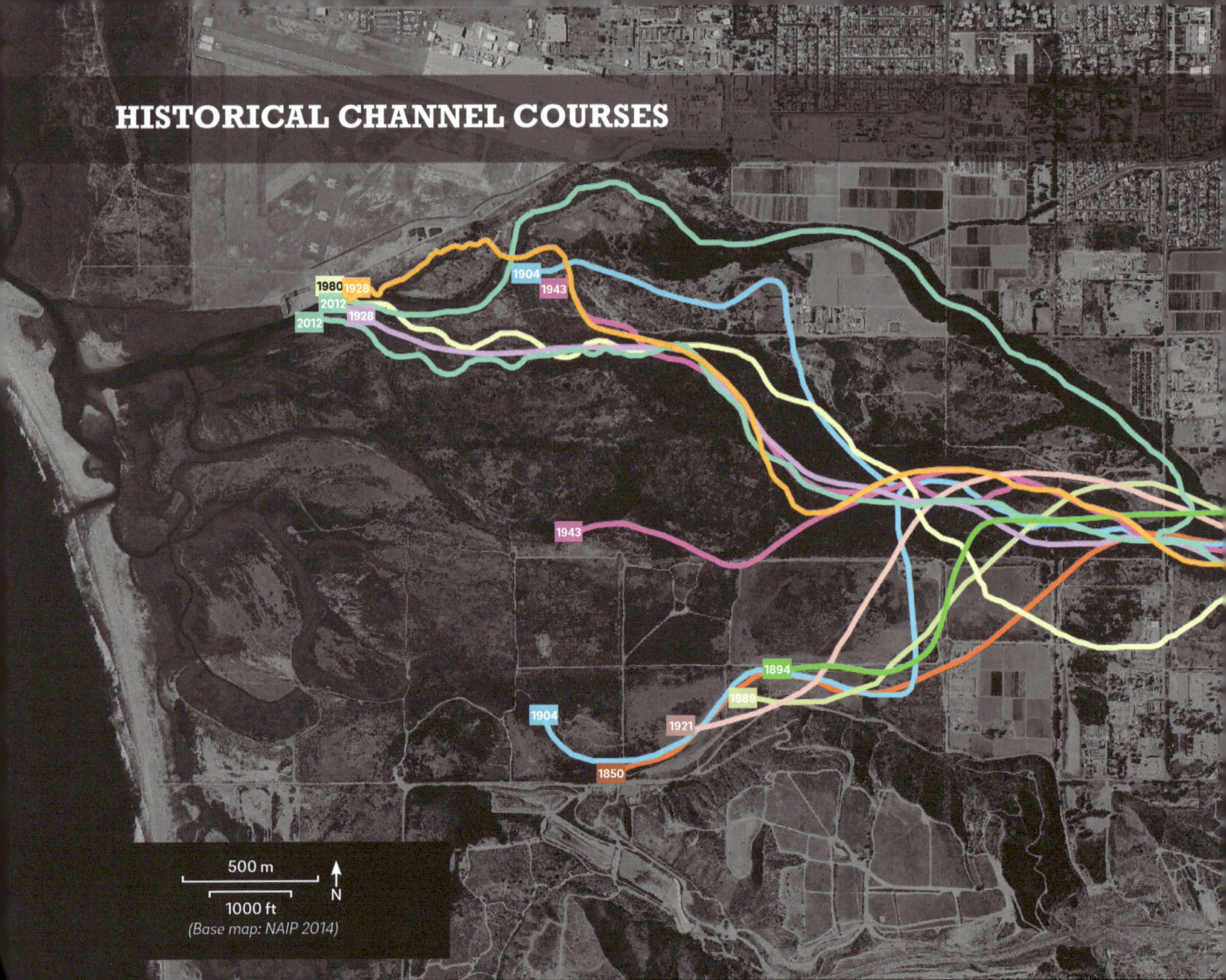

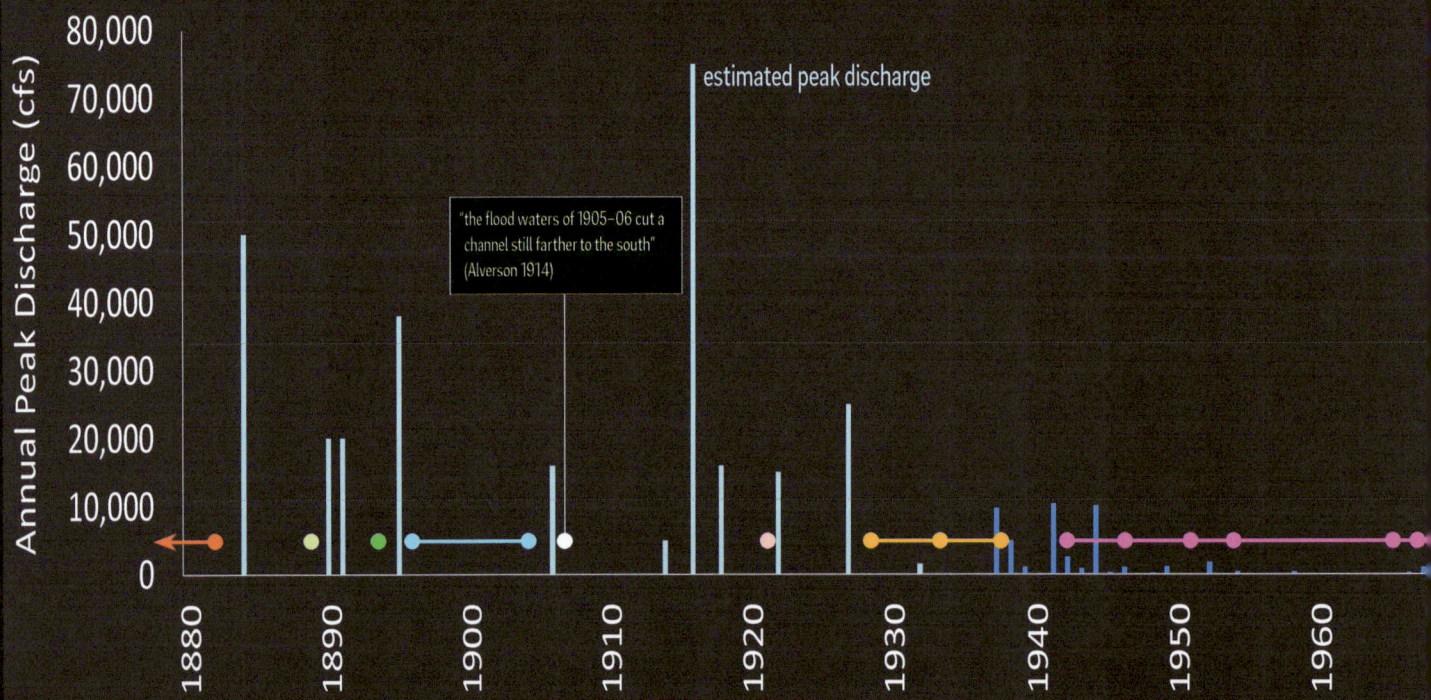

500 m
1000 ft
N
(Base map: NAIP 2014)

1980 1928
2012
2012 1928
1904
1943
1943
1894
1889
1904
1921
1850

Annual Peak Discharge (cfs)

estimated peak discharge

"the flood waters of 1905–06 cut a
channel still farther to the south"
(Alverson 1914)

80,000
70,000
60,000
50,000
40,000
30,000
20,000
10,000
0

1880   1890   1900   1910   1920   1930   1940   1950   1960

Legend (map):
- 1850
- 1889
- 1894
- 1904
- 1921
- 1928
- 1943
- 1980
- 1989
- 2012

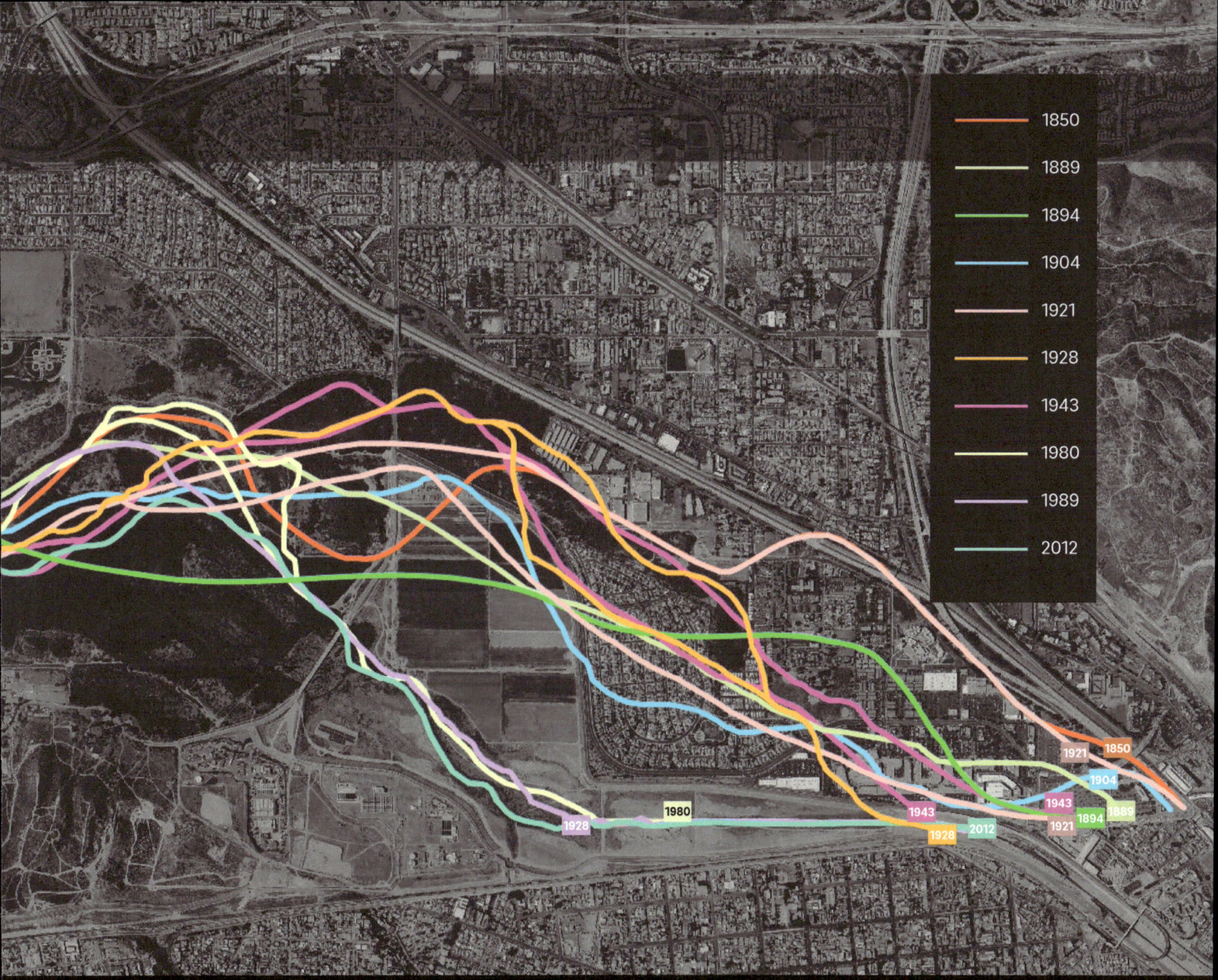

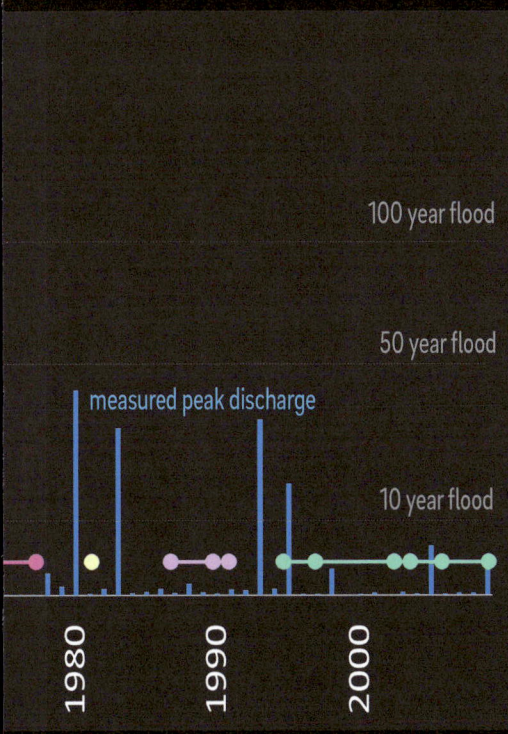

This map depicts many of the major courses the Tijuana River has taken through the U.S. portion of the valley since 1848. Plotting the courses onto a chart of peak annual discharge helps illustrate the relationship between river movement and high flow events. With few exceptions, floods above a certain size (approximately 10,000 cfs) have caused the channel to change course, whereas smaller events (less than approximately 5,000 cfs) have not led to documented large-scale changes in the position of the river.

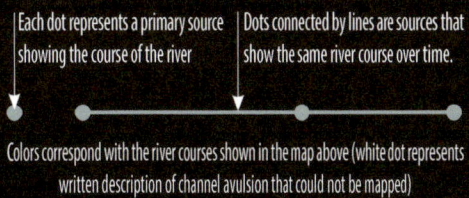

100 year flood

50 year flood

measured peak discharge

10 year flood

Each dot represents a primary source showing the course of the river. Dots connected by lines are sources that show the same river course over time.

Colors correspond with the river courses shown in the map above (white dot represents written description of channel avulsion that could not be mapped)

See p. 110 for peak discharge data sources.

1980

1990

2000

PATTERNS and CONTROLS OF CHANNEL MOVEMENT

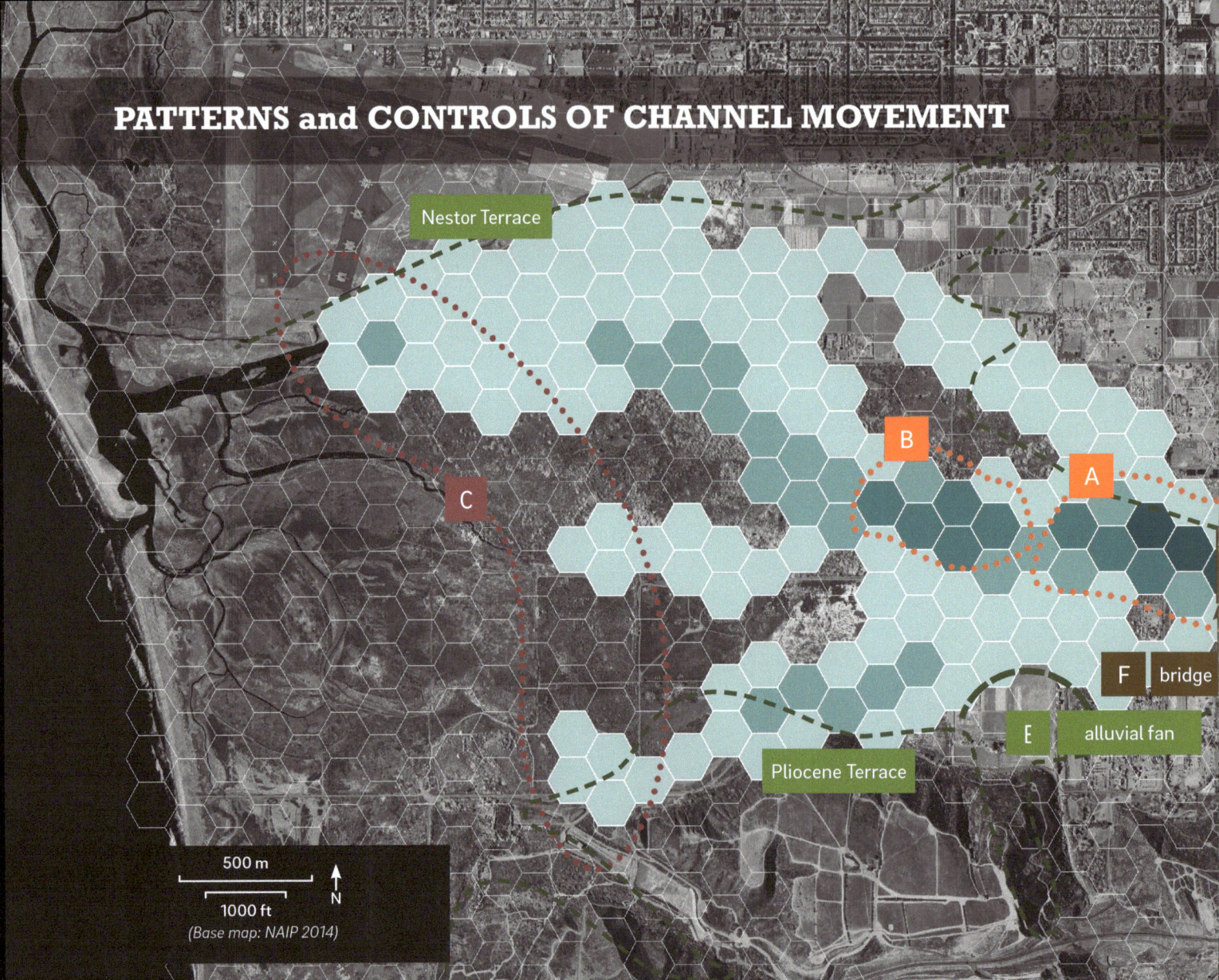

Nestor Terrace

B

A

C

F | bridge

E | alluvial fan

Pliocene Terrace

500 m

1000 ft

N

(Base map: NAIP 2014)

A   **Primary convergence zone:** At the present-day site of the Hollister Street Bridge, the position of the Tijuana River has been relatively stable over time, with all ten of the documented historical courses since 1849 passing within a zone less than 200 m (700 ft) wide. The convergence zone seems to be a product of geological controls: the historical courses are constricted between a projection of the low terrace to the north and the high mesas to the south.

B   **Secondary convergence zones:** High-probability convergence zones often create secondary zones of convergence where bends in the channel's meander sequence are accommodated (Graf 2000). Likely secondary convergence zones are apparent 1 km (0.6 mi) upstream and downstream of the primary zone.

C   **Downstream divergence zone:** During high-flow events, the downstream reach of the Tijuana River frequently shifted between one of three tidal sloughs, which created a large divergence zone just upstream of the estuary. Here, at its maximum, the historical migration zone is more than 2 km (1 mi) wide.

D   **Upstream divergence zone:** A second divergence zone is located downstream of the international border. Here, the floor of the valley is relatively wide and gently sloped, factors that promote channel movement.

E   **Geologic controls:** To the north, most of the historical river courses abut against the southern boundary of the Nestor terrace. The terrace is often slightly elevated above the floodplain and consists of less-erodible mid-Pleistocene siltstone, sandstone, and conglomerate deposits. We hypothesize that southern movement has been limited by colluvium deposited at the base of the steep mesas, the alluvial fan of Smuggler's Gulch, and the northern trend in the valley as it enters the U.S.

F   **Human infrastructure:** The primary area of convergence is located immediately upstream and downstream of the Hollister St. Bridge, where a structure has been situated since at least 1904. The bridge, which confines and directs the flow of the river, likely reinforces the underlying geological controls. Since 1977, the river has also been affected by the energy dissipater, a large flood control structure that limits channel movement near the international boundary.

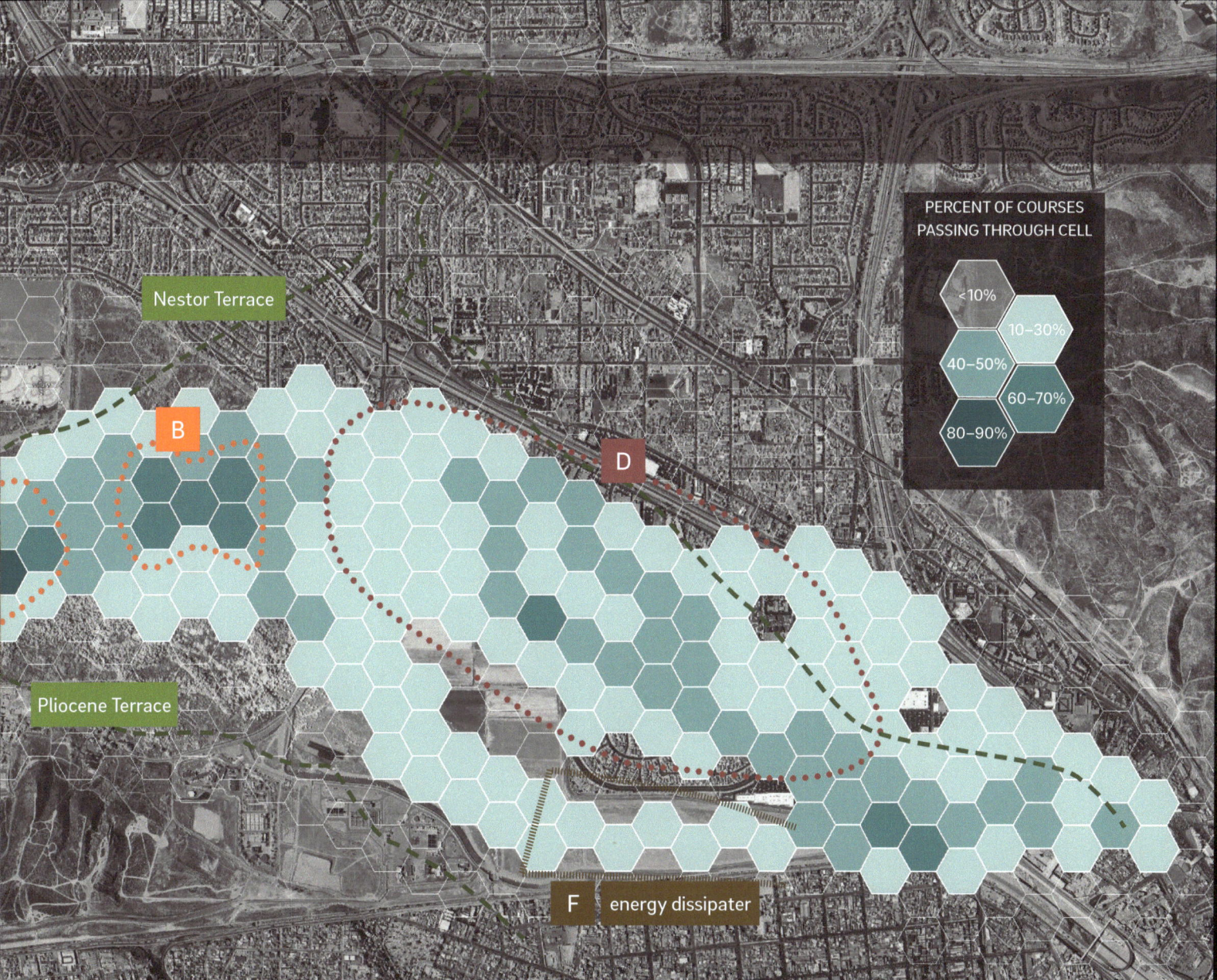

PERCENT OF COURSES
PASSING THROUGH CELL

<10%
10–30%
40–50%
60–70%
80–90%

Nestor Terrace

B

D

Pliocene Terrace

F   energy dissipater

Although early observers characterized the entire lower five miles of river channel as "unstable" and subject to course-changes during floods (San Diego County Flood Control 1937), our analysis reveals a degree of spatial variability in channel stability across the river valley. To assess and visualize channel stability, we adapted the methods of Graf (1981, 1983, 2000), who utilizes superimposed historical channel courses to generate maps of locational probability. The locational probability map of the Tijuana River indicates that along some reaches (where the cells are darker), the low-flow channel is persistently located within a relatively small area (a *convergence* zone). Along other reaches (where the cells are lighter), the location of the channel has little persistence over time (a *divergence* zone). These spatio-temporal patterns —along with some of the factors that might be controlling them—are highlighted here. Note that the map shows the percent of unique courses passing through each cell, not the percent of time the river has occupied each cell (see p. 120 for a description of the methods to this analysis).

(continued
from p. 113) Channel instability was not uniform across the full length of the river. Instead, a variety of physical controls created patterns in how the river moved and drove spatial variations in stability. While some reaches shifted dramatically during most floods and had little persistence over time (areas of "divergence"), others were more stable and consistently located within a relatively confined area (areas of "convergence"; see p. 116).

In the U.S., the primary area of convergence was centered at the present-day location of the Hollister Street Bridge. At this location, all ten of the mapped historical courses pass within a zone less than 200 m wide. Channel movement was likely limited in this area by geology: the northern edge of the convergence zone is defined by the Nestor terrace, which, although not strongly pronounced in the local topography, is made up of older and less-erodible mid-Pleistocene deposits (Kennedy and Tan 2007). Although the southern edge of the convergence zone is less defined, it coincides with a fault line, the alluvial fan of Smuggler's Gulch, and the northern edge of the Pliocene San Diego Formation, which forms the steep mesa at the southern edge of the valley (Storie and Carpenter 1930, USAED 1974, Kennedy and Tan 2008). The constriction point is situated within a low point in what Ellis and Lee 1919 term the "Tia Juana Terrace" (Fig. 5.26; but note that this unit is not indicated in contemporary maps of the region's geology). The convergence zone is also the location of the Hollister Street Bridge, which confines and directs the flow of the river and has likely reinforced underlying geological controls. A bridge has existed at the location since at least 1904 (USGS 1904).

Two weaker convergence zones are also located approximately 1 km (0.6 mi) upstream and downstream of the primary convergence zone (see p. 116). These areas, which we term "secondary convergence zones," seem related to the pattern described by Graf (2000), wherein constriction points that limit channel movement often create additional convergence zones at neighboring bends in a channel's meander sequence.

In addition to the convergence zones, analyses of historical channel positions reveal two major divergence zones (areas where the location of the active channel has relatively low persistence over time; p. 116). The first is located downstream of the international border, where the width of the valley floor is relatively large and the slope relatively low (factors that promote channel movement). The second is found at the downstream end of the study extent, where the river meets the estuary. Here, we see evidence for a series of possible "nodal" avulsions, which are recurring avulsion events that originate from a relatively fixed area of a floodplain (Slingerland and Smith 2004). In this case, the node is positioned just upstream of the estuary proper, approximately 3 km east of the shoreline. From this location, each historical course steers towards one of the

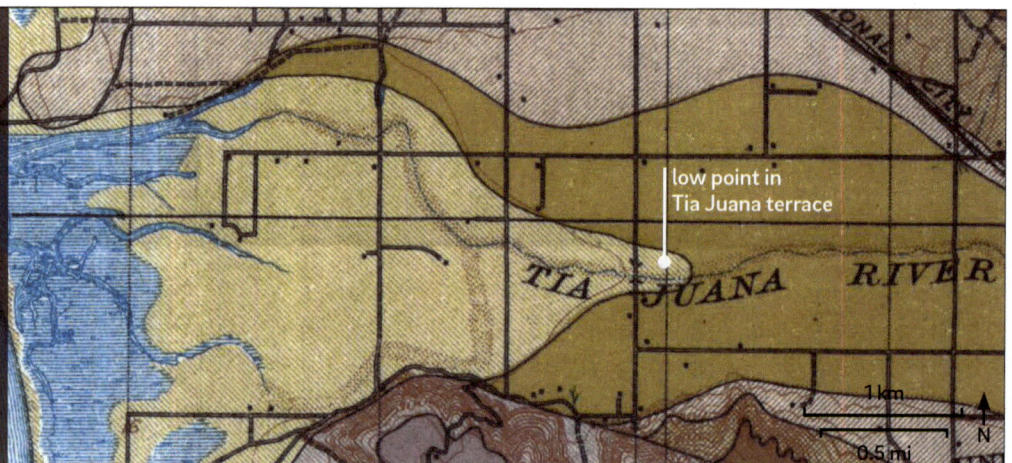

Figure 5.26. Geologic map showing the Nestor Bridge and Tijuana River aligned with a low point in the "Tia Juana terrace." This geologic feature is one possible explanation for the high degree of channel stability observed at the location over time. (Ellis and Lee 1919)

"Tia Juana terrace"

"Coastal flats"

"Salt-marsh deposits"

low point in
Tia Juana terrace

TIA JUANA RIVER

1 km

0.5 mi

estuary's main tidal sloughs (see Fig. 5.25), creating a migration zone more than 2.1 km (1.3 mi) wide. The geomorphic processes driving this divergence zone seem related to the process of "delta switching," whereby the locus of deltaic deposition repeatedly shifts (in this case between the estuary's major tidal sloughs) due to aggradation of the active channel. This divergence zone is thus associated with the river's "fluvial fan," a term referring to a sediment body that results from frequent nodal avulsions (North and Warwick 2007). With the exception of Oneonta Slough, each of the estuary's major tidal channels served as the primary outlet of the river at some point during the 20th-century.

The analysis of river movement serves as a second line of evidence for the historical location and extent of the river corridor. Though the river corridor was mapped using historical soil data independently from the channel courses, the two datasets are closely aligned and exhibit a high degree of overlap (Fig. 5.27). This pattern—river movement confined to the extent of the historical river corridor—is to be expected and reflects important feedbacks between flow, channel avulsions, sediment deposition, and the establishment of riparian vegetation (c.f. Gran et al. 2015). In recent years, however, the pattern has broken down with the river travelling well outside the historical river corridor since 1980. This change can be largely attributed to the construction of the energy dissipater structure at the international boundary in 1978, which directs the flow of the river due west. The new northern course of the river, cut during the flood of 1993, also exits the historical river corridor. This particular change was likely related to the filling of land within the floodplain west of the Hollister Street bridge for the development of a horse paddock during the late 1980s (J. Zedler, personal communication).

This said, human activities also directly affected the course of the Tijuana River during the 19th and early 20th century. In 1891, for example, a Mr. Teavan successfully altered the river's course near the international boundary after constructing a wing wall of brush and rock (Alverson 1914). Similarly, after the river cut a new northward course near the estuary during the flood of 1895, residents of the lower valley constructed a dam of sand and brush to turn the river back towards its former southern path (Perry 1936). This plan seems to have backfired, since it ultimately resulted in the old channel "filling...up level." Another small dam within the high-flow channel, presumably built after the flood of 1927, is apparent in 1928 aerial photographs. Altering the course of the channel was thought to be relatively simple task and did not necessarily require a wall or dam: "If one property owners [sic] happens to get up enough willows and trees on his land to push the river on to his neighbor's good cultivated soil, he could do so" (Bruhlmeier in Wyman 1937b).

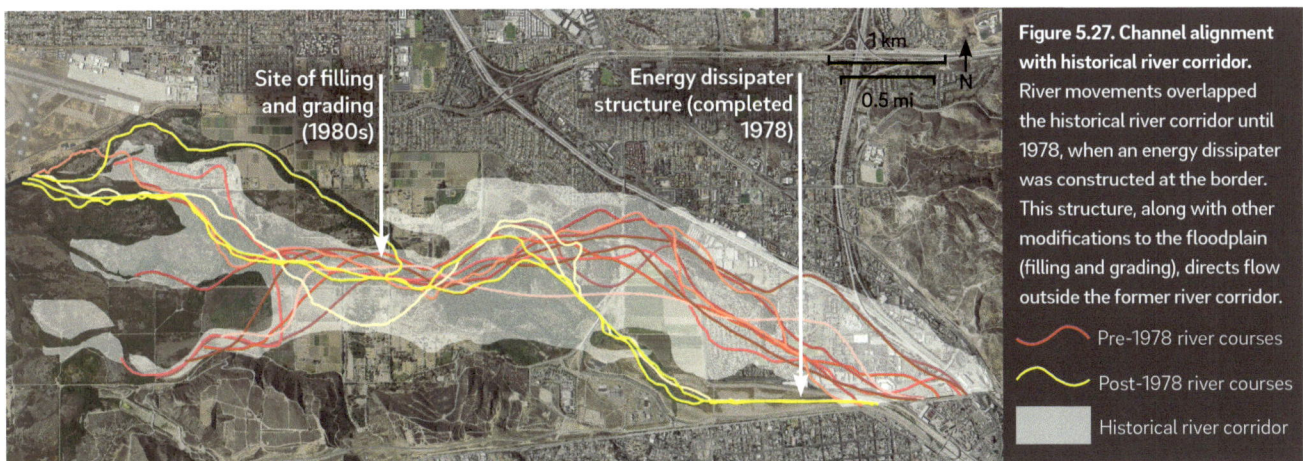

Figure 5.27. Channel alignment with historical river corridor. River movements overlapped the historical river corridor until 1978, when an energy dissipater was constructed at the border. This structure, along with other modifications to the floodplain (filling and grading), directs flow outside the former river corridor.

— Pre-1978 river courses
— Post-1978 river courses
Historical river corridor

Site of filling and grading (1980s)

Energy dissipater structure (completed 1978)

3 km
0.5 mi
N

HISTORICAL CHANNEL MOVEMENT ANALYSIS: METHODS

Our analysis was designed to assess the spatial and temporal components of river movement within the lower Tijuana River valley and to help identify the primary controls on these patterns. Ultimately, we used two related approaches to analyze and visualize channel movement. First, using maps and aerial photographs of the lower river valley from 1849–2015, we developed a spatial database and map of historical channel courses. From this map, we developed a second map of locational probability, which measures spatial variability in locational stability. The methods for and assumptions inherent in each of these approaches are discussed here. Due to data availability, both analyses were limited to the U.S. portion of the study extent.

To generate the map of historical channel courses, we first reviewed available historical data for maps and orthophotographs depicting the location of the Tijuana River. Next, we grouped contemporaneous sources that depicted equivalent channel locations into "unique courses." Finally, we digitized the center lines of each unique course from the most spatially accurate representative source. A critical component of the analysis was determining and consistently applying a threshold for what constitutes a new or "unique" channel course. Since the scale and spatial accuracy of the early historical data was not appropriate for assessing intermittent lateral migrations or changes in the width of existing low-flow channel banks, we looked for evidence of shifts in channel locations at a relatively large scale (approximately 1:20,000). This scale seemed appropriate for ignoring differences in location due to mapping and other relatively small changes in channel position and allowed us to instead identify avulsion events and "event-driven change episodes" (Graf 2000). The sources used to map each of the historical channel courses are listed in Table 5.3. For courses mapped from aerial photographs, we digitized the low-flow channel centerline at a scale of 1:12,500. For maps depicting the river as a polygonal feature, we digitized the polygon's centerline. We mapped channel braids only when explicitly depicted by an original map. We clipped channel center lines at the inland edge of the estuary.

The locational probability analysis is based on the methods originally developed by William Graf (1981) and entails superimposing a cell grid on the map of individual channel courses to determine how many have occupied each individual cell. As described by Graf (1983), the analysis is useful because it "reduces a lengthy and complex record to a single, easily interpreted map with areas of stability and instability clearly defined." To generate the map seen on pp. 116–17, we created a grid of regular hexagons (each with a side length of 100 m) and intersected these features with the historical courses map in ArcGIS. Channels were counted as belonging to cells that they intersected or if they came within 25 m (to account for approximate channel low-flow channel widths). For each cell, locational probability equals the number of courses it contains divided by the total number of courses—channel courses and locational probability were not weighted by time (see Graf 2000).

Table 5.3. Sources used to map historical river courses.

Sources depicting course	Known date range
Gray 1849 [this report] (ca. 1850 course) Hardcastle & Gray 1850 Ilarregui and de Chavero 1850 Poole 1854 Freeman 1854 (partial)	1849–1854
Beasley and Schuyler 1889	1889
International Boundary Commission 1901 (1894 course)	1894
Perry 1936 (1895 course; partial) USGS 1904	1904
Ervast 1921	1921
San Diego County 1928 Knox 1933 Cruse 1937 Lee 1937	1928–1937
USGS 1943 (1941 course) Unknown 1946 USGS 1953 (1950 course) Nationwide Environmental Title Research 2016 (1953 course) USAED 1964 Nationwide Environmental Title Research 2016 (1964 course) USCGS 1974 (1966 course) Nationwide Environmental Title Research 2016 (1966 course) USGS 1967 Nationwide Environmental Title Research 2016 (1968 course) Nationwide Environmental Title Research 2016 (1971 course) PWA 1987 (1977 course)	1941–1977
USGS 1980a USGS 1980b	1980
PWA 1987 (1986 course) USGS 1989 USGS 1990	1986–1990
USGS 1994 USGS 1996 Nationwide Environmental Title Research 2016 (2002 course) Nationwide Environmental Title Research 2016 (2003 course) Nationwide Environmental Title Research 2016 (2005 course) USGS 2012 NAIP 2012	1994–2015

RIPARIAN HABITATS & BIODIVERSITY

River wash and riparian scrub were the primary riparian habitats found within and along the river and supported dozens of native plant and animal species.

Today, the Tijuana River supports broad willow riparian woodlands that cover approximately 3 km² (700 acres; see p. 184) and regularly exceed 15 m (50 ft) in height (Boland 2014). Although some have hypothesized that the pre-Euro-American landscape also supported extensive riparian woodlands (Cleisz et al. 1989), analysis of the historical record suggests otherwise. From the late 18th through much of 20th century the valley was largely treeless, with riparian vegetation dominated not by willow woodlands or forests, but by willow scrub interspersed with less densely vegetated swaths of river wash. In total, riparian habitats covered approximately 18 km² (4,400 acres), or 40% of the valley floor, in a corridor often more than 1 km (0.6 mi) wide.

The landscape the first Europeans found upon entering the valley in 1769 was devoid of trees (see p. 66), providing evidence for the absence of extensive riparian woodland or forest during the late 18th century. Camped near the river, members of Juan Crespí's expedition—desperate for wood with which to build a fire—resorted to burning their tent poles (Cañizares et al. 1952). The lack of firewood, coupled with descriptions of a grass-covered valley (Cañizares et al. 1952, Crespí and Brown 2001), raise the possibility that riparian vegetation at the time was largely herbaceous. Native landscape management practices could be one possible explanation for herbaceous cover — though they did not cut live trees, in valleys across the region aboriginal Kumeyaay did plant and gather a semi-domesticated native grain-grass that was maintained with fire (Shipek 1989, 1993). Fields were planted via broadcast seeding, harvested in June or July, then intentionally burned before being re-sowed. The Kumeyaay also quickly re-seeded areas scoured by floods (Shipek 1993). Since various leafy greens and other annual plants were also broadcast seeded with the grain, the fields would not necessarily have been recognized for what they were by Europeans such as Crespí, who were accustomed to monocultures of wheat or row crops.

During the mid-19th century, after the displacement of Kumeyaay from the valley, willow-dominated riparian scrub was the primary plant community along the river. The earliest evidence for this comes from a ca. 1840 Mexican *diseño* map, which notes *"monte de saus,"* or willow scrub, extending eastward from the approximate location of Goat Canyon along both sides of the Tijuana River (see Fig. 5.3; U.S. District Court California,

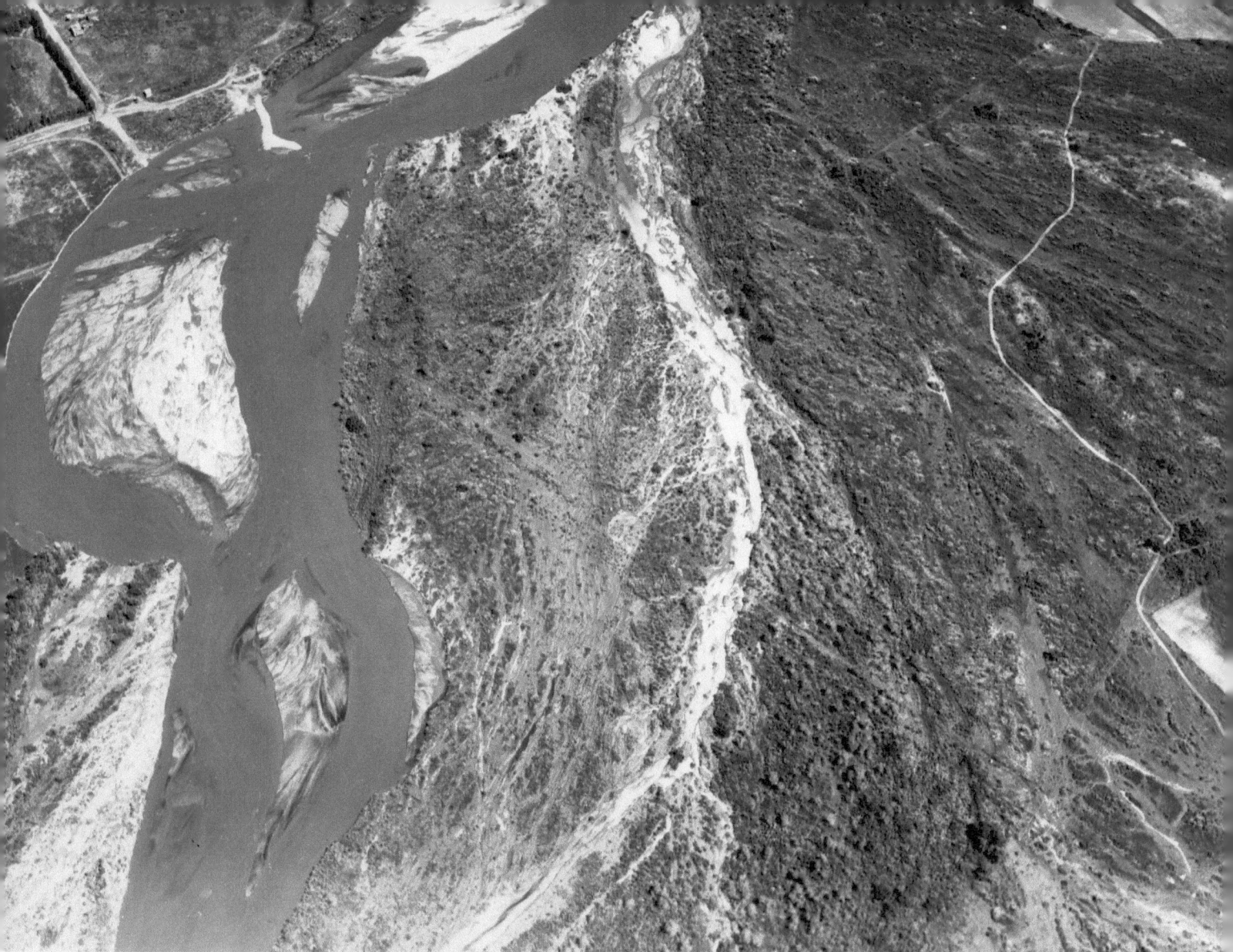

Southern District ca. 1840). Other sources from the period describe the river corridor as "weeds brush and little grass" (Freeman 1854) and depict the river as lined with vegetation (Fig. 5.28; Gray 1849, Hardcastle and Gray 1850). The lack of riparian forests around the time of California statehood is affirmed by the earliest GLO survey (Freeman 1854), which recorded only six bearing trees, each extremely small (12–46 cm [5–18 in] in diameter) and located dozens of meters from the survey points. Since GLO surveyors preferred to use bearing trees that were large and close (White 1983), this suggests that even small trees were few and far between. Surveyor J. Pascoe (1869b), walking the same section line 15 years later, recorded no bearing trees "as none came within reach of the corners." This may reflect a difference in surveying technique, or alternatively may reflect a difference in vegetative cover before and after intervening floods (see p. 132 for descriptions of temporal variation in riparian vegetation).

Since riparian willows can be a source of livestock forage (e.g., Kovalchik and Elmore 1992) and cattle are known to have grazed within the river corridor (e.g. Unknown 1890), it is also possible that grazing in the 19th century could have impacted the structure or composition of riparian habitat, though the effects of these practices are unknown (see pp. 46–47). The prevalence and impact of harvesting wood for fuel, another activity that could have affected riparian vegetation during mid-1800s, are also unknown.

Figure 5.28. (facing page, bottom) Riparian vegetation as depicted on an early map. Maps of the river valley dating from the time of statehood, like this boundary survey from 1848, depict vegetation along the river. Though the map seems to indicate the presence of trees, we are confident this depiction is largely diagrammatic; more detailed sources from the period corroborate the presence of riparian scrub, but not riparian woodland. *(Gray 1849, courtesy Coronado Public Library)*

Figure 5.29. (above) Extensive areas of riparian scrub are evident in this photograph dated May 1, 1941. Different areas of riparian scrub had different densities. The boundaries between these areas were likely defined by the layering of flood events over time. The photo looks upstream from the present-day location of Dairy Mart Rd. *(Photo #79:741-897, courtesy San Diego History Center)*

Figure 5.30. (below) Riparian vegetation seen in subsequent historical aerial photographs. The earlier set of aerials (top), were flown during the later months of 1928 and show the valley approximately two years after the major flood of February 1927. From these photos it is clear that the low-flow channel is the most sparsely vegetated surface within the river corridor, with varying—but still relatively low—riparian plant densities across other river wash surfaces. The area of river wash is bounded by dense riparian scrub. Aerials flown in 1946 (18 years and four >5,000 cfs flood peaks later) show that much of the river wash habitat remained sparsely vegetated, with extensive swaths of sand still apparent. *(1928: County of San Diego, 1928; 1946: Unknown 1946, Photo #19496-CGE-7221-A-01, courtesy Mapoteca Manuel Orozco y Berra, Servicio de Información Agroalimentaria y Pesquera)*

Figure 5.31. "Channel of Tia Juana," ca. 1920. A view of the sandy, sparsely vegetated Tijuana River channel showing the riparian plant community typical of river wash habitat. The river wash in the foreground grades into areas of denser riparian scrub in the background. Location unknown. *(LIPP, Box 85, Photo #3069, courtesy Water Resources Collections and Archives, UC Riverside)*

1928

1946

Figure 5.32. "In the bed of the Tijuana River," June, 26, 1903. This early photograph highlights both the river's dry, sandy substrate and its relatively sparse and short-statured riparian vegetation. *(F869.S22 P46, Photo VII, courtesy Special Collections & Archives, UC San Diego Library)*

By the early 20th century, numerous accounts explicitly described extensive areas of willow riparian scrub (Storie and Carpenter 1923, Huey 1931; also see *Daily Alta California* 1890b, Stephens 1908, and Wyman 1937a, which describe generic "brush"; Fig. 5.29). Common species included sandbar willow *(Salix exigua),* which was frequently documented along the river corridor by naturalists beginning in 1882; Goodding's willow *(S. gooddingii),* red willow *(S. laevigata)* and arroyo willow *(S. lasiolepis)* were also present during the early 20th century (see Table 5.4). The prevalence of willow scrub as the dominant early 20th century riparian habitat type is confirmed by the T-sheet resurvey (Knox 1933), which shows over 1,000 acres of "willow brush" within the river's floodplain, and corroborated by early aerial and landscape photographs (San Diego County 1928; Fig. 5.30).

Interspersed within the extensive areas of riparian scrub were sandy bands of river wash. Riparian scrub and river wash were broadly associated with the different geomorphic features of the historical river corridor (described on p. 90): dense willow scrub was primarily found on the floodplain, while river wash was primarily found on the younger, flood-scoured surfaces of active and recently abandoned channels (Fig. 5.30). Areas of river wash were more sparsely vegetated than areas of riparian scrub, with exposed sand, scattered willows, and seasonal grass cover (Fig. 5.31; Unknown 1890, Storie and Carpenter 1923). That said, the two habitat types were closely related and might best be thought of as two points along one cyclical ecological gradient, with boundaries that shifted over time and in response to floods. Indeed, to those on the ground the boundary between riparian scrub and river wash were somewhat indistinct. Harwood, describing conditions in 1931, noted how, because "there are [also] willow trees in the river bottom… there can be no definite separation" between the willow thickets and sandy clearings. Small sandy clearings were also present amidst areas of otherwise dense scrub (von Bloeker 1931).

Table 5.4. Historical plants of the river corridor. A partial list of native plants historically present in the Tijuana River corridor, drawn from early accounts and pre-1950 herbarium records. The records suggest a mix of both dryland and wetland species, as indicated by the brown and blue dots, respectively (see Fig. 5.33).

Species were selected for inclusion based on a combination of the locality information provided in the herbarium records and known associations with riparian habitat types. All data were provided by the participants of the Consortium of California Herbaria, except for additional citations listed in the Notes field. Harwood (1931) notes that his list includes "the commonest species only," and that rare "does not mean very few found but few found in accordance to the other species mentioned."

While willows were the most consistently documented riparian plant, the Tijuana River corridor also supported many additional species (Table 5.4). Taken together, these species are indicative of a wide range of hydrological conditions and a high degree of historical habitat heterogeneity. Plants historically found in the river corridor include many species associated with drier coastal sage and alluvial scrub, as well as facultative and obligate wetland species (Figure 5.33). The river corridor also supported numerous riparian plant species that are now considered rare or threatened: early naturalists gathered specimens of Matilija poppy *(Romneya coulteri,* which was described as "abundant"; Jepson 1907), slender woolly-heads *(Nemacaulis denudata var. gracilis),* Nuttall's lotus *(Lotus nuttallianus),* and southwestern spiny rush *(Juncus acutus subsp. leopoldii),* which are all listed in the California Rare Plant inventory. These findings are consistent with some contemporary research that suggests intermittent streams may have higher long-term riparian species diversity than those with perennial or strictly ephemeral flow (a "Goldilocks hypothesis"; Katz et al. 2012).

Riparian biodiversity was also supported by disturbance events that periodically altered the structure of the environment and maintained habitat heterogeneity at the landscape scale. Tules *(Schoenoplectus acutus),* for example, were only documented around groundwater-filled pools carved within the river corridor by floods (see p. 104). The river's riparian habitats also supported a range of wildlife. Although not explored in depth in this report, we highlight some of the riparian species historically documented within the valley on pp. 128–29.

Common name	Scientific name	Relevant excerpts	Years	Notes
Trees				
sandbar willow	*Salix exigua*		1882, 1895, 1902, 1903, 1913, 1919, 1936, 1938, 1949	Schneider 1919, Higgins 1949; includes records for *S. sessilifolia*
Goodding's willow	*S. gooddingii*		1903	
red willow	*S. laevigata*		1938	
arroyo willow	*S. lasiolepis*		1938	
cottonwood	*Populus sp.*	"rare" (Harwood)	1883, 1931	Wilson 1883, Harwood 1931
California sycamore	*Platanus racemosa*		1894	
Shrubs				
mulefat	*Baccharis salicifolia*	"abundant"	1902, 1913, 1931	Harwood 1931
black sage	*Salvia mellifera*	"common"	1931	Harwood 1931
white sage	*S. apiana*	"common"	1931	Harwood 1931
common sagebrush	*Artemisia tridentata*	"rare"	1931	Harwood 1931
arrowweed	*Pluchea sericea*		1903, 1949	Higgins 1949
bush senecio	*Senecio douglasii*		1903	
fourwing saltbush	*Atriplex canescens*	"banks of Tia Juana River"	1903, 1935	
chaparral mallow	*Malacothamnus fasciculatus*		1935	
California fagonia	*Fagonia laevis*		1903	
blue elderberry	*Sambucus nigra subsp. caerulea*	"Tia Juana River near ocean"	1902	

Common name	Scientific name	Relevant excerpts	Years	Notes
Herbs				
nightshade	*Solanum sp.*	"common"	1931	Harwood 1931
clematis	*Clematis ligusticifolia*	"common"	1931	Harwood 1931
branching phacelia	*Phacelia ramosissima*	"rare"	1931	Harwood 1931
Chinese parsley	*Heliotropium curassavicum*	"common"	1931	Harwood 1931
rigid bird's beak	*Cordylanthus rigidus*	"common"	1931	Harwood 1931
bladderpod	*Peritoma arborea*	"common"	1931	Harwood 1931
skunkbush	*Navarretia squarrosa*	"common"	1931	Harwood 1931
Matilija poppy	*Romneya coulteri*	"edges of dried streams" (Parry); "abundant in the Tia Juana River bed. It is blooming now. Also at High School Commencement time, since they use it in decorating" (Jepson, December 13, 1907)	1849, 1907	Parry 1849, Jepson 1907
California evening primrose	*Oenothera californica*		1893, 1903	
spiny rush	*Juncus acutus*		1902, 1903	
southwestern spiny rush	*Juncus acutus subsp. leopoldi*		1903	
California croton	*Croton californicus*		1903	
Heermann's lotus	*Acmispon heermannii*		1903	
Nuttall's lotus	*Lotus nuttallianus*	"Tia Juana Wash"	1903	
Beardless wild rye	*Elymus triticoides*		1903	
spiny goldenbush/ spiny chloracantha	*Chloracantha spinosa, C. spinosa var. spinosa*		1902, 1938	
bush seepweed	*Suaeda nigra*	"Tia Juana River near Monument School"	1903, 1936	
slender woolly-heads	*Nemacaulis denudata var. gracilis*		1903	
scarlet lupine	*Lupinus concinnus*		1903	
California sealavender	*Limonium californicum*		1935	
Indian hemp	*Apocynum cannabinum*	"Tia Juana River at Monument School"	1938	
mugwort	*Artemisia douglasiana*		1935	
wide throated yellow monkeyflower	*Mimulus brevipes*		1913, 1894	
volcanic gilia	*Gilia ochroleuca ssp. Exilis*	"Tia Juana River"	1903	
ropevine clematis	*Clematis pauciflora*		1913	
tule	*Schoenoplectus acutus*	"Tia Juana River...pot holes"	1931	Harwood 1931

Drier, coastal sage and alluvial scrub species (e.g. fourwing saltbush)

Wetter, obligate and facultative wetland species (e.g. tule)

Figure 5.33. Drier and wetter species in the river corridor. The Tijuana River corridor supported a wide range of riparian plant species that, taken together, are indicative of a high degree of habitat heterogeneity. Many of the plants, like the fourwing saltbush (*Atriplex canescens*), are associated with drier coastal sage and alluvial scrub communities. Others, like tule (*Schoenoplectus acutus*), are facultative or obligate wetland species. Note that these photographs are not from the Tijuana River valley. *(Tule: Andrey Zharkikh, 2013, CC BY 2.0; Saltbush: Steven Perkins, 2006)*

RIPARIAN WILDLIFE

In southern California and beyond, riparian corridors support a diverse array of fauna requiring close proximity to a water source and dense vegetation. Of these, a handful of riparian birds are often used as focal species for conservation planning, and are thought to be good indicators of certain riparian habitat attributes (Chase and Geupel 2005, RHJV 2004). A wide array of these focal species were historically found in the Tijuana River's riparian zone, including riparian vegetation generalists, ground-dwelling species, wetland species, and species with preference for dense shrub thickets.

The **San Diego song sparrow** (*Melospiza melodia cooperi*) is a common riparian focal species primarily found in riparian habitats and wetlands (Humple and Geupel 2004). Recorded in the lower Tijuana River region by Frank Stephens as early as 1908 (VertNet specimen record), the song sparrow thrives in both fresh and brackish marshes, and prefers "tall rank growth of cattails and bulrushes" and "early successional riparian habitat" (Humple and Geupel 2004). Song sparrow abundance is positively correlated with willow presence, but the birds are generally absent in riparian habitat with a dense forested canopy (Marshall 1948, Sanders and Edge 1998). Historical presence of the species thus suggests the presence of open willow scrub and/or marshy riparian habitat.

A second focal species, the **blue grosbeak** (*Passerina caerulea salicaria*)—known for its vibrant colors and large beak—was also historically observed in the valley (the first known record is from M. Canfield in 1927; VertNet specimen record). Blue grosbeaks are considered a riparian edge species and prefer breeding habitat with "herbaceous annuals and young, shrubby willows and cottonwoods, such as those regenerating after a flood" (White 1998; the structure of the vegetation is considered to be more important than its species composition). The historical presence of grosbeak could suggest the presence of this early successional vegetation.

The federally endangered **least Bell's vireo** (*Vireo bellii pusillus*) was once considered one of the most common birds in California's riparian habitats. However, the widespread destruction of woody riparian habitat and extreme vulnerability to cowbird parasitism led to precipitous population declines during the later half of the 20th century (Kus 2002). In 1986, only 300 territorial males remained in southern California, which prompted the federal government to list the sub-species as endangered (Kus et al. 2010). The species was first documented in the Tijuana River valley in 1920 (VertNet specimen record). Early successional riparian stands of five to ten years of age characterized by dense, shrubby vegetation are most suitable for least Bell's vireo nesting, which generally occurs only one meter off the ground (Kus 2002).

While each of the above species was documented in the valley prior to 1950, some focal species present today have only recently been recorded, including a few that favor dense riparian woodlands (a habitat type that established in the valley after 1980; see pp. 184–87). The federally endangered **southwestern willow flycatcher** (*Empidonax traillii extimus*), for example, breeds in dense stands of riparian vegetation dominated by willow shrubs and trees (Craig and Williams 1998) and generally requires a developed canopy for successful breeding (Craig and Williams 1998, Sogge et al. 1997) Although early records confirm southwestern willow flycatcher nested in nearby regions (like the San Diego River as early 1898 and the Sweetwater River as early as in 1910), territorial birds were not documented in the Tijuana River valley until 1981, just after riparian woodlands first developed along the river (Unitt 1987). This pattern—the establishment of willow flycatcher colonies only after woodlands recently developed—is also known to have occurred in at least two other locations in San Diego County (Haas and Unitt 2004). Although the absence of historical records is not definitive evidence that the species itself was absent, it does raise the possibility that the valley, which was dominated by riparian scrub, did not historically provide suitable breeding habitat for this endangered species, even prior to major European-American landscape modification.

Other riparian indicator species that were not documented in the valley during the breeding season prior to the establishment of riparian woodlands (ca. 1980), but which have since established breeding populations, include **Swainson's thrush** (*Catharus ustulatus*, which typically nests in riparian woodland with a rather closed canopy; Unitt 1987), **tree swallow** (*Tachycineta bicolor*, which nests in tree cavities in snags at the edge of openings in riparian woodland [Unitt et al. 2004]), and **yellow warbler** (*Setophaga petechia*, which "symbolizes mature riparian woodland" [Unitt et al. 2004]). Conversely, at least one species that relies on the sparsely vegetated, sandy alluvium that characterized the historical riparian corridor—the endangered **Pacific pocket mouse** (*Perognathus longimembris pacificus*; USFWS 1994a)—has been extirpated from the valley (though this did occur prior to the establishment of riparian forests).

Together, these findings highlight potential management trade-offs. Managing for an intermittent flow regime, as is done now, might preclude the long-term persistence of the locally novel riparian forests that now serve as breeding habitat for a variety of riparian indicator species, including an iconic endangered species. On the other hand, reestablishing historical riparian conditions could support a different suite of endangered species. The recent outbreak of an invasive ambrosia beetle (*Euwallacea* sp.), which has led to widespread tree mortality (Boland 2016), further expedites the need for discussions around the future of riparian habitats in the valley and larger region (see p. 195).

RIPARIAN SPATIAL VARIABILITY

Though mostly riparian scrub, riparian habitat differed in a few locations.

Reach-scale hydrologic variability in variables such as flow permanence (see p. 100) often
underpins reach-scale differences in the structure, composition, and diversity of riparian plant communities (e.g., Stromberg et al. 2005, Beller et al. 2011, Katz et al. 2012). Due to idiosyncratic data sources and the dynamic nature of the Tijuana River itself, very few strong longitudinal patterns emerge from the available historical information. However, some historical reach-level variability is apparent.

At the broadest scale, willow scrub was less prevalent in the lowest reaches of the river as it approached the estuary: Stephens (1912) noted how the willows "gradually grow fewer and smaller as the river bottom merges into the salt marshes near the sea." This gradient was likely controlled, at least in part, by increasing downstream salinities, which would have limited growth of salt-intolerant plants. Indeed, Stephens also noted that willows and mulefat were found "higher up [after] the soil becomes less saline." Stephens' observations are supported by Harwood (1931), who found that "willow thickets" were first found approximately 2 km (1 mile) inland, just past the historical inland margin of salt marsh. Curiously, the willow thickets, which he described as "very dense," were said to only extend inland along the banks of the river for another 2.4 km east (1.5 mi), or to the approximate present-day location of Hollister Street. Since contemporaneous sources (e.g. Knox 1933) confirm the presence of willow scrub much further east (to the international border and beyond), Harwood's "willow thickets" could refer to a more densely vegetated (or somehow otherwise distinct) portion of the floodplain.

A second longitudinal pattern concerns the presence of large riparian trees, which were more prevalent in the upper portions of the Tijuana River watershed and grew less frequent further downstream. *The Daily Alta California* (1868) noted that "willow, cottonwood, and sycamore are to be had in considerable quantity higher up the stream, but become very scarce near the mouth." The newspaper goes on to describe how settlers in the lower valley used driftwood

Figure 5.34. (below) Views of herbaceous wetlands within the river corridor at Agua Caliente, August 13th, 1921. The herbaceous cover (labeled with "*cienega* [marsh]") is likely related to the perennial surface water found at the location. *(Photos #AS-723-10513-1-83 (1) & #AS-723-10513-1-83 (2), both courtesy Archivo Historico de Agua)*

Figure 5.35. (facing page, bottom) Large trees along the bank of the river at Agua Caliente. This photograph, taken on July 29th, 1910, shows the "first train" of the SD&A Railway unloading passengers at the site of the Agua Caliente springs. A grove of trees (probably cottonwoods) are visible on the raised land adjacent to the river. This area is the only known location in the lower valley to have supported large trees. Reflections under the small footbridge suggest there is water in the channel bed, even in late July. The presence of the riparian trees at this site is very likely related to the presence of perennial stream flow (and a high/stable groundwater table). *(Unknown 1910, courtesy Coleccionista De Tijuana)*

MANANTIAL DEL AGUA CALIENTE SOBRE LA MARGEN IZQUIERDA Y DENTRO DEL CAUCE DEL RIO DE TIJUANA.

F.C. TIJUANA A TECATE. S. DIEGO ARIZONA. MANANTIAL BORDO. CIENEGA DENTRO DEL CAUCE DEL RIO DE TIJUANA

and scrub from the hills for fuel in lieu of timber. Another early observer noted "plenty of wood...cottonwood and willow" along Cottonwood Creek, which also suggests riparian trees were more prevalent upstream of our study extent (Andrews 1853).

Although riparian scrub was the primary riparian plant community, a short and narrow reach of the river at the Agua Caliente hot springs was instead dominated, at least during some years, by herbaceous freshwater wetlands (Fig. 5.34). Photographs from the summer of 1921 label a portion of the river bed as "*cienega* [marsh]" and the visible vegetation is low and herbaceous. This unique plant community was likely sustained by water from the springs, which contributed to year-round flow at this location(see p. 100). This finding is broadly consistent with contemporary research of other semi-arid rivers in the southwest, which has found that reaches with saturated soils often support herbaceous wetlands that are absent (or at least decrease in seasonal cover) along non-perennial reaches (Stromberg et al. 2005).

Additionally, one report from 1883 suggests that a grove of cottonwood trees grew in the immediate vicinity of the Agua Caliente springs (Wilson 1883), which is notable since the valley was mostly treeless. A grove of trees is visible on the raised left bank of the river in a number of photographs from the early 20th century (e.g. Fig. 5.35). A slightly earlier photograph, likely taken during the summer of 1894, clearly shows mature cottonwoods (*Populus fremontii*) at the site (Blanco 1901). Cottonwood is considered an obligate phreatophyte sensitive to fluctuations in water table; dense, multi-aged stands are only sustained where groundwater under the floodplain is shallow and stable (Stromberg et al. 2007). These are conditions one would expect to find at a location with year round surface flows.

RIPARIAN TEMPORAL VARIABILITY

Floods created natural variability in riparian habitats over time.

In the Tijuana River corridor, floods drove significant interannual variation in the structure and composition of riparian vegetation. Floods affected riparian habitats through processes such as physical scour, sediment deposition, groundwater recharge, and propagule transport. The effects of floods on decadal-scale variability in bottomland morphology and associated riparian vegetation are especially pronounced in dryland rivers with high flow variability, since the geomorphic effects of floods persist longer in these systems (Baker 1977, Friedman and Lee 2002). In the Tijuana River, comparison of historical photographs from the frequently photographed international border show the scour, regrowth, and subsequent re-scouring of riparian scrub along the route between San Ysidro and Tijuana during the late 19th and early 20th centuries before and after major floods (Fig. 5.36). A similar sequence can be seen in photographs of Matanuco Canyon, the present-day site of the Rodríguez Dam (Fig. 5.37).

Seasonally, grasses within the river corridor responded to intra-annual fluctuations in precipitation and runoff. Herbaceous vegetation was said to "spring up after the high-water stage," which provided forage for grazing cattle and other primary consumers (Storie and Carpenter 1923).

Figure 5.36. Temporal variation in riparian vegetation at the international boundary. In the top photo, likely taken after floods in 1891 or 1895, the foreground appears scoured and largely devoid of woody vegetation. By 1910 (middle), this area had largely re-vegetated with dense riparian scrub. In both 1916 and 1918, major floods swept through the valley. In a photo from after these floods (bottom) much of the vegetation present in 1910 has again been scoured out. Note that livestock are visible grazing in the river corridor in both the first and second photographs; this activity also could have impacted the structure and composition of riparian vegetation over time.

(ca. 1895: Photo #FEP 836, courtesy San Diego History Center; 1910: Photo #1113, courtesy San Diego History Center; ca. 1918: F1391.T36 T5532, courtesy Special Collections & Archives, UC San Diego Library)

ca. 1895
Tijuana
newly scoured
crossing

1910
re-vegetated with willow scrub
crossing

ca. 1918-1920
Tijuana Mexico.
re-scoured
crossing
Tijuana, Calif

Figure 5.37. Temporal variation in riparian vegetation at Matanuco Canyon. Each of these photos look upstream from the Matanuco Canyon railroad bridge.

ca. 1912: At the time this photo was taken, approximately seven years had elapsed since the last significant flood, and dense riparian scrub was growing within the channel corridor. A road cuts though the riparian vegetation in the bottom-right hand portion of the frame. *(Bonillas and Urbina 1912, Lam. IV, courtesy HathiTrust Digital Library and The University of Michigan)*

1920: Much of the riparian scrub growing within the channel corridor ca. 1912 (top) was absent in 1920. The vegetation was presumably washed out by the floods of 1916 and 1918. The photo shows a sandy point bar partially recolonized with short plants. There is no evidence of the road seen in the first image. *(Photo by Hiram Savage, July 1920, MS 76/16, Box 2 Folder 44, #387, courtesy Water Resources Collections and Archives, UC Riverside)*

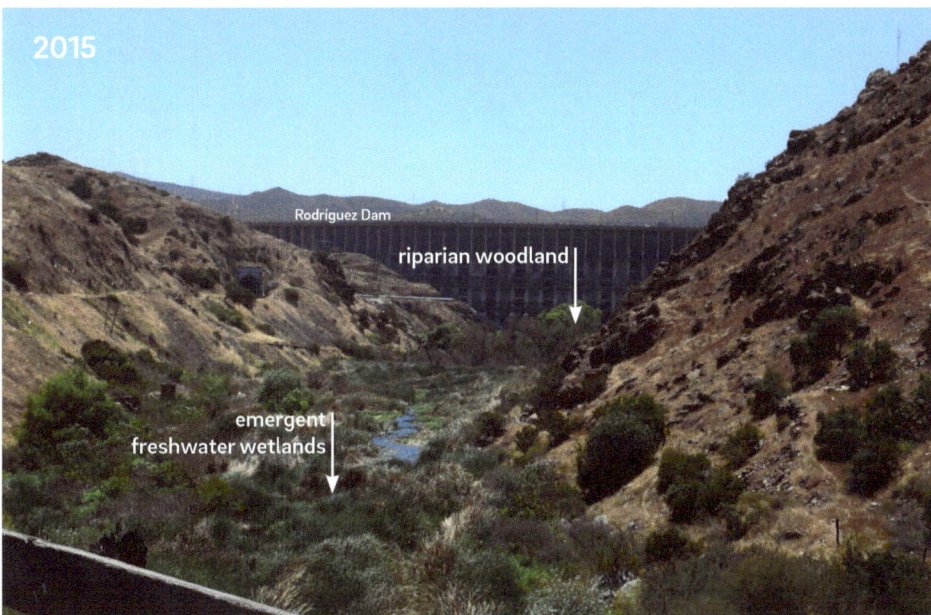

2015: The canyon is now the site of Rodríguez Dam, which was completed in 1936. Regular dam releases have increased the water supply to the canyon, which is now the site of emergent wetlands and a handful of large riparian trees, neither of which appear to have been found historically along this reach. *(Photo by Samuel Safran, April 2015)*

6 TIJUANA RIVER ESTUARY

Perhaps nowhere in the Tijuana River valley are the dynamic fluctuations and physical processes that shape the landscape so clearly visible as in the Tijuana Estuary. Situated on the westernmost end of the valley, the estuary occupies a transitional zone between the terrestrial, riparian, and wetland habitats of the Tijuana River valley and the Pacific Ocean. As such, the estuary is profoundly influenced by the interplay between the freshwater and sediment inputs from the watershed and waves, tides, and littoral sand movement from the ocean.

This chapter describes the physical processes and habitat mosaics that characterized the estuary during the mid-19th century. Analysis of historical data provides insights into key physical processes – streamflow, tidal flux, and inlet dynamics – that shaped the estuary and generated physical gradients in hydrology, salinity, and elevation. Those gradients are reflected in the distribution of habitat types that existed in the estuary historically, including Salt Marsh, Mudflat/Sandflat, Salt Flat / Open Water, Subtidal Water, Beach/Dune, and High Marsh Transition Zone (mapped together with Alkali Meadow Complex).

(Douglas Inman Papers, SMC 57, Box 169, Folder USA C1.1, Tijuana Slough, Regional Photos, courtesy Special Collections & Archives, UC San Diego Library)

TIJUANA ESTUARY, ca. 1850

Historical habitat types & key findings

p. 138

Tides reached more than 2.5 km inland and were a major driver of the historical morphology and ecology of the estuary. Approximately 3.5 km² of the river valley was subject to tidal inundation, creating a potential diurnal tidal prism volume on the order of 680,000 m³.

p. 156

A barrier system supporting Beach and Dune habitats separated the estuary from the Pacific Ocean. This undulating landscape was characterized by small-scale habitat heterogeneity and supported a range of unique plant and animal species.

p. 142

Early sources suggest that the dominant condition of the Tijuana Estuary inlet was either fully or partially open (i.e., closed in the subtidal or intertidal).

Mid-valley Slough

Tijuana River Slough

Oneonta Slough

Legend

- Dune
- Beach
- Subtidal Water
- Mudflat/Sandflat
- Salt Flat / Open Water
- Salt Marsh
- Alkali Meadow Complex / High Marsh Transition Zone
- River Channel
- River Wash / Riparian Scrub
- Grassland / Coastal Sage Scrub
- Perennial Freshwater Wetland
- Pond
- Vernal Pool

500 m
← N
1000 ft

p. 160
At higher elevations, estuarine habitats types graded into terrestrial habitat types, creating an array of different ecotones. These transition zones provided a wide range of unique ecological functions.

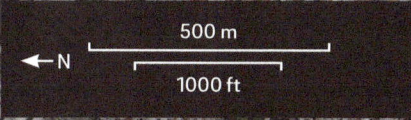

p. 152
Salt Marsh was the predominant estuarine habitat type, covering three times more area than unvegetated tidal habitats. Although dominated by pickleweed, the salt marsh plain also featured supratidal islands and large salt flats, which contributed to habitat complexity.

p. 146
More than 21 km of tidal channels conveyed tides and river water through the estuary. At low tide most of these channels were exposed, creating large swaths of Mudflat and Sandflat.

Tijuana River

Old River Slough

South Slough

ESTUARY MORPHOLOGY & HYDROLOGY

Dynamic physical drivers such as streamflow, tidal flux, wave action, and sediment flux shaped and maintained the morphology of the Tijuana Estuary in the mid-19th century, which in turn influenced the distribution of estuarine habitats. This section focuses primarily on tidal influence, with a brief discussion of river dynamics as they impacted estuarine morphology (river dynamics are described in detail in Chapter 5).

The Tijuana River delivered both freshwater and sediment to the estuary. The volume of these inputs was highly variable spatially and temporally, with the vast majority of freshwater discharge and sediment flux occurring during rare episodic flood events (see p. 96; Haltiner and Swanson 1987). The effects of these flood events on estuarine morphology, and in particular on the sediment balance within the estuary, depended on flood magnitude and other factors, and likely varied spatially throughout the estuary (Moffatt and Nichol Engineers 1987, Zedler et al. 1992, Callaway 2001). During moderate floods, sediment would have accumulated in some portions of the estuary, raising mudflat elevations and reducing intertidal area. The largest floods, however, scoured out tidal channels and mudflats —or created entirely new channels —and transported large volumes of sediment out into the ocean (Morin 1916, Perry and Hull in Wyman 1937b, Swanson 1987a, Zedler et al. 1992, Callaway in Zedler 2001). While the net effects of flooding on the sediment balance within the estuary historically is unknown, Jacobs et al. (2011) suggest that "the current sub and intertidal space in the estuary is all or nearly all hydraulic space created by floods... [likely including] large early 19th and 17th century events."

Sediment deposition and scour within the estuary were also regulated by tidal flux and wave action. Flood tidal deltas, formed by sediment transported into the estuary from the beach, are evident in historical maps (e.g., Harrison 1852; Fig. 6.1), and were likely a primary source of sediment to the estuary in times

Figure 6.1. The central tidal channel (now known as Tijuana River Slough) shown on the 1852 T-sheet extends nearly 2.5 km (1.5 mi) inland, suggesting that tidal inundation extended at least this far east. Tidal channels also extend over 1.5 km (1 mi) to the north and south of the inlet, which provides a minimum estimate for the extent of tidal flooding in these other arms of the estuary. A flood tidal delta is visible near the mouth of the estuary. *(Harrison 1852, courtesy NOAA)*

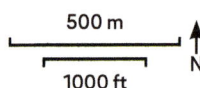

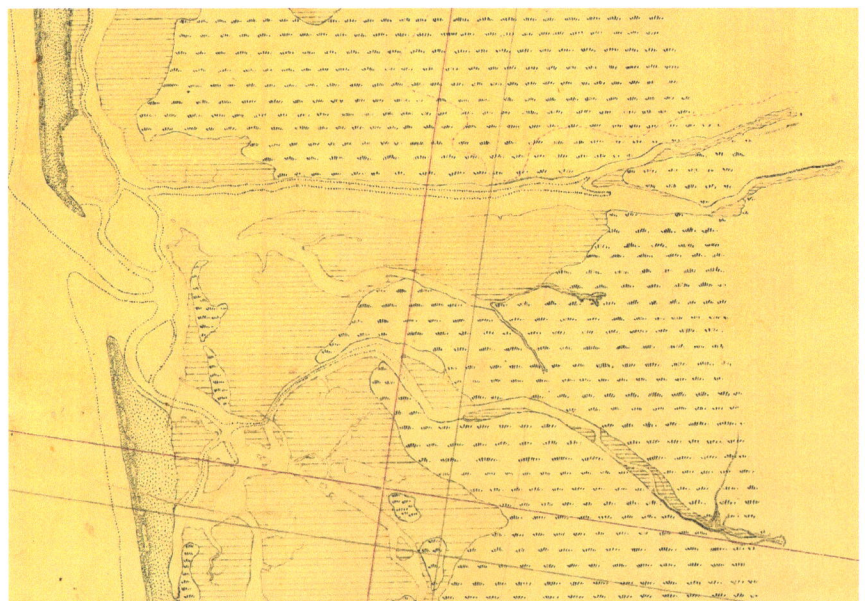

500 m

1000 ft

N

of low or no streamflow (Mayer 1987, Jacobs et al. 2011). An observer in the 1930s, for instance, noted that "the tides built long sandbars from the beach far into the slough" (Richards 2002). Conversely, the erosional shear stress created by tidal flows, which was a function of tidal prism volume, resulted in the export of sediment from the estuary (Williams and Swanson 1987). Based on the historical habitat types mapped for this report, as well as the work of others, we estimate that the potential diurnal tidal prism at the Tijuana Estuary was between 680,000 m³ (24,000,000 ft³) and 2,000,000 m³ (72,000,000 ft³) during the mid-19th century (see box text pp. 140–41; Swanson 1987a). As discussed below, this range is much higher than estimates of contemporary tidal prism volume.

Tidal influence varied throughout the estuary, with lower-elevation areas close to the ocean exposed to regular tidal inundation and higher-elevation areas further inland only exposed to tidal flooding during extreme events. Early sources indicate that tides extended inland for a substantial distance along tidal channels and across surrounding marshes. Both the 1852 T-sheet (Harrison 1852; Fig. 6.1) and the 1849 boundary survey map (Gray 1849; Fig. 6.2) and show a region of tidal influence extending approximately 2.5 km (1.5 mi) inland along the central estuarine channel. Large areas to the north along Oneonta Slough and to the south and southeast along tidal channels extending towards present day Border Field State Park were also exposed to tidal influence. GLO Surveyor Pascoe, traveling west about one mile south of the estuary mouth in September 1869, reported that the land was "salt marsh but easily reclaimed," implying that he was likely near the upper limit of tidal influence (Pascoe 1869). During storms and spring tides, tidal flooding would have spread over a larger area: Purer (1942) stated that storm tides could "carry in debris over a stretch of several miles."

By the late 20th century, both tidal prism volume and the inland extent of tidal influence had decreased substantially as a result of multiple factors, including conversion of Mudflat to Salt Marsh (especially in the northern arm of the estuary; see p. 180), Beach/Dune transgression (up to 200 m [650 ft] in some areas; see p. 183), and sediment deposition from the Tijuana River and tributaries (Williams and Swanson 1987, Swanson 1987a). Swanson (1987a) estimated that by 1986 tidal prism had decreased to just 300,000 m³ (10,890,000 ft³), a 55–85% reduction from the estimated historical range. The decrease in tidal prism volume has potential implications for tidal channel morphology, inlet cross-sectional area, and inlet closure frequency (see pp. 142–45; Williams and Swanson 1987).

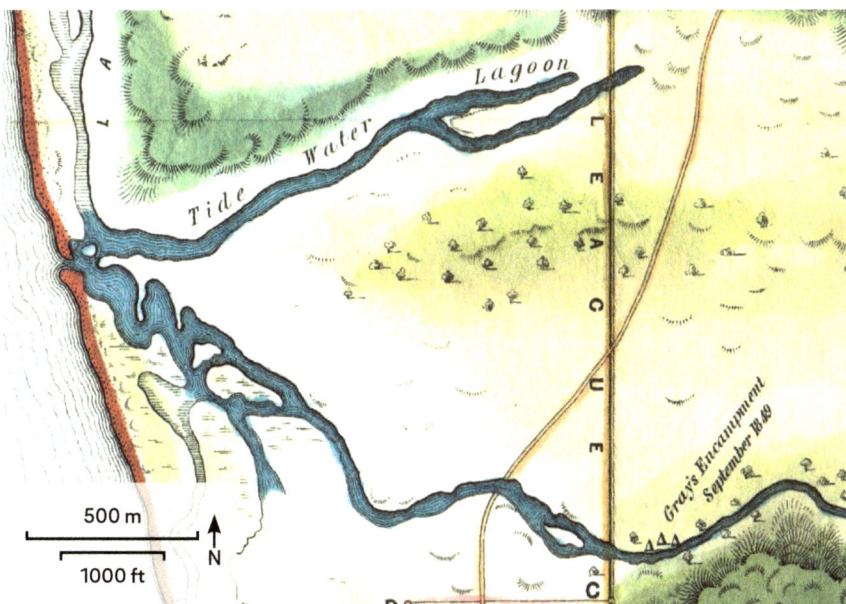

Figure 6.2. The central estuarine channel (now known as Tijuana River Slough) is labeled as a "Tide Water Lagoon" in this 1849 survey, which suggests the feature was tidal along most of its length and provides a sense for the inland extent of tidal inundation. The label and symbology also raises the possibility that, at the time they were mapped, tidal influence was stronger in Tijuana River Slough than in Oneonta or South sloughs (which were not colored blue or labeled explicitly as tidal). *(Gray 1849, courtesy Coronado Public Library)*

TIDAL PRISM CALCULATION

We used the historical synthesis mapping and estimated elevations for each habitat type to estimate potential diurnal tidal prism volume at the Tijuana Estuary in the mid-19th century. Potential diurnal tidal prism refers to the volume of water between Mean Higher High Water (MHHW) and Mean Lower Low Water (MLLW). In contrast to actual tidal prism (not calculated), which takes into account tidal attenuation within the estuary and is dependent on inlet conditions (Coats et al. 1995), potential tidal prism uses the full offshore tidal range, which averages 1.64 m (5.37 ft) at Imperial Beach (just north of the estuary mouth; NOAA gage #9410120).

Though a lack of detailed information regarding the historical elevations of estuarine habitat types precludes a precise calculation of the historical tidal prism volume, a reasonable estimate can be calculated from the available data with some basic assumptions. Table 6.1 shows the historical area, estimated elevation, and relative contribution to tidal prism volume of each of the estuarine habitat types, and describes the methodology and assumptions used in the calculations. Based on these values, we estimate that the potential diurnal tidal prism of the Tijuana Estuary was approximately 680,000 m³ (24,000,000 ft³) during the mid-19th century. This value is substantially lower than the value obtained by Swanson 1987a, who estimated an 1852 tidal prism volume of approximately 2,000,000 m³ (72,000,000 ft³) using a slightly different set of assumptions about estuarine habitat elevations.

The discrepancy between the two estimates appears to be primarily due to differing assumptions about the historical elevation of the Salt Marsh. In our analysis we assumed that the marsh plain had an average elevation of 0.014 m (0.04 ft) below MHHW based on contemporary elevation surveys conducted by Takekawa et al. 2013, while Swanson (1987a) used a conic calculation for marsh volume assuming a lower elevation of MHHW minus 0.58 m (1.9 ft) and an upper elevation of MHHW. The contribution of the Salt Marsh to historical tidal prism volume would have depended greatly on the relative extent of low marsh, which today occurs at elevations ranging from -0.241 to -0.534 m (-0.8 to -1.75 ft) relative to MHHW at the Tijuana Estuary (Thorne et al. 2016), to marsh plain, which is typically found at elevations between MHW and MHHW (Zedler et al. 1999, Sullivan 2001). We did not differentiate marsh types in our reconstruction of historical habitat types at the Tijuana Estuary, but the evidence available suggests that pickleweed (*Sarcocornia pacifica*)-dominated marsh plain was much more widespread than cordgrass (*Spartina foliosa*)-dominated low marsh (see discussion of marsh heterogeneity on pp. 152–55). In the absence of reliable information on historical marsh elevations, it seems reasonable to consider these two estimates of historical tidal prism volume (680,000 to 2,000,000 m³) as approximate upper and lower bounds on the true value. Both estimates of historical tidal prism volume for the Tijuana Estuary are substantially (2–13 times) larger than the historical tidal prism volume estimates for other coastal wetland systems in northern San Diego County (Fig. 6.3; Beller et al. 2014).

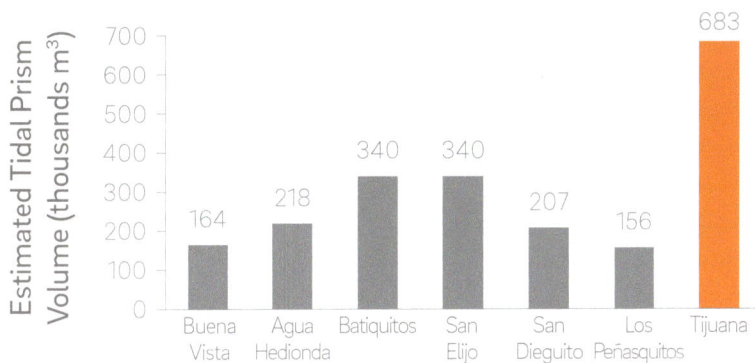

Figure 6.3. Estimated historical tidal prism volumes for multiple San Diego County systems.
The Tijuana Estuary had an estimated diurnal tidal prism volume of approximately 680,000 m³ (24,000,000 ft³) during the mid-19th century. Other estuarine systems in northern San Diego County had much smaller tidal prism volumes, ranging historically from 160,000 to 340,000 m³ (5,800,000 to 12,000,000 ft³; Beller et al. 2014). The watersheds of these North County lagoons are 5–80 times smaller than the Tijuana River watershed.

Habitat type	Area (m²)	Elevation	Contribution to tidal prism volume (m³)
Subtidal Water	91,000	MLLW (1.64 m below MHHW)	149,700
Mudflat	733,000	*Lower elevation:* 1.64 m below MHHW *Upper elevation:* MHHW	499,800
Salt Marsh	2,478,000	0.014 m below MHHW	33,800
Salt Flat / Open Water	169,000	MHHW	0
Total	**3,471,600**	-	**683,300**

Table 6.1. Values used to estimate potential diurnal tidal prism volume at the Tijuana Estuary during the mid-19th century. For each habitat type except Mudflat, area (derived from the historical synthesis mapping) was multiplied by elevation to obtain the estimated contribution to total tidal prism volume.

Since reliable historical data on estuarine elevations were not available, elevations were assigned based on a combination of contemporary data and some simplifying assumptions. The Salt Marsh elevation was derived from elevation surveys conducted in 2010, which found the average elevation of the marsh platform at the Tijuana Estuary to be 1.55 m (5.09 ft) NAVD88, which converts to 0.014 m (0.04 ft) below MHHW (Takekawa et al. 2013). Salt Flat / Open Water was assumed to be at or above MHHW, which is consistent with contemporary measurements of Salt Flat / Open Water elevation in other systems such as Carpinteria Salt Marsh (Callaway et al. 1990). The Subtidal Water habitat type was assigned an elevation of MLLW (any portion of a subtidal channel below MLLW by definition does not contribute to diurnal tidal prism volume).

Mudflats were assumed to be have a conical shape with base elevations equal to MLLW and maximum elevations equal to MHHW. To calculate the tidal prism volume contribution from mudflats, we first calculated the combined contribution from the Mudflat and the Subtidal Water habitat types using the formula $V = h/3(\sqrt{(ab)} + a + b)$, where h is the elevation difference between the top and bottom of the channels, a is the area of the Subtidal Water habitat type, and b is the combined area of the Subtidal Water and Mudflat habitat types. We then subtracted the tidal prism volume contribution from the Subtidal Water habitat type to obtain the contribution from Mudflats alone.

INLET DYNAMICS

The estuary's inlet was a dynamic feature that shifted along a short stretch of coast and experienced varying degrees of closure.

Inlet dynamics have profound effects on physical and ecological conditions within an estuary, influencing estuarine morphology, hydrology, salinity gradients, and vegetation patterns (Zedler et al. 1992, Jacobs et al. 2011). In general, inlet closure is more likely to occur under conditions of small tidal prism volume, low freshwater inflow, and high wave energy (Battalio et al. 2006, Jacobs et al. 2011).

To assess historical inlet closure dynamics at the Tijuana Estuary, we examined observations of inlet condition in historical maps, photographs, and textual sources. Of the 55 maps we reviewed that offer unique depictions the Tijuana Estuary's mouth between 1849 and 1960 (at least 24 of which were likely based on field observations or surveys), 53 show an open inlet (i.e., a direct connection between the estuary and the ocean), one (a coarse-scale county map from 1935) shows a closed inlet, and one is ambiguous (Fig. 6.4, Table 6.2). Since inlet depiction in historical maps is generally binary (open or closed), we interpret depictions of "open" conditions to include any state in which ocean water could enter the estuary, even if circulation was limited to a narrow tidal range (i.e., all conditions with closure anywhere in the intertidal or subtidal zone). We interpret a "closed" state to mean the estuary was closed to tidal influence at high tide, corresponding to the conditions described in Jacobs et al. (2011) as "dune dammed," "perched," and "closed."

Textual accounts (along with a limited number of maps) offer a more nuanced view of inlet dynamics, and suggest that the inlet opening was often either partial (i.e., closed in the intertidal) or shallow. For example, one of the few 19th century textual descriptions of inlet dynamics at the Tijuana Estuary states that "the bar [at the mouth of the river] is very shoal, and always breaks," which we interpret to mean that the inlet channel was generally open but shallow (i.e., closed in the intertidal or high in the subtidal; Mendenhall 1890). Another observer described how in the 1930s, during periods of no streamflow, sand accumulated in the entrance channel and resulted in partial inlet closure (i.e., closure in the intertidal):

> [The mouth] is partially plugged immediately after the river ceases to flow, and even while it is still flowing the bar is built up gradually and yet, I think there was some communication with the ocean— perhaps one feet or two deep, or even three feet deep, during the summer of 1935. My son waded across a little place there in a bathing suit. At low tide it might be completely plugged and yet at high tide might be considerable flow inward. (Perry 1936)

At other times, the beach barrier completely severed the tidal connection. Mary Louise Richards, who as a child in the 1930s lived with her family near the mouth of the Tijuana River, recalled that "overnight sand could be deposited until the mouth was completely closed" (Richards 2002). Though inlet closure could occur abruptly, as Richards describes,

One high tide could change the whole topography at the mouth of the slough. Where yesterday there was a sand bar, today not a sign of one.

— RICHARDS 2002, DESCRIBING DYNAMIC CONDITIONS AT THE MOUTH OF THE TIJUANA RIVER IN THE 1930s other observers recounted how the formation of the beach barrier could occur over a period of over several years (Perry 1936; Richards in Wyman 1937b).

During periods of inlet closure, streamflow into the estuary could cause water to back up behind the sandbar, resulting in widespread flooding within the estuary, changes in water chemistry, and altered habitat conditions for marsh wildlife (Bruhlmeier in Wyman 1937b, Richards 2002). In at least one case the flood water formed "a lake which extends back from the ocean from a mile to a mile and half"; during these periods water in the estuary "became stagnant and seaweed grew rampant" (Richards 2002; Perry 1936).

Because of the flooding and other impacts caused by inlet closure, early residents and local officials made repeated attempts to reopen the estuary mouth following closure events (Chancellor in Wyman 1937b, Richards 2002). The effects of these manipulations could be dramatic, though the results were not always lasting, as Richards (2002) describes:

> The county road department came and made a couple of cuts across the barrier with a grader to start the water flowing out to the ocean. Once it got started, the bank washed away almost as fast as the grader could travel. When the bulk of the water had flowed out, the mouth of the slough was about a mile wide, and we had a good view of the ocean for a while. It didn't stay that way very long before the sand piled up again.

Our analysis of historical maps, photographs, and textual documents suggests that the dominant inlet condition at the Tijuana Estuary between the mid-19th and mid-20th centuries was at least partially open (i.e., closed in the intertidal or subtidal), and that complete (i.e., supratidal) inlet closure was relatively infrequent. This result suggests that inlet closure was potentially less frequent than predicted by Jacobs et al. (2011), who used a combination of coastal setting, coastal exposure, watershed characteristics, and formation process to assess the likely closure frequency of estuary inlets along the California coast. As a hydraulically formed (flood-generated) estuary with a medium-sized, intermediate-gradient watershed in a high exposure, progradational setting, Jacobs et al. (2011) predicted that the Tijuana Estuary would be either perched above high tide or closed at high tide approximately 60% of the time, closed in the

INLET MIGRATION

An examination of 27 georeferenced maps and aerial photos dating from the mid-19th through late 20th centuries reveals that the Tijuana Estuary inlet has migrated within an approximately 1,000 m (3,200 ft) zone over the past 150 years. Given that the estuary spans 4.5 km (2.8 mi) of coastline, this degree of movement represents relatively small shifts in inlet location. The stability of the inlet may be attributable to the shape of the offshore delta formed by sediment discharge from the Tijuana River, which creates a wave convergence zone that prevents large-scale northward migration of the inlet (Dingler and Clifton 1994).

Small-scale inlet migration seems to have been a constant and relatively rapid process, at least during the early 20th century. While camped in the estuary during the summer of 1931, for example, Harwood (1931) found that the mouth of the river moved 22.9 m (75 ft) south in just six weeks (an average rate of 0.5 m/day). He attributed this movement to the "eternally... restless waves and tides [that] are ripping away the south side of the river mouth and redepositing it on the north side." As a result of the "continually shifting" banks of the river mouth, he found that, compared to other parts of the beach, there was "little life to be found in the sand" at the inlet (Harwood 1931).

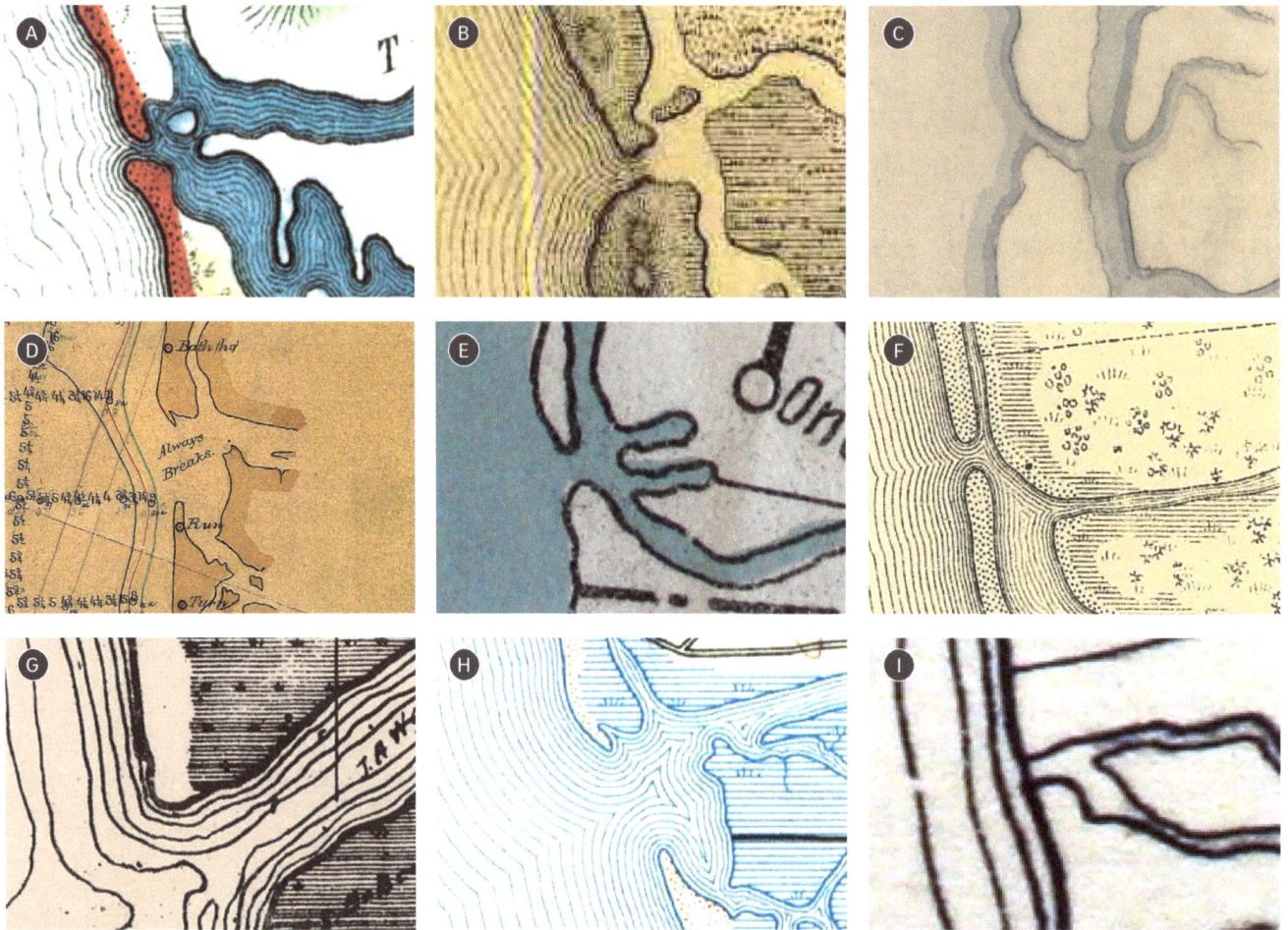

Figure 6.4. Historical maps depicting the mouth of the Tijuana Estuary. Early maps consistently depict the mouth with an open inlet. Results from our review of 55 historical maps for information on inlet dynamics are shown on the facing page in Table 6.2 (refer to this table for image sources).

intertidal (i.e., partially open) approximately 20% of the time, and in a transitional state the remaining 20% of the time. The consistent depiction of fully or partially open inlet conditions at the Tijuana Estuary in historical maps is also distinct from historical inlet depictions and observations for smaller estuarine systems in northern San Diego County (Buena Vista, Agua Hedionda, Batiquitos, San Elijo, San Dieguito, and Los Peñasquitos lagoons), which are shown as both open and closed by multiple sources (see Beller et al. 2014), suggesting that supratidal inlet closure was less frequent at the Tijuana Estuary than for these smaller lagoon systems.

While the sources used in this analysis provide valuable insights to historical inlet dynamics, there are a number of limitations that should be kept in mind, particularly in the case of cartographic data. First, as previously mentioned, inlet depiction in the maps tends to be binary (i.e., either open or closed) and typically does not reflect different degrees of inlet closure (i.e., closure in the subtidal, intertidal, or supratidal). Second, because maps provide only a static "snapshot" of inlet condition, infrequent or short-duration closure events would likely not be reflected in the dataset. Third, inlet condition likely varied on a seasonal basis, but the season in which the system was surveyed is often unknown, and thus was not factored in to the analysis. Finally, many of the maps postdate major anthropogenic modifications to the watershed, which may have altered inlet dynamics and influenced the depiction of inlet condition. Some of these modifications, such as mechanical opening

	Year	Source	Inlet condition		Year	Source	Inlet condition	
A	1849	**Gray 1849**	O		1918	Savage 1918	O	
	1850	**Hardcastle and Gray 1850**	O		1919	Scolam 1919	O	
B	1850	**Ilarregui and de Chavero 1850**	O		1920	Rodney Stokes Co. Inc. 1920	O	
	1850	**Ilarregui 1850**	O		1921	**Ervast 1921**	O	
	1852	**Harrison 1852**	O		1922	San Diego & Arizona Railway 1922	O	
C	1854	**Poole 1854**	O		1928	Mora 1928	O	
	1875	Denton and Lauteren 1875	O		1929	Adams and Favela 1929	O	
	ca. 1880	San Diego Land & Town Company ca. 1880	O		1929	**Department of Public Works 1929**	O	
	1881	Unknown 1881	O		1930	Automobile Club of Southern CA ca. 1930	O	
	1883	Fox and Willey 1883	O		1930	**USGS 1930**	O	**H**
	1886	Clark 1886	A		1931	Blackburn 1931	O	
	1887	**San Diego Land and Town Company 1887**	O		1931	**Harwood 1931**	O	
	1889	Beasley and Schuyler 1889	O		1933	**Knox 1933**	O	
D	1889	**Mansfield 1889**	O		1935	City of San Diego 1935	C	**I**
E	1889	Ryan and Humphreys 1889	O		1935	Klare 1935	O	
	ca. 1900	Burbeck ca. 1900	O		1935	Rand McNally 1935	O	
	ca. 1900	Knight ca. 1900	O		1935	**Sipe and McBean 1935**	O	
	1900	Denton 1900	O		1937	**Lee 1937**	O	
	1900	Ruhlen 1900	O		1937	**Cruse 1937**	O	
F	1901	**International Boundary Commission 1901**	O		1937	Barreto 1937	O	
	1904	**USGS 1904**	O		1943	**USGS 1943**	O	
	1906	Crowell 1906	O		1944	Quayle 1944	O	
	1910	Bedford and Cromwell 1910	O		1950	Thomas 1950	O	
G	1912	**Alexander 1912**	O		ca. 1950	Metsker Maps ca. 1950	O	
	1914	**Alverson 1914**	O		1953	**Nichols 1953**	O	
	1915	Harris and Cromwell 1915	O		1953	**USGS 1953**	O	
	1917	Automobile Club of Southern CA 1917	O		n.d.	SCMWC n.d.	O	
	1917	Guldbaum 1917	O					

Inlet condition	
O	open
C	closed
A	ambiguous

Table 6.2. Tijuana Estuary inlet conditions as depicted by maps (1849–1960). The inlet was classified as "open" if a source depicted any direct connection between the water of the estuary and ocean. All but two of the 55 unique maps analyzed for this project depict the Tijuana Estuary with an open inlet: one map shows the estuary with a closed inlet, and one shows it in an ambiguous state. Bold font is used for maps that were likely drawn from direct observations/surveys of the Tijuana Estuary. Circled letters relate sources with the maps shown in Figure 6.4.

of the inlet (documented as early as the 1930s; Chancellor in Wyman 1937b, Richards 2002), may have skewed the depiction of inlet condition towards "open." Other modifications, such as the construction of dams within the watershed (the first of which was completed in 1912), may have skewed the depiction towards "closed" (as a result of decreased frequency and intensity of flood scour events). While prior studies have suggested that inlet closure frequency has increased since the mid-19th century (e.g., Williams and Swanson 1987, Goodwin and Kamman 2001), the uncertainties described above make it difficult to draw quantitative conclusions about the frequency or duration of closure events during the historical period, or how inlet dynamics might have changed over time.

TIDAL CHANNELS & FLATS

More than 21 kilometers of tidal channels conveyed salt water from the ocean and freshwater from the river through the Tijuana Estuary.

> During a low tide there are exposed numerous jetties, points and islands, but at high tide all is changed. The channels have expanded till the islands of sand have been covered, the estuaries are deeper and wider, and the water laps at the very edges of one's habitation.
>
> —HARWOOD 1931

> The pattern the sloughs take can roughly be compared to the right hand outstretched, with the wrist representing the mouth or channel through which the tides come and go and runoff from the rains finds its way to the ocean.
>
> —RICHARDS 2002, DESCRIBING CONDITIONS IN THE 1930s

From its mouth, the Tijuana Estuary split into a series of tidal channels. These sloughs, which generally filled and drained with the tides twice daily, covered an area of more than 80 ha (200 acres). Close to 90% of this surface area was intertidal and exposed at low tide, creating extensive swaths of mudflat and sandflat. The remaining 10% of the tidal channel surface area was subtidal, remaining submerged at low tide (though the ratio between intertidal and subtidal habitat would have varied with the position of mouth within the tidal frame and, to a lesser degree, the volume of freshwater inflows). Some tidal channels connected upstream to the Tijuana River, while others dead-ended within the marsh. Although early residents do not seem to have had unique names for individual sloughs, (they were primarily referred to using cardinal directions, e.g. the "north slough"; Richards 2002), the sloughs were nonetheless considered distinct features. One early observer even considered each of the primary sloughs an individual estuary and referred to the larger system with the plural "Tia Juana Estuar*ies*" (Harwood 1931; emphasis added). This section uses contemporary names for the individual sloughs (see map on p. 136–37).

The primary tidal sloughs each extended more than 2 km from the mouth, one pointing north and the others arranged at various angles between east and south. In total, the historical estuary featured approximately 21 km of channels (measured at a scale of 1:5,000 and disregarding islands less than 0.5 ha in size). Channel density was thus relatively low, with approximately 6 km of channels per square kilometer of tidal habitat (channel density within salt marshes in other parts of California frequently exceed 15 km/km²; Collins and Grossinger 2004). The low channel density could be related to the estuary's intermediate tidal prism (see p. 141) and relatively high sediment load.

The estuary's tidal channels varied in size, generally becoming narrower with distance from the mouth (Harwood 1931). At their widest, the primary channels were between 30–130 m across. One exception was Oneonta Slough (or "lagoon," as it was commonly called in early sources), which grew wider as it moved further from the mouth and ultimately reached a width twice that of any other channel (270 m). This slough's unique morphology is likely a product of a distinct formation process.

Figure 6.5. A tidal slough just inside the sand dunes at the western edge of the estuary, possibly south of the estuary mouth. At the time of this photograph, the tide is relatively high and the slough is largely filled with water. The photo looks north-east. *("Classification L. Living bird, animal, insect subjects, including nests and eggs of birds not in museum groups," Photo #L5721, courtesy San Diego Natural History Museum)*

While Tijuana River, Mid-Valley, and Old River sloughs likely formed as courses of the Tijuana River and were all active river channels during the 18th and 19th centuries (see p. 114), Oneonta Slough is thought to have been a remnant portion of a larger ancestral lagoon that occupied the lower end of the valley during the mid to late Holocene (Swanson 1987a; see p. 36) and was not an outlet of the River during the historical period. The area occupied by the lagoon would have been slower to fill with river-derived sediments due to its sheltered position behind the Nestor Terrace (Swanson 1987a).

All of the estuary's channels were relatively shallow. Although there are no direct accounts of channel depth before the mid-1900s, reports from that period describe how most channels were less than 0.6 m (2 ft) deep (San Diego Regional Water Quality Control Board 1967) and how only boats with very low draft could navigate the subtidal habitat (Moffatt and Nichol, Inc. 1957). Channel depth during the historical period can be inferred by the extensive swaths of intertidal habitat, which could not have been any deeper than the maximum tidal range (~2.6 m [8.6 ft] in the open ocean, likely much less within the estuary; NOAA 2016). On the subject of channel geometry, Harwood (1931) describes how channels were not symmetrical in cross-section, noting that in each slough "one bank rises gradually like a sloping shelf, while the other bank rises at a steeper angle, more like a broken shelf." This asymmetry is ubiquitous in natural tidal channels (J. Collins, personal communication). Asymmetry appears to have been especially pronounced at the landward

> Near its mouth this river splits into four estuaries, one of which curls off to the north while the other three strike off at various directions between south and east.... One may follow these estuaries for a mile and a half before the water is exhausted."
>
> —HARWOOD 1931

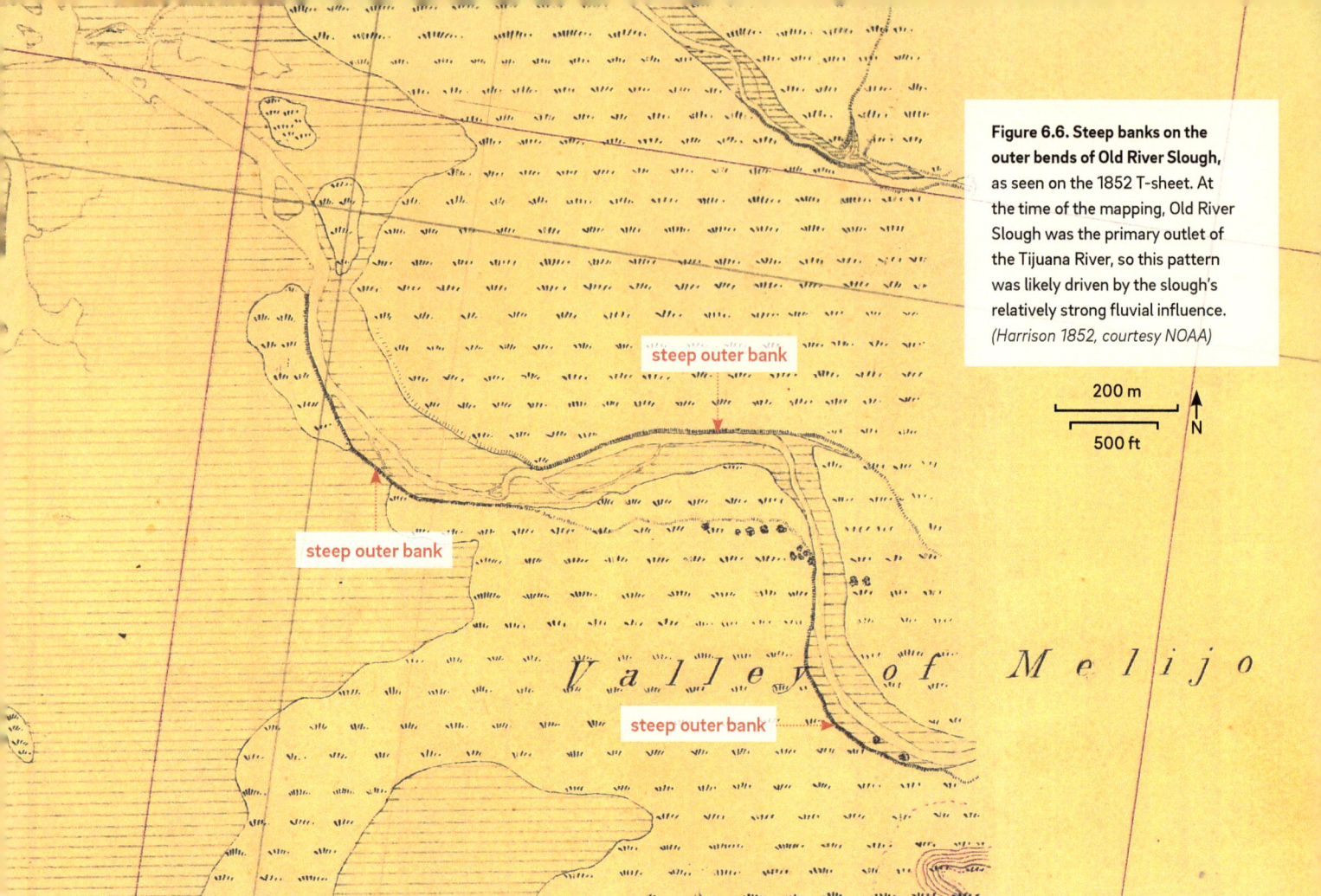

V a l l e y   o f   M e l i j o

Figure 6.6. Steep banks on the outer bends of Old River Slough, as seen on the 1852 T-sheet. At the time of the mapping, Old River Slough was the primary outlet of the Tijuana River, so this pattern was likely driven by the slough's relatively strong fluvial influence. *(Harrison 1852, courtesy NOAA)*

200 m

500 ft

N

> At the mouth of the river the bottoms of the four estuaries are clean, white, and sandy....Farther back from the mouth...the white sand is replaced by brown sandy mud....At the upper end of these-estuaries where the decomposing algae is deposited by the tides, the brown sandy mud is replaced by a thick black gummy ooze with all the characteristics of quicksand combined with those of rotten eggs.
>
> —HARWOOD 1931

margins of the major sloughs, which are depicted on the 1852 T-sheet with steep outer banks, a pattern that could be related to greater fluvial influence further inland (Harrison 1852; Fig. 6.6).

Tidal channel substrate varied with distance from the mouth of the estuary. Submerged channels and intertidal flats were sandy closest to the mouth, where waves deposited coarse sediment. Farther inland, the substrate grew finer and was generally characterized as "muddy" (Harwood 1931). The organic component of the substrate was especially high at the upper ends of the sloughs, where the tides deposited algae which decomposed to create a "thick black gummy ooze" (Harwood 1931).

Although the estuary's tidal flats might have appeared relatively barren, they served a variety of important ecological functions. Today, these areas are recognized as important habitat for a range of organisms, including microalgae, phytoplankton, invertebrates, fish, and birds (see Zedler et al. 1992). This is reflected in historical descriptions of the plants and animals that inhabited the network of sloughs: over a six week period during the summer of 1931, Harwood recorded 22 species of mollusks, 22 species of crustaceans, 17 species of fish, 10 species of echinoderms, 3 species of sponges, and 2 species of bryzoa. Benthic invertebrates were quite abundant; at two study sites, Harwood

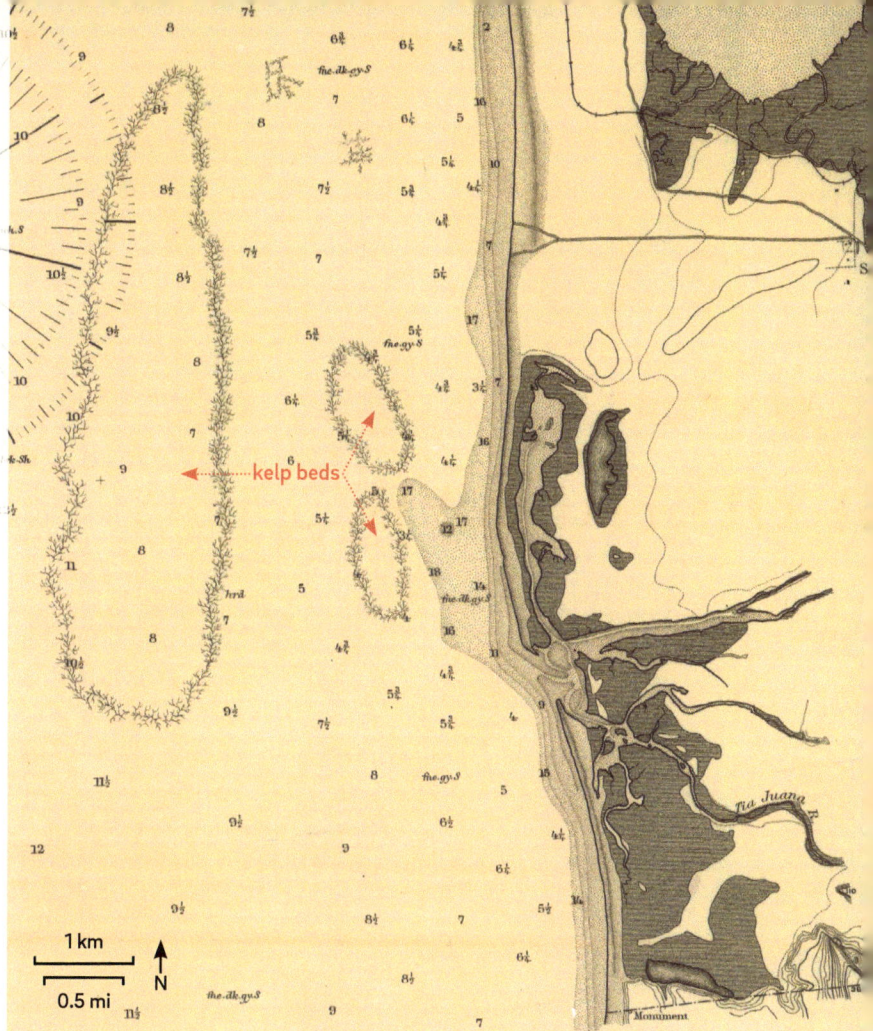

Figure 6.7. "Thick kelp beds" covering 4.7 km² just outside the Tijuana Estuary in 1895. Historically, a substantial amount of kelp detritus routinely washed into the estuary (Harwood 1931, Richards 2002). In this manner, kelp beds subsidized the amount of organic carbon available to consumers within the estuary. For example, beached kelp, which harbors abundant insect prey, is known to help support the endangered Belding's savannah sparrow (Zedler et al. 1992). Additionally, historical sources describe how detached kelp "roots"—known as holdfasts—served as transport vectors into the estuary for an array of attached invertebrates, including annelid worms, chitons, tube mollusks, brittle stars, sea urchins, starfish, and sea cucumbers (Harwood 1931).

Approximately 8 km² (3 mi²) of kelp forests disappeared from the estuary's nearshore environment in the mid-20th century (Limbaugh 1955) and were absent until 1979 (North et al. 1993), which likely decreased the amount of detritus and associated organisms washed into the estuary. Though the precise cause of this decline is unknown, there is significant natural temporal variability in the local extent of kelp forests (Steneck et al. 2002). Kelp beds have been present at this location in most years since 1979, reaching a maximum size of 1.9 km² in 2008 (MBC 2015). It is possible that the initial recovery of the kelp beds in 1979 can be traced to the flood of 1978, which could have decreased sea urchin abundance and relieved grazing pressure on kelp stipes (J. Zedler, personal communication). Many other changes in estuarine invertebrate populations were caused by the sudden and prolonged low-salinity conditions in 1978 (c.f. Zedler et al. 1992) and sea urchins are known to be sensitive to salinity changes (Irlandi et al. 1997). *(USCGS 1895, courtesy NOAA)*

recorded densities of 25 and 50 individual organisms per cubic foot (0.03 m³), the most abundant of which were the California horn snail *(Cerithideopsis californica)* and bent nosed clam *(Macoma nasuta).* It should be noted, however, that since benthic communities are affected by substrate type, soil salinities, and physical disturbance, the density and composition of invertebrates would have varied over time and space. This was illustrated during the early 1930s when a flood washed "all the clams out to the ocean," after which it was "a long time" before they reestablished (Richards 2002).

Although the tidal channels were largely unvegetated, Harwood (1931) did document the presence of eelgrass, an important community-structuring aquatic plant found in some subtidal habitats. The plant does not grow within the estuary today (Zedler et al. 1992, Bernstein et al. 2011) and it is unclear if the eelgrass recorded by Harwood was rooted and established within the estuary or was simply loose material that washed in with the tides. Tides frequently carried in other outside material, including kelp, which would have been a source of organic carbon and a variety of invertebrate organisms in the estuary *(Macrocystis pyrifera;* see Figure 6.7). Macroalgae such as *Enteromorpha* (sea lettuce) and other seaweeds supplemented tidal channel plant life (Harwood 1931, Zedler et al. 1992).

> After rough seas the kelp and seaweed were unusually bad, and...Bill had to go out and clear it [their bait box] off several times during the incoming and outgoing tides.
>
> —RICHARDS 2002, DESCRIBING THE ESTUARY IN THE 1930s

PACIFIC POCKET MOUSE:
"Rarest of mammals"

This exceedingly small Pocket-Mouse is one of the rarest of mammals yet, though some one may find them plentiful unexpectedly.

— STEPHENS AND FENN 1906, DESCRIBING THE PACIFIC POCKET MOUSE

The federally endangered Pacific pocket mouse *(Perognathus longimembris pacificus),* originally described by Edgar Mearns according to a type specimen collected in the lower Tijuana River valley in 1894, is the smallest and most narrowly distributed subspecies of the little pocket mouse *(Perognathus longimembris;* Mearns 1898, USGS 2012). Similar to other pocket mice, it is a nocturnal granivore with external cheek pouches used to store seeds (USGS 2012, USFWS 2010). The Pacific pocket mouse is exceptionally long-lived for its size, and one of the smallest hibernating mammals (USFWS 2010).

Endemic to the southwestern California coastline, the Pacific pocket mouse was historically rare and patchily distributed. It was typically found in sandy substrates in coastal dune, coastal strand, river alluvium, and open coastal sage scrub habitats, usually confined to areas within 4 kilometers (2 mi) of the ocean (USFWS 2010, Brylski et al. 1998). Mearns (1898) described the type locality as "a flat, often submerged by high ocean tides, at the mouth of the Tijuana River, where it appeared to be abundant." Von Bloeker (1931) discovered several specimens in the Tijuana River bed, in an area "covered by a dense growth of weeds and brush" but with "small open spaces of alluvial river-bottom." At least 132 specimens were taken from the lower Tijuana River valley between 1894 and 1932, the last year in which the species was observed in the valley (Spencer 2005, Brylski et al. 1998).

The mouse was thought to be extinct until its rediscovery at Dana Point in 1993 (Brylski et al. 1998). Today, there are only four known occupied sites (400 hectares of habitat in total), three of which are within the Marine Corps Base Camp Pendleton (USGS 2012, USFWS 2010, Brylski et al. 1998). The species continues to face serious threats from habitat destruction and fragmentation, native and domestic predators, and human disturbance, making it the target of ongoing protection efforts. Although the extent of the river corridor and sandy alluvium has been decreased over time due to agricultural development and other land uses, the Tijuana River valley was included in a list of seven prospective Pacific pocket mouse translocation sites in 2010 (USFWS 2010). Changes since the extirpation of the mouse— specifically the conversion of the river's riparian habitats from a mosaic of willow scrub and sandy river wash to dense riparian forest in the 1980s (p. 184)—have probably contributed to a decrease in the amount of appropriate habitat available today.

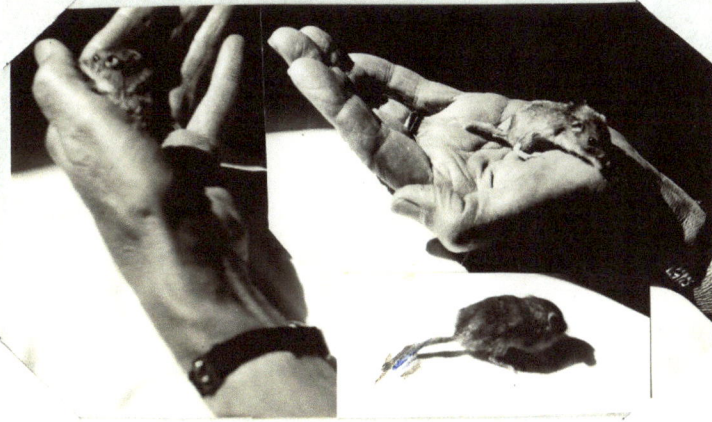

Bobbity liked nothing better than to sit in one's [your] hand and if the light was kept out of his eyes, he would go to sleep there.

BOBBITY

Dear little mouse with shiny coat
    Bright black eyes and dainty hands
Watching us with a wistful look
    And a far away gaze that understands
More than we think of our intent
    And more than we know of distant lands.
Deserts and wastes of sandy soil
    Where treasures of seeds in a cool deep cell,
The rich rewards of nights of toil
    Were [Are] the dainty foods that please[d] him well.
But now [when] he has [came] come to share our life
    And freely to us his valued trust to give
To accept from our hands (relief from strife) protection and care
    And teach us how his people live
It is [was] only for us to understand
    And write his life with a friendly [knowing] hand.

                                Vernon Bailey

Figure 6.8. Archival materials from the manuscripts of mammologist Vernon Bailey about "Bobbity". Bobbity was a Pacific pocket mouse collected alive from the sandy river bottom of the Tijuana River in 1931 by Jack Von Bloeker and given to Bailey for study (Von Bloeker 1931, Bailey 1939).

Bailey advocated for reducing unnecessary suffering of the animals collected for study and devoted himself to the design and manufacture of more humane mammal traps (Mathias 2011).

(Photo & poem: SIA RU 007267, Box 7, Folder 46, # SIA2011-1522 & #SIA2011-1523, both courtesy Smithsonian Institution Archives)

MARSH PLAIN & SALT FLATS

Salt marsh was the estuary's most extensive habitat type, with substantial within-marsh heterogeneity.

> On the whole, the area covered by marsh plants is more extensive than the open water.
>
> —PURER 1942

In the 1850s, Salt Marsh covered approximately 250 ha (610 acres) of the Tijuana Estuary. Marshland spanned the full north-south extent of the estuary and reached up to 2.2 km inland (1.2 mi) inland. Although approximately 25% of the Salt Marsh was situated in the northern part of the estuary (along Oneonta Slough), the majority of the marshland was found south of Tijuana River Slough (the opposite is true today, see p. 181). In total, the Tijuana Estuary's marshes constituted nearly 8% of total extent of vegetated estuarine wetlands along the Southern California Bight (between Point Conception and the U.S.-Mexico border; Stein et al. 2014). As is still the case, the area of vegetated Salt Marsh in the estuary exceeded the area of unvegetated tidal habitats (Subtidal Water, Mudflat, and Sandflat) by a factor of three (see also Purer 1942). The vast majority of the Salt Marsh habitat was subject to inundation during the highest tides, which occasionally made the estuary appear like a large expanse of open water (Purer 1942).

Salt marsh vegetation was dominated by *Sarcocornia pacifica* (pickleweed; Figure 6.10), but included a range of other species (Stephens 1908, Holmes and Pendleton 1918, Harwood 1931, Purer 1942; Table 6.3). Species distribution within the marsh would have been determined by a variety of interrelated factors, including elevation, inundation frequency, salinity, proximity to channels, and interspecific competition (Pennings and Callaway 1992, Zedler et al. 1999, Pennings et al. 2005, Bonin and Zedler 2008, Varty and Zedler 2008). Additionally, the species composition of the Tijuana Estuary is known to shift in response to major disturbance events, such as floods and mouth closure, and was likely somewhat dynamic over time (Zedler et al. 1992). Currently rare species that were recorded in the salt marshes of the Tijuana Estuary include aphinasma (*Aphanisma blitoides*), salt marsh bird's beak (*Cordylanthus maritimus* subsp. *maritimus*), Leopold's spiny rush (*Juncus acutus* subsp. *leopoldii*), estuary seablite (*Suaeda esteroa*), and bush seepweed (*Suaeda taxifolia;* specimen records were obtained from the Consortium of California Herbaria).

Although generally mapped as homogenous area of undifferentiated marsh, historical data suggest the marsh plain had notable aspects of heterogeneity. Much of this complexity occurred over the estuary's subtle elevation gradient, which created variability in qualities like inundation frequency, salinity, and species composition. At low elevations, marsh patches within wide tidal sloughs—likely dominated by *Spartina foliosa*—were inundated with each high tide. This contrasts with higher areas of salt marsh at the periphery of the estuary. The southern arm of the estuary, for example, was characterized as "salt marsh but easily reclaimable," suggesting a lower magnitude and frequency of inundation (Freeman 1852). At its upper reaches, Salt Marsh habitat transitioned into a number of other terrestrial and wetland habitat types, including Alkali Meadow Complex, Riparian Scrub, Grassland / Coastal Sage Scrub, and the steep cliffs of the mesas to the south (see pp. 160–65). Additional aspects of marsh plain heterogeneity, including islands in the marsh and non-tidal salt marsh, are explored in Figure 6.9.

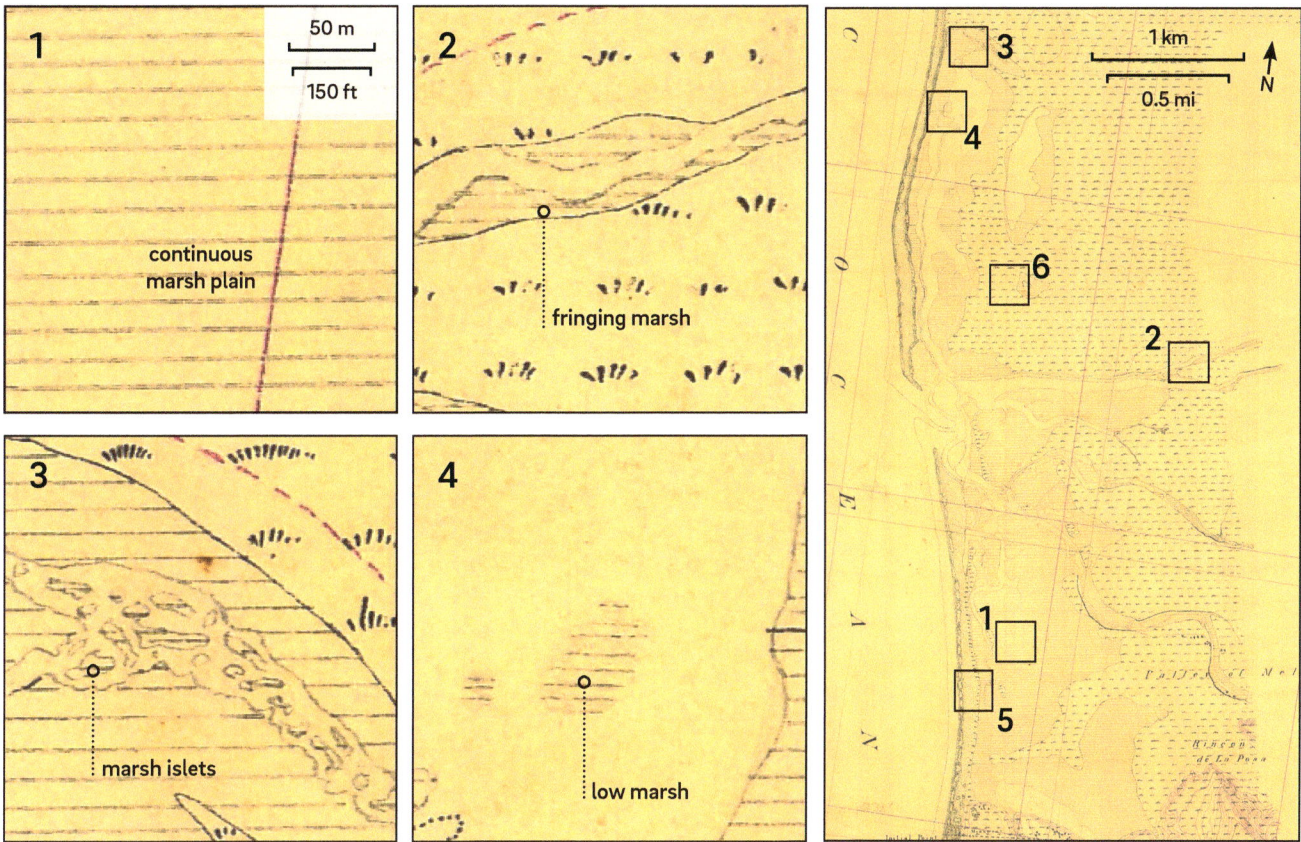

Figure 6.9. Snapshots of marsh plain heterogeneity evident on the 1852 T-sheet. *(Harrison 1852)*

1) Continuous marsh plain. The majority of the vegetated marsh surface is thought to have been found at an elevation of approximately MHHW and would have only been inundated during the highest monthly tides.

2) Fringing marshes. Narrow bands of marsh vegetation lined the upper reaches of the estuary's tidal sloughs.

3) Islets of marsh vegetation, each ranging from 10–400 m^2 in size, at the north end of Oneonta Slough. Although the processes driving this pattern are uncertain, mudflats in the slough rapidly vegetated between the mid-19th and mid-20th centuries (see pp. 180). It is possible that these marsh features are pioneering clumps of vegetation that indicate active marsh progradation (c.f., Johannessen 1964). The clumps could also be related to the presence of mima mounds, though they do not overlap the known location of the mound field mapped by Cox and Zedler (1986).

4) Low marshes within tidal sloughs. These "submerged marshes," which were drawn without an outside border, were largely flooded during normal high tides. The largest of these patches approach 1 acre in size.

5) Islands in the marsh. Within the marsh plain, a handful of "islands" rose above the reach of the tides. These elevated patches would have served as high tide refuge for a variety of marsh species.

6) Non-tidal marsh. A few areas of salt marsh did not have a direct tidal connection. Some of these isolated marshes, such as the one depicted in this snippet, surrounded salt flats.

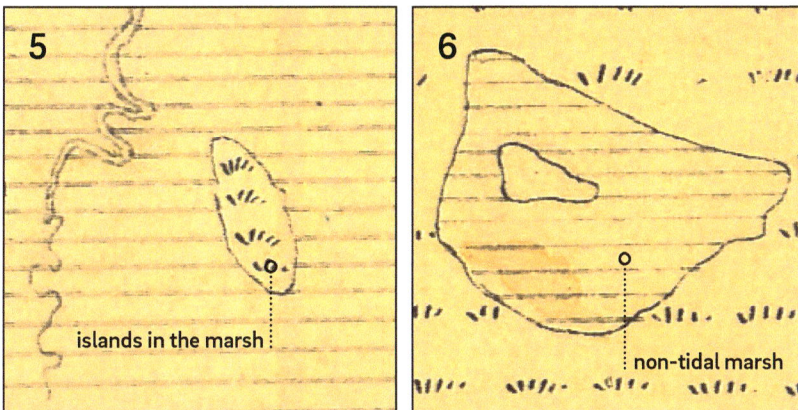

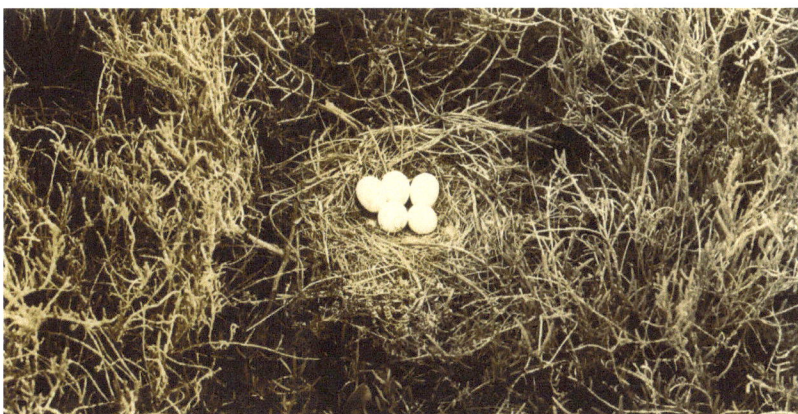

Figure 6.10. The pickleweed nest of a northern harrier (*Circus cyaneus*; also known as "marsh hawk") in the salt marshes of the Tijuana Estuary, 1924. *("Classification L. Living bird, animal, insect subjects, including nests and eggs of birds not in museum groups," Photo #L5722, courtesy San Diego Natural History Museum)*

Table 6.3. Salt marsh plant list. Partial list of native plants that were historically present in the valley and were likely associated with the estuary's salt marshes, drawn from early accounts and pre-1950 herbarium records. Data were provided by the participants of the Consortium of California Herbaria (ucjeps.berkeley.edu/consortium/), unless another source is listed in the notes field. Species were selected for inclusion in the table based on a combination of the locality information provided in the herbarium records and known associations with the Salt Marsh habitat type. All data were provided by the participants of the Consortium of California Herbaria, except for additional citations listed in the Notes field.

Common name	Scientific name	Locality	Years	Notes
pineapple weed	Amblyopappus pusillus	"along borders of salt marsh," "Imperial Beach," "mouth Tia Juana Valley"	1902, 1913, 1918, 1923, 1936, 1937, 1938	
aphinasma	Aphanisma blitoides	"alkali flat near ocean"	1903, 1941	CNPS Rare Plant Rank 1B.2
Parish's glasswort	Arthrocnemum subterminale	"Sea beach, mouth of Tia Juana River," "Imperial Beach"	1935, 1936	
Watson's saltbush	Atriplex watsonii	"Imperial Beach, edge of salt marsh"	1903, 1923, 1936, 1938, 1942	Purer 1942
saltwort	Batis maritima	"Tijuana River slough" (Purer 1942)	1942	Purer 1942
salt marsh bird's beak	Cordylanthus maritimus subsp. maritimus	"Tia Juana Slough," "upper edge of salt marsh, Imperial Beach"	1935, 1938, 1942	Purer 1942; CNPS Rare Plant Rank 1B.2
salt grass	Distichlis spicata	"Imperial Beach"	1895, 1938, 1942	Purer 1942
alkali seaheath	Frankenia salina	"Mouth of Tia Juana River," "sandspit south of Imperial Beach"	1913, 1935, 1938, 1942, 1950	Purer 1942
salt heliotrope	Heliotropium curassavicum	"Tiajuana R. Estuary"	1931	
alkali heliotrope	Heliotropium curassavicum var. oculatum	"Salt marsh"	1936	
spiny rush	Juncus acutus	"Tia Juana River"	1902, 1903, 1913	
Leopold's spiny rush	Juncus acutus subsp. leopoldii	"Tiajuana River Estuary"	1903, 1931	CNPS Rare Plant Rank 4.2
California sealavender	Limonium californicum	"Tia Juana slough," "Tijuana Estuary"	1935, 1941, 1942	Purer 1942
California sealavender	Limonium californicum var. mexicanum	"Tia Juana slough"	1935	
shore grass	Monanthochloe littoralis	"Edge of the [Tijuana] estuaries" (Harwood), "Imperial Beach"	1931, 1938, 1942	Harwood 1931, Purer 1942
pickleweed	Salicornia pacifica (Sarcocornia pacifica)	"Tia Juana Valley near ocean," "marsh near Mexican Boundary," "Tiajuana Est."	1935, 1939, 1941, 1942	Purer 1942
California cordgrass	Spartina foliosa	"Salt marsh, in water, Tia Juana River Slough"	1938, 1942	Purer 1942
California seablite	Suaeda californica	"Tijuana Estuary," "Imperial Beach"	1918, 1938, 1941, 1942	Purer 1942; CNPS Rare Plant Rank 1B.1
estuary seablite	Suaeda esteroa	"Imperial beach, salt marsh," "Near beach 1 mile north of Monument 258"	1936, 1938	CNPS Rare Plant Rank 1B.2
bush seepweed	Suaeda nigra	"Mouth Tia Juana River," "Salt marsh," "Slough south of Imperial Beach"	1935, 1936, 1939, 1950	Includes records for *S. torreyana*, S. moquinni
woolly seablite	Suaeda taxifolia (Suaeda californica var. pubescens)	"Imperial Beach"	1903, 1918, 1936, 1938	CNPS Rare Plant Rank 4.2

Salt flats were also a distinctive feature of the marsh plain. At both the northern and southern ends of the estuary, sizeable salt flats—together totaling 17 ha (42 acres)—were situated in depressions at the marsh's upper edge. When Harrison (1852) surveyed the southern salt flat during the winter of 1851–2, the feature was inundated; he labeled it a "pond." Additional surveys confirm the feature was seasonally dry and covered with a salt crust (Gray 1849, Pascoe 1869; Fig. 6.11, also see Figure 6.14). The northern salt flat, which was put to light industrial use as a salt source from the mid-1800s until the flood of 1916, was known as the "Salt Works" (Pascoe 1869, Richards 2002). These features were largely unvegetated, since the high soil salinities associated with salt flats preclude significant plant growth (Pennings and Bertness 1999, Noe and Zedler 2001a).

Salt flats are often found in low-latitude estuaries with low rainfall, high evaporation rates relative to inflow, strong seasonal variation in precipitation, and/or irregular tidal inundation (Pennings and Callaway 1992, Largier et al. 1997, Pennings and Bertness 1999). Although now relatively rare, large salt flats were once a common feature of coastal wetlands in southern California and performed a number of important ecological functions (e.g., Callaway et al. 1990, Beller et al. 2014). Salt flats are important for a variety of insects, including tiger beetles; the Tijuana Estuary supports a higher diversity and abundance of tiger beetles than any other coastal locality in southern California (Nagano 1982 in Zedler et al. 1992). Salt flats also are used by a number of endangered bird species, including Belding's savannah sparrow (foraging habitat), California least tern (nesting habitat), and western snowy plover (foraging habitat when they are inundated and nesting habitat when they are dry; Zedler et al. 1992).

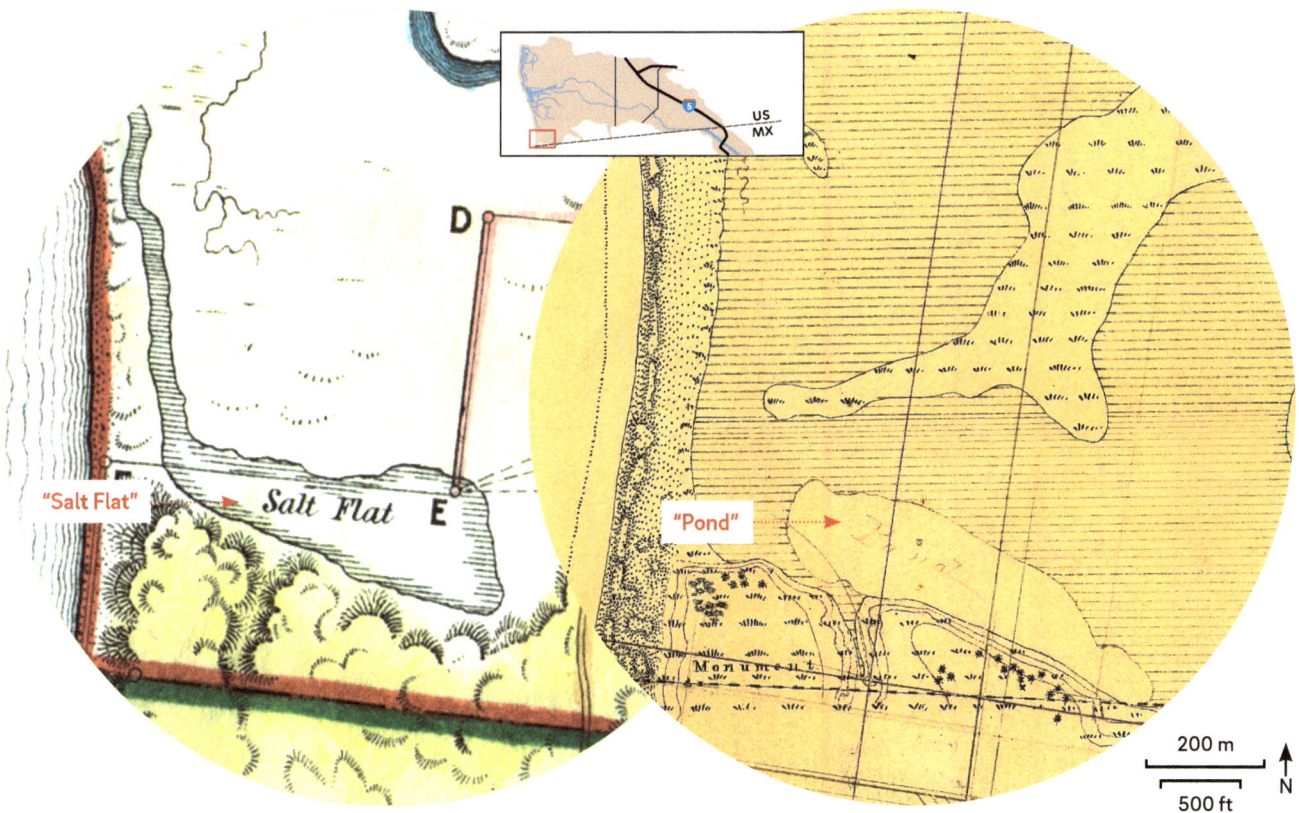

Figure 6.11. The two lives of a salt flat. During winter and spring, salt flats accumulate rainfall and seawater from extreme tides, which then evaporate during the summer and leave behind a concentrated crust of salt. This seasonal variability is captured by these two maps: during the summer of 1849 (left) the feature was dry and labeled a "salt flat," but during the winter of 1851–1852 (right) the feature was wet and labelled a "pond." Since the dry season is longer than the winter inundation period, present-day researchers refer to these features as salt flats (rather than temporary tidal ponds; Zedler et al. 1992). *(Left: Gray 1849, courtesy Coronado Public Library; Right: Harrison 1852, courtesy NOAA)*

Complex beach and dune habitats formed the western boundary of the estuary.

The western edge of the estuary is bounded by a long barrier system or "sand spit" that formed 4–5,000 years ago during the last marine transgression (Dingler and Clifton 1994; see p. 36). This elevated land spanned the full width of Tijuana River valley, only interrupted at the outlet of the Tijuana River (see p. 142). Shaped by the interplay of wind, waves, and tides, the barrier system was quite dynamic and featured an array of interrelated beach and dune habitats.

In total, the beach-dune barrier system covered approximately 45 ha (110 ac) of land above the high water line. Its topography varied along the north-south axis of the estuary, but generally consisted of two parallel dune ridges spaced 20–60 m (70–200 ft) apart (Harrison 1852; see Fig. 6.12). As mapped by Harrison (1852), the individual ridges each ranged from 15–50 m (49–164 ft) wide, while the complete supratidal barrier beach and dune system was generally 70–160 m (230–520 ft) in width. When the T-sheet was surveyed in 1852, the dunes bounding the Tijuana Estuary were all less than 3 m (10 ft) above sea level, making them shorter than some of the dunes farther north along the Silver Strand (which, at the same point in time, occasionally exceeded 6 m [20 ft] in height; Harrison 1852, Storie and Carpenter 1923).

It is likely that the outer (seaward) ridge of dunes was a "foredune ridge"—a partially vegetated, less coherent, and relatively dynamic ridge of sand parallel to the beach above ordinary high tides—while the inner (landward) string of dunes was more of a classical "dune ridge"— a taller, more vegetated, coherent, and stable ridge of sand (Pickart and Barbour 2007). This distinction is supported by the accounts of William Cooper (1967), who surveyed the estuary's dunes in 1919 and described how the "dune belt" consisted of a strip of "foredune hillocks" and then an "inner dune ridge, with much blowout activity of the windward side." The distinct ridges are also visible in at least one early landscape

Figure 6.12. Components of the historical barrier beach system. The 1852 T-sheet depicts a variety of geomorphic features, including the beach, a foredune ridge, dune swale, and dune ridge. These features created heterogeneity within the beach-dune environment. (*Harrison 1852, courtesy NOAA*)

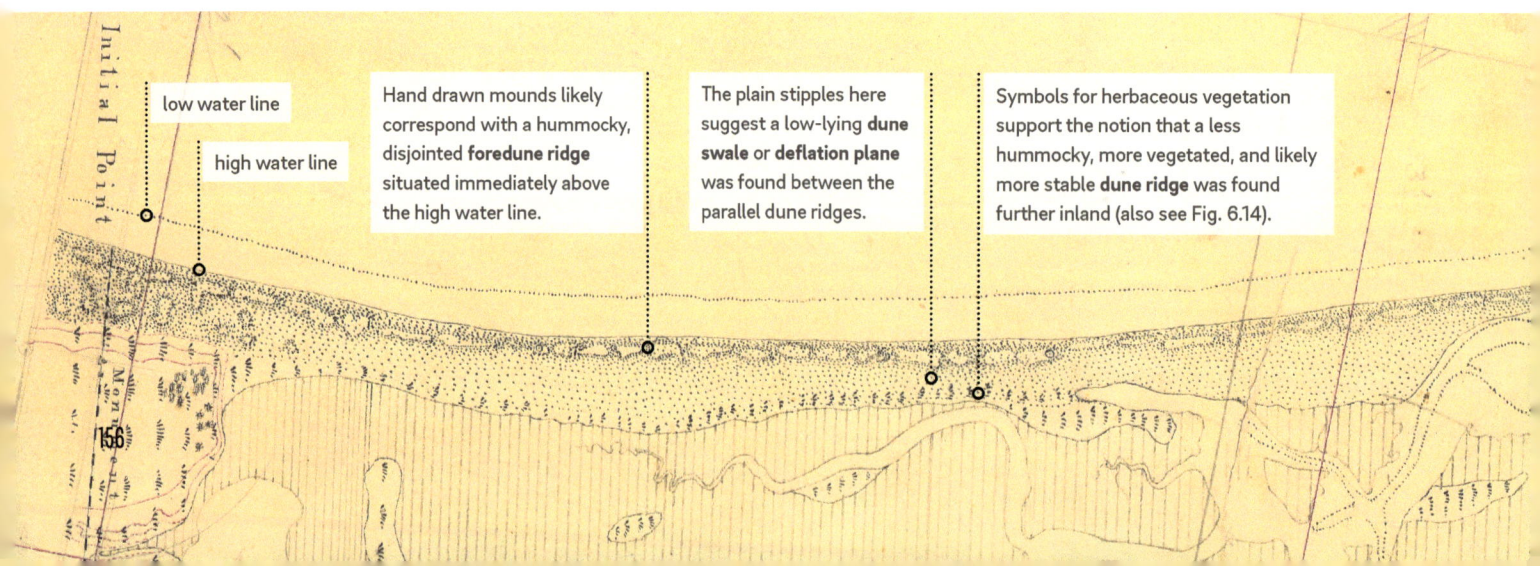

low water line

high water line

Hand drawn mounds likely correspond with a hummocky, disjointed **foredune ridge** situated immediately above the high water line.

The plain stipples here suggest a low-lying **dune swale** or **deflation plane** was found between the parallel dune ridges.

Symbols for herbaceous vegetation support the notion that a less hummocky, more vegetated, and likely more stable **dune ridge** was found further inland (also see Fig. 6.14).

Initial Point

photograph, with the inner ridge appearing less hummocky and more vegetated (see Figure 6.14). Cooper's description of foredune "hillocks" reflects the characteristic discontinuous and hummocky form of foredunes in southern California (Pickart and Barbour 2007; also see Harrison 1852, Storie and Carpenter 1923; Fig. 6.13). At the northern end of the estuary, the foredune ridge was less pronounced (Harrison 1852), and might have instead constituted "coastal strand" habitat (Dugan and Hubbard 2010). Regardless of how these various geomorphic features are classified, together they suggest the historical complexity and heterogeneity of beach-dune habitats at the western margins of the estuary.

Figure 6.13. Silver Strand dunes, ca. 1929, at a location 3.5 km north of the Tijuana Estuary (San Diego Bay is visible to the right). This dune system continued largely uninterrupted south to the international boundary. (Photo #UT 1982, courtesy San Diego History Center).

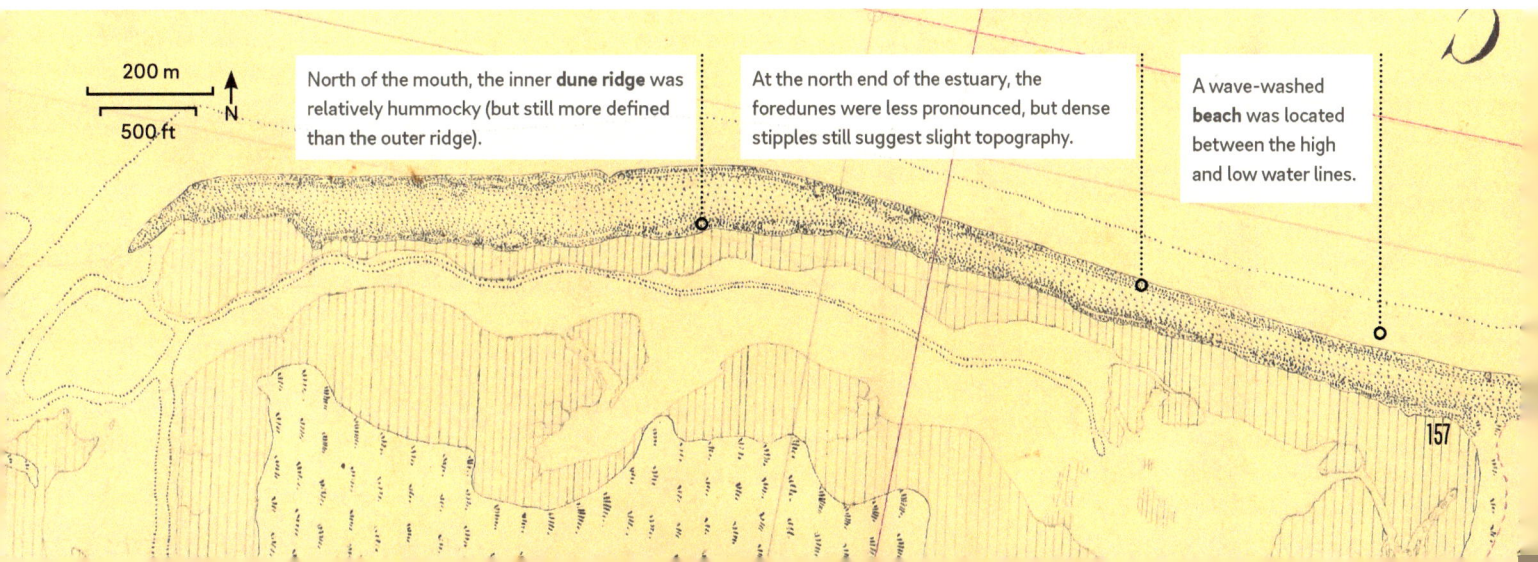

200 m
500 ft
N

North of the mouth, the inner **dune ridge** was relatively hummocky (but still more defined than the outer ridge).

At the north end of the estuary, the foredunes were less pronounced, but dense stipples still suggest slight topography.

A wave-washed **beach** was located between the high and low water lines.

Figure 6.15. Species records for California fagonia (*Fagonia laevis*). Most records of the plant are from the Mojave and Sonoran deserts, but a disjunct coastal population was found in the Tijuana River valley, possibly associated with the estuary's dune system. *(Map courtesy CalFlora)*

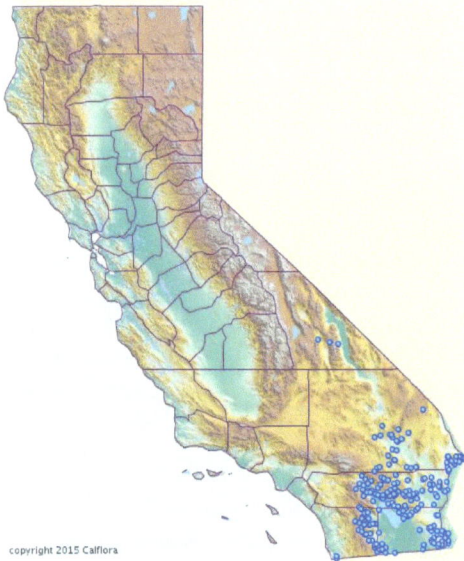

copyright 2015 Calflora

Figure 6.14. (below) View of the southern portion of the estuary, looking north (ca. 1912). Though roads and other developments have affected the native land cover by this time, the elevated beach-dune habitats at the western edge of the estuary are still largely intact. The image shows both the hummocky outer foredune ridge and more densely vegetated inner dune ridge extending north towards the river mouth. The ridges are separated by a lower sandy swale. On its landward side, the dune ridge grades down into the high marsh transition zone (see p. 160), areas of salt marsh (see p. 152), and a large remnant salt flat (see p. 155). Tidal sloughs are visible in the distance (see p. 146). *(Photos #80:4250 & #80:4249, both courtesy San Diego History Center)*

The historical beach-dune system featured a wide range of plant species, mostly low herbs and subshrubs, but also some larger shrubs and species generally associated with other habitat types. Cooper (1967), reporting on the condition of the dunes in 1919, states that dune vegetation was of the "foredune pioneer type, with a few representatives of the dune shrub community." These would likely have included species such as sticky sand verbena (*Abronia maritima*), pink sand verbena (*Abronia umbellata*), dune ragweed (*Ambrosia chamissonis*), and salt bush (*Atriplex leucophylla*; Purer 1936, Zedler et al. 1999). Large clumps of lemonade sumac (*Rhus integrifolia*) and laurel sumac (*Malosma laurina*), which are generally associated with chaparral and alluvial scrub, were noted as occurring on lee slopes and in sheltered hollows (Cooper 1967). Where the dunes bordered salt marsh, certain species—specifically Parish's glasswort (*Arthrocnemum subterminale*)—were noted for their ability to keep pace with deepening sand and were found growing from invading dune slipfaces at a height of 2 m (7 ft).

Cooper also noted the appearance of plants "belonging to the desert"—which, when moving north to south along the California coast, first became conspicuous parts of the dune plant community south of the entrance to San Diego Bay. These included at least one species of prickly pear (*Opuntia prolifera*), yucca (*Yucca schidigera*), Mexican tea (*Ephedra californica*), and several other desert shrubs. An examination of the species found in the Tijuana River valley reveals a number of species with desert affinities that have disjunct local populations. One notable example of this is the shrub California fagonia (*Fagonia laevis*); prior to 1950, the nearest records of the plant in the U.S. outside of the Tijuana River valley were from the Colorado Desert approximately 100 km inland (Fig. 6.15).

A snapshot of the wildlife that historically utilized the dunes was provided by the zoologist Robert Harwood, who surveyed the area during the summer of 1931. He found one important species, the (now) endangered California Least Tern (*Sterna antillarum browni*), in "great abundance" on the dunes near the estuary mouth (Harwood 1931; Fig. 6.16). There were "many nests," each built on the sand with bits of shell. A set of new nests appeared

outer foredune ridge

inner dune ridge

during late July, indicating that nesting at the estuary took place in at least two waves. Other dune species noted were the common side-blotched lizard (*Uta stansburiana*) and cottontail rabbit (*Sylvilagus auduboni*); cottontails were found in burrows on the backside of the dunes near the marsh edge. Finally, Harwood found that the insect life of the sand dunes "varies to a great extent"; he collected a total of ten different species, including six species of beetles.

On the beach, many early observers described catching California grunion (*Leurethes tenuis*). The fish "came in swarms" during high tides in the spring and summer to spawn in the wet sand (Bolla n.d.) and were a source of food for those living near the shore (Meigs 1925, Cuero and Shipek 1968).

In the evening, Hermann took us to the beach to catch smelt, which came up with the waves to spawn. Between 10 P.M. & 11 P.M. we caught 18—with our hands....[The] beach was lined with the fires of smelt-hunting parties.

—PEREVIL MEIGS 1925, ON CATCHING GRUNION

remnant salt flat

The estuarine-terrestrial transition zone is the area of interactions between tidal and terrestrial or tidal and fluvial processes that result in mosaics of habitat types, unique assemblages of plants and animals, and sets of ecosystem services that are distinct from those of adjoining estuarine, riverine, or terrestrial ecosystems alone (Zedler et al. 1992, Uyeda et al. 2013, Goals Project 2015). Historically, the Tijuana Estuary transitioned into multiple terrestrial habitat types, creating an array of transition zone types, along the estuary edge. These included transitions between (in decreasing order of their linear extent): estuarine habitat types and Alkali Meadow Complex (17.3 km), River Wash / Riparian Scrub (5.2 km), Beach/Dune (4.2 km), Grassland / Coastal Sage Scrub (3.6 km), and the steep mesa cliffs (1.5 km; Fig. 6.17). Each of these transition zone types had certain unique characteristics, differing in aspects such as width, slope, and species composition.

The earliest transition zone descriptions come from local biologists, who studied the estuary during the early to mid-20th century. Robert Harwood, in a 1931 report on the estuary's plant and animal life, noted "a gradual and almost imperceptible slope from the wet muddy land near the estuaries to the high and dry soil farther away" (Harwood 1931). Along this gradient, Harwood observed four plant species — *Salicornia ambigua, Distichlis littoralis, Juncus acutus,* and *Frankenia salina. Salicornia* was said to be most abundant in the lowest areas, while the latter three plants were associated with drier, sandier soils further up slope. Harwood cautioned, however, that dividing lines between the four species were "indiscernible"; the plants of the estuary were said to "blend in with the plants of higher ground." A similar ecotone (in terms of plants composition and progression) was observed by Edith Purer, who found that *S. pacifica*— although growing in pure stands in the lower levels of the estuary—was "at higher levels associated with *Distichlis,* and where still higher with *Distichlis* and *Frankenia [salina]"* (Purer 1942).

These historical accounts suggest that the divide between high marsh and salt grass-dominated alkali meadow was generally broad and indistinct. This quality is reflected in our historical synthesis map, which combines the High Marsh Transition Zone and Alkali Meadow Complex habitat types. The broad ecotone between these estuarine and terrestrial habitats is also supported by the historical soils map, which shows how the soil type dominated by "pickleweed and salt grass" (Alviso very fine sandy loam) extended up to 650 m beyond the mapped edge of Salt Marsh (Fig. 6.18). These soils then generally transitioned into a soil type (Foster very fine sandy loam) characterized by purer stands of salt grass without pickleweed. The spatial continuity of salt grass—from marsh to meadow—was explicitly described by Robert Harwood (1931): the plant was "very plentiful" within the estuary, but also had "the ability to grow back from the [estuary] where the soil is dry."

Elsewhere, the estuary transitioned into the sandy bed and scrubby riparian habitat of the Tijuana River (see Figure 6.20). A handful of historical sources described this estuarine-riverine transition zone and noted how willows "gradually grow fewer and smaller as the river bottom merges into the salt marshes near the sea" (Stephens 1912). The 1852 T-sheet depicts scattered shrubs or trees along the upper margins of

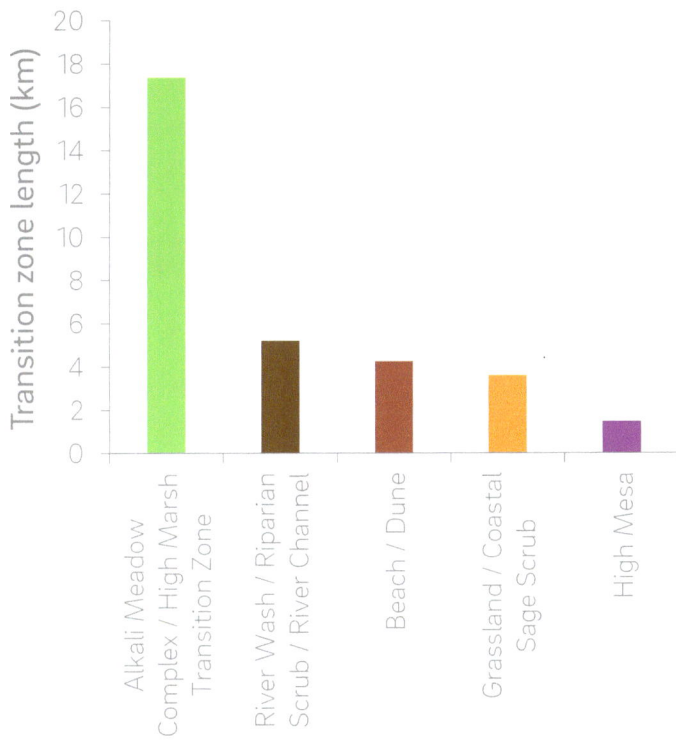

Figure 6.17. The historical estuarine-terrestrial transition zone overlaid on contemporary aerial imagery. At least five kinds of transition zones were found around the estuary, though transitions to Alkali Meadow Complex were by far the most prevalent. *(Base map: NAIP 2014)*

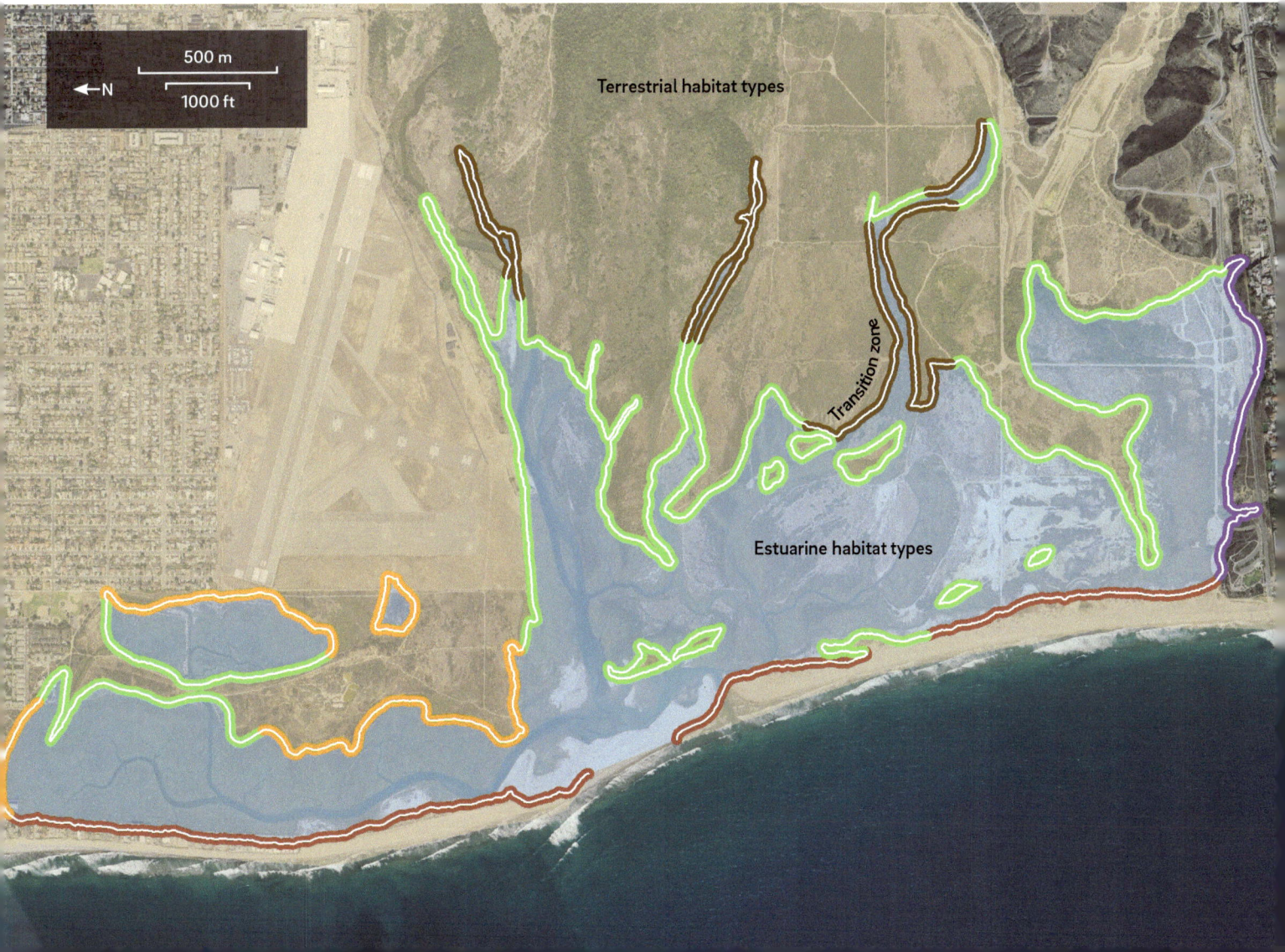

Fig. 6.18. A broad ecotone between high marsh and salt grass-dominated alkali meadows is suggested by this historical soils map. The extent of Alviso very fine sandy loam ("Av")—a soil type with a high salt content that supported a cover of "pickleweed and salt grass"—extended up to 650 m (0.4 mi) beyond the mapped inland edge of Salt Marsh (shown on the map with a dark blue overlay). We mapped this area as part of the High Marsh Transition Zone. At its upper margins, Alviso very fine sandy loam often gave way to Foster very fine sandy loam ("Fv"), a soil type associated with purer stands of salt grass (without mention of pickleweed), which we mapped as Alkali Meadow Complex (see p. 70). *(Storie and Carpenter 1930, courtesy University of Alabama)*

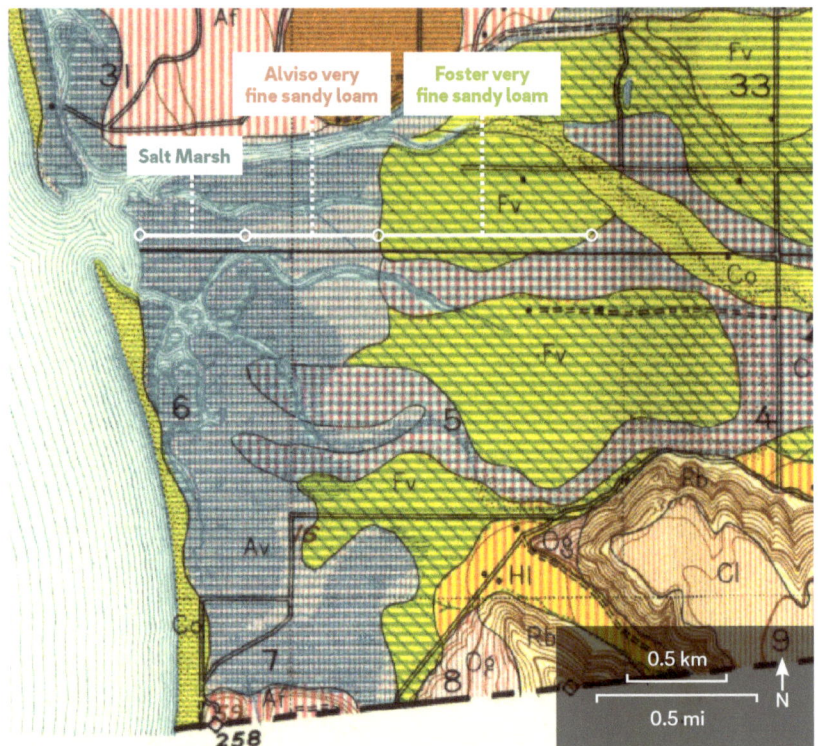

Old River Slough, some growing within what is mapped as Salt Marsh (Fig. 6.19). In oblique aerial photos taken nearly 100 years later, the same site supports a distinct thicket of relatively tall and dense riparian vegetation.

Unlike the transitions to alkali meadow and riverine habitats, the transitions to other terrestrial habitats—Beach/Dune, Grassland / Coastal Sage Scrub, and the high mesas—occurred over relatively steep topographic gradients, and were often narrower as a result. Contemporary research supports this inference: Zedler et al. (1992) found an abrupt transition and narrow band of overlap between remnant salt marsh and coastal sage scrub communities, which they attributed to the steep slope leading up to the low mesa where the scrub community grew.

 Contemporary research also notes a concentration of mima mounds in the estuary's high marsh transition zone. These earthen mounds were found to be up to approximately 0.5 m in height 18.5 m in diameter (Cox and Zedler 1986), and are thought to be the product of long-term soil movement by pocket gophers (but several other hypotheses for their formation exist; Cox 1984). Although no mention of these mounds was found in the historical record, the features studied in the 1980s seem to be apparent in the historical orthophotographs taken in 1928 (San Diego County 1928). These raised features would have added an element of local topographic complexity to the transition zone and supported terrestrial vegetation found further upland elsewhere (Cox and Zedler 1986). One possible explanation for their concentration in the transition zone relates

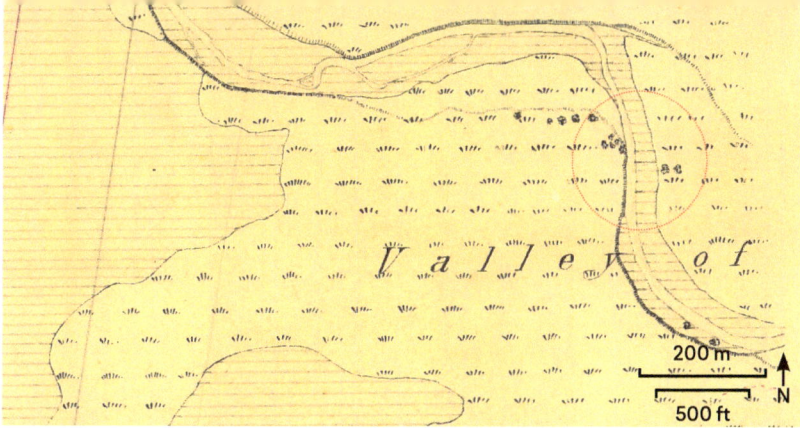

Figure 6.19. Woody vegetation at the estuarine-terrestrial transition zone. On the 1852 T-sheet, symbols for woody vegetation are drawn at the upper margin of Old River Slough, within and adjacent to the estuarine habitat types. Here, at the fluvial-tidal interface, intertidal flat and salt marsh transitioned into river wash and riparian scrub. The feature seems to have persisted into the mid-20th century: large shrubs or trees are visible in aerial photographs taken in 1927 (middle) and 1944 (bottom). *(Top: Harrison 1852, courtesy NOAA; Middle: San Diego County 1928; Bottom: Photo #79:744-890, courtesy San Diego History Center)*

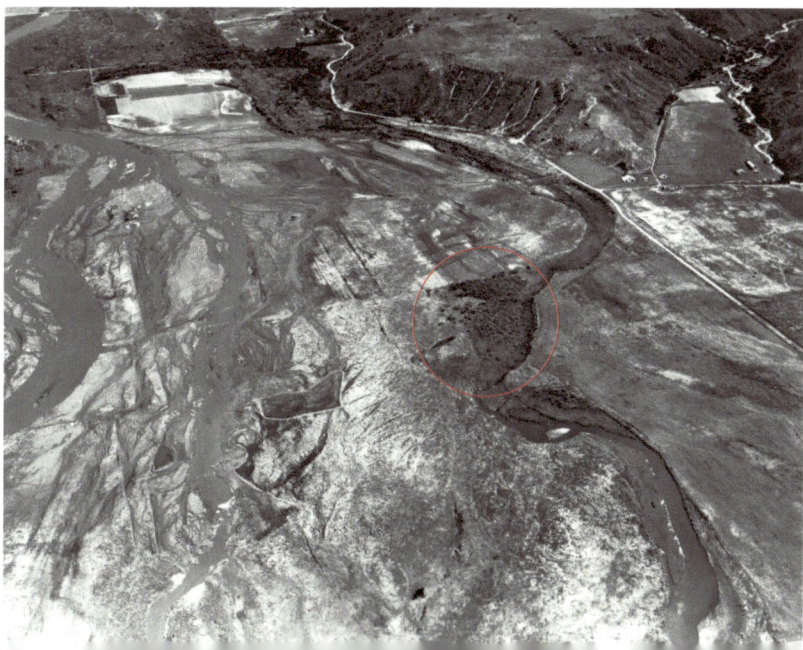

to flood frequency; flooding in the transition zone is hypothesized to be frequent enough to force gophers to locate their nest chambers in elevated areas, yet infrequent enough to allow extensive tunneling into lower surrounding areas (the result being a strong displacement of soil to mounded areas over time; Cox 1984, Cox and Zedler 1986).

Driven by gradients in elevation, moisture, inundation frequency, and salinity, the estuary's transition zones provided a wide range of ecological functions that together supported numerous species of plants and animals. Due to highly variable, seasonally fluctuating soil salinity, transition zones can support greater numbers and densities of winter annual plant species, which are able to survive the hypersalinity and low moisture of the dry season as seeds (Callaway et al. 1990, Noe and Zedler 2001a, Noe and Zedler 2001b). For many animals, including the endangered Ridgway's rail *(Rallus obsoletus levipes)*, the transition zone is crucial as high tide refuge. During extremely high tides and storm events, rails and other tidal marsh wildlife tend to concentrate in the transition zone, which also makes it an important foraging area for various predators, including bobcats *(Lynx rufus)*, coyotes *(Canis latrans)*, long-tailed weasels *(Mustela frenata)*, short-eared owls *(Asio flammeus)*, northern harriers *(Circus cyaneus)*, and white-tailed kites *(Elanus leucurus;* Zedler et al. 1992; Richards 2002). Many species utilize the transition zone for seasonal migrations: California meadow voles *(Microtus californicus)*, for example, spend the dry season within wetlands (including tidal marsh) before moving upland during the wet season to forage on new terrestrial plant growth (Collins et al. 2007). Other species—like the California ground squirrel *(Spermophilus beecheyi)*—make daily migrations across the transition zone. The squirrels construct their nest chambers within the transition zone and terrestrial habitats but often move down into the tidal environment to feed (Collins et al. 2007). Finally, the supratidal transition zone provides habitat for many insects, including ground-nesting bees (which pollinate marsh plants like the endangered salt marsh bird's beak [*Chloropyron maritimum* subsp. *maritimum*]), rove beetles (an important food source for many birds), and a variety of rare tiger beetles (J. Zedler, personal communication).

Figure 6.20. The estuarine-riverine transition zone, looking east from the head of Tijuana River Slough, 1944. An overflow channel of the Tijuana River extends from the upper-right hand corner of the image towards the lower-left hand corner of the image, where it meets a tidal slough. At this location riparian scrub associated with the overflow channel merges into tidal habitats like salt marsh. *(Photo #79:741-893, courtesy San Diego History Center)*

7
HABITAT TYPE CHANGE ANALYSIS

Overview

The Tijuana River valley has witnessed dramatic land cover changes since the mid-19th century stemming from urban and agricultural development, river channelization, groundwater pumping, and a range of other land and water use changes. Cumulatively, these land cover changes have resulted in substantial losses of most native habitat types, shifts in habitat distribution, and changes in habitat quality. Nevertheless, despite the significant habitat losses and alterations that have occurred over the past 150 years, contiguous undeveloped areas still occupy large portions of the estuary and valley floor.

Understanding how landscapes have changed over time is useful for setting restoration and management priorities and for envisioning how they might become more ecologically resilient in the future. This chapter contributes to such an understanding for the Tijuana River valley by examining changes in land cover since ca. 1850.

The chapter begins with maps of the past and present valley and then describes the methods we used to develop the contemporary habitat type map, as well as the methods and assumptions used in our analysis of landscape change over time. The remainder of the chapter is devoted to the results of this analysis, with a focus on changes in habitat type extent, distribution, and composition. These results set the stage for Chapter 8, which articulates management implications and recommended avenues for future research.

(Photo by Samuel Safran, April 2015)

to San Diego Bay

Tijuana Estuary

Tijuana River

Yogurt Canyon

Goat Canyon

Smuggler's Gulch

HISTORICAL CONDITIONS | Tijuana River valley, circa 1850

UNITED STATES OF AMERICA

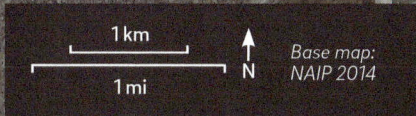

MEXICO

	Dune
	Beach
	Subtidal Water & Mudflat/Sandflat
	Salt Flat / Open Water
	Salt Marsh
	Alkali Meadow Complex / High Marsh Transition Zone
	River Channel
	River Wash / Riparian Scrub
	Grassland / Coastal Sage Scrub
	Perennial Freshwater Wetland
	Pond
	Vernal Pool

1 km

1 mi

N

Base map: NAIP 2014

Río Alamar

Cerro Colorado ▲

Presa Rodríguez

to San Diego Bay

Tijuana Estuary

Tijuana River

Yogurt Canyon

Goat Canyon

Smuggler's Gulch

MODERN CONDITIONS | Tijuana River valley, circa 2012

UNITED STATES OF AMERICA

MEXICO

■	Dune
■	Beach
■	Subtidal Water & Mudflat/Sandflat
■	Salt Flat / Open Water
■	Salt Marsh
■	Alkali Meadow Complex / High Marsh Transition Zone
■	River Channel
■	River Wash / Riparian Scrub
■	Grassland / Coastal Sage Scrub
■	Perennial Freshwater Wetland
■	Pond
■	Riparian Forest
■	Concrete Channel
■	Agriculture
■	Developed/Disturbed

1 km
1 mi

N

Base map:
NAIP 2014

Río Alamar

Cerro Colorado

Presa Rodríguez

Methods

We used contemporary land cover and vegetation mapping to evaluate how habitat type extent and distribution have changed in the lower Tijuana River valley over the past 165 years. Since no recent effort to map modern natural communities in the Tijuana River valley covers the full project study extent, modern habitat type data were compiled from multiple sources (Fig. 7.1). The classification systems for each of the compiled datasets were then crosswalked to the habitat types used to map the historical river valley (Table 7.1).

The Western San Diego County Vegetation dataset developed by the San Diego Association of Governments (SANDAG) in 2012 served as the primary source for the classification of vegetated habitats in the United States. Because this layer did not map estuarine wetland features, we supplemented the SANDAG map with data from the Southern California Wetland Mapping Project (SCWMP 2012). Data for developed and agricultural areas in the U.S. were derived from the National Land Cover Dataset (NLCD) of 2011, while habitat types in Mexico were derived from the Tijuana River Watershed Digital Vegetation File published by the Center for Earth Systems Analysis Research (CESAR) in 2000. Although the CESAR dataset was derived from imagery gathered between 1994 and 1995, it is still the most comprehensive and detailed we were able to acquire for Mexico. Certain features, including some reaches of the Tijuana River in the U.S. and the concrete flood control channel in Mexico were mapped or updated by SFEI staff from recent aerial imagery. Additional details concerning the development of the modern habitat type map are provided in the layer's GIS metadata.

One of the fundamental challenges to assessing changes in habitat type distribution is that complex habitat types must be combined into a limited number of classifications that (1) reflect the resolution of available data and (2) can be applied to both the historical and modern landscape. Reflecting these challenges, not all changes in habitat type distribution are captured by the change analysis. For example, though there is very little area of the River Wash habitat type left in the valley, the decrease in River Wash is not captured quantitatively because historical data resolution required us to lump the River Wash class with the Riparian Scrub class. Similarly, because the modern boundary between Subtidal Water and Mudflat/Sandflat could not be accurately demarcated with the available data, we combined these two habitat types when comparing them over time. Since we were unable to reliably demarcate the contemporary boundary between Beach and ocean in a manner consistent with the historical Beach classification, we used the classification Beach/Ocean. These compromises were necessary to ensure that the habitat types being compared in the historical and modern landscape are as comparable as possible.

Finally, despite our best efforts to make faithful comparisons of the extent of each habitat type over time, it is important to note that the species composition, ecological functions, and habitat quality of historical and modern habitat types are not necessarily equivalent. For example, the contemporary River Wash / Riparian Scrub class includes areas dominated by *Arundo donax* and *Tamarix* spp., non-native species that would not have been found in the river corridor prior to European-American colonization. Similarly, the contemporary Grassland / Coastal Sage Scrub class includes areas dominated by the non-native plant *Glebionis coronaria*. These differences were important to consider when assessing change over time and are addressed qualitatively in this chapter.

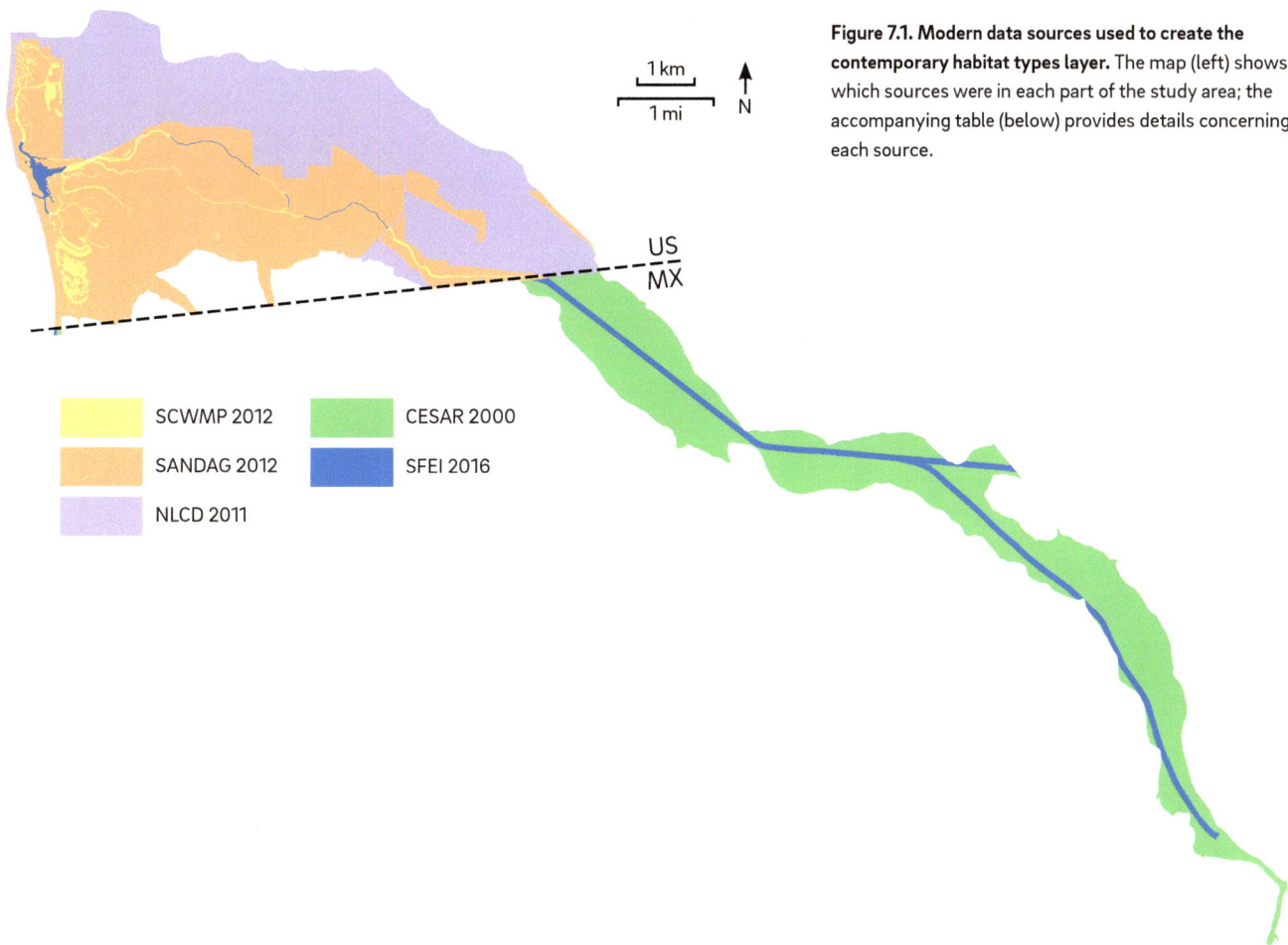

Figure 7.1. Modern data sources used to create the contemporary habitat types layer. The map (left) shows which sources were in each part of the study area; the accompanying table (below) provides details concerning each source.

Legend:
- SCWMP 2012
- SANDAG 2012
- NLCD 2011
- CESAR 2000
- SFEI 2016

Title	Source	Citation	Year depicted	Classification scheme	Minimum mapping unit of original source	Incorporated area (ha)	Study extent coverage
Southern California Wetlands Mapping Project	Southern California Wetlands Mapping Project	SCWMP 2012	2005	Cowardin (modified)	0.5 acres (0.2 ha)	65	1.5%
Western San Diego County Vegetation	San Diego Association of Governments	SANDAG 2012	2012	National Vegetation Classification Standard	Terrestrial systems- 1 ha Wetlands- 0.5 ha	1,432	33.6%
National Land Cover Dataset (2011 Edition)	Multi-Resolution Land Characteristics Consortium	NLCD 2011	ca. 2011	Anderson Land Cover Classification System (modified)	900 m² (0.09 ha)	1,437	33.7%
Tijuana River Watershed Digital Vegetation File	Center for Earth Systems Analysis Research	CESAR 2000	1994–1995	Holland (modified)	Riparian areas- 0.5 acres (0.2 ha) Non-riparian areas- 5 acres (2.0 ha)	1,147	26.9%
SFEI Supplemental Mapping	San Francisco Estuary Institute	SFEI 2016	2008–2010	N/A	N/A	185	4.3%

Table 7.1. Crosswalk between the original source classifications used in the contemporary datasets and the habitat types used to map the historical river valley. Several additional habitat types were incorporated into the modern classification system when analogues to historical classifications were unavailable (e.g., Concrete Channel, Riparian Forest, Developed/Disturbed). Some important exceptions to this crosswalk (for classifications marked with an asterisk) are described in the modern habitat layer's GIS metadata.

Source	Original source classification	Crosswalk classification
SCWMP 2013 System-Class (Code/s)	Estuarine subtidal-Unconsolidated bottom (E1UBL)	Subtidal Water*
	Estuarine intertidal-Aquatic Bed (E2ABM, E2ABN)	Mudflat/Sandflat*
	Estuarine intertidal-Streambed (E2SBM,E2SBN)	Mudflat/Sandflat*
	Estuarine intertidal-Unconsolidated Shore/Aquatic Bed (E2US/ABN)	Mudflat/Sandflat*
	Estuarine intertidal-Unconsolidated Shore/Emergent (E2US/EMN, E2US/EMP)	Mudflat/Sandflat*
	Estuarine intertidal-Unconsolidated Shore (E2USM, E2USN, E2USP)	Mudflat/Sandflat*
	Riverine lower perennial-Unconsolidated Bottom (R2UBF)	River Channel
	Riverine intermittent-Streambed (R4SBA, R4SBC, R4SBCx)	River Channel
SANDAG 2012 Alliance	Abronia latifolia-Ambrosia chamissonis Alliance	Dune
	Agriculture	Agriculture
	Artemisia californica-Eriogonum fasciculatum Alliance	Grassland / Coastal Sage Scrub
	Arthrocnemum subterminale Alliance	Salt Marsh*
	Arundo donax Semi-Natural Stands	River Wash / Riparian Scrub
	Atriplex lentiformis Alliance	Alkali Meadow Complex / High Marsh Transition Zone
	Baccharis pilularis Alliance	Grassland / Coastal Sage Scrub
	Baccharis salicifolia Alliance	River Wash / Riparian Scrub*
	Baccharis sarothroides Provisional Alliance	Grassland / Coastal Sage Scrub
	Bahiopsis laciniata Alliance	Grassland / Coastal Sage Scrub
	Beach	Beach / Ocean
	Bolboschoenus maritimus Alliance	Perennial Freshwater Wetland
	Brassica (nigra) and Other Mustards Semi-Natural Stands	Grassland / Coastal Sage Scrub
	Bromus (diandrus, hordeaceus)-Brachypodium distachyon Semi-Natural Stands	Grassland / Coastal Sage Scrub
	Cressa truxillensis-Distichlis spicata Alliance	Alkali Meadow Complex / High Marsh Transition Zone
	Developed	Developed/Disturbed
	Distichlis spicata Alliance	Alkali Meadow Complex / High Marsh Transition Zone
	Encelia californica Alliance	Grassland / Coastal Sage Scrub
	Eucalyptus (globulus, camaldulensis) Semi-Natural Stands	Developed/Disturbed
	Frankenia salina Alliance	Alkali Meadow Complex / High Marsh Transition Zone
	Glebionis coronaria semi natural stands	Grassland / Coastal Sage Scrub
	Graded/Scraped/Maintained	Developed/Disturbed
	Isocoma menziesii Alliance	Grassland / Coastal Sage Scrub*
	Juncus acutus Provisional Alliance	Alkali Meadow Complex / High Marsh Transition Zone
	Lycium californicum Provisional Alliance	Grassland / Coastal Sage Scrub
	Mediterranean California Naturalized Annual and Perennial Grassland Semi-Natural Stands	Grassland / Coastal Sage Scrub
	Naturalized Warm-Temperate Riparian and Wetland Semi-Natural Stands	River Wash / Riparian Scrub
	Open Water	Pond*
	Ornamental	Developed/Disturbed
	Rhus integrifolia Alliance	Grassland / Coastal Sage Scrub
	Salix gooddingii Alliance	Riparian Forest
	Salix laevigata Alliance	Riparian Forest
	Salix lasiolepis Alliance	Riparian Forest
	Sarcocornia pacifica (Salicornia depressa) Alliance	Salt Marsh*
	Spartina foliosa Alliance	Salt Marsh
	Tamarix spp. Semi-Natural Stands	River Wash / Riparian Scrub
	Tidal/Mudfat	Mudflat/Sandflat*
	Typha (angustifolia, domingensis, latifolia) Alliance	Perennial Freshwater Wetland
NLCD 2011 Land cover class	Barren Land	Developed/Disturbed
	Cultivated Crops	Agriculture*
	Developed, High Intensity	Developed/Disturbed
	Developed, Low Intensity	Developed/Disturbed *
	Developed, Medium Intensity	Developed/Disturbed *
	Developed, Open Space	Developed/Disturbed *
	Emergent Herbaceous Wetlands	Developed/Disturbed
	Evergreen Forest	Developed/Disturbed
	Hay/Pasture	Agriculture*
	Herbaceuous	Developed/Disturbed *
	Open Water	Developed/Disturbed
	Shrub/Scrub	Developed/Disturbed
	Woody Wetlands	Developed/Disturbed
CESAR 2000 Description	Beach	Beach / Ocean
	Coastal Sage Scrub	Grassland / Coastal Sage Scrub*
	Coastal Sage Scrub - Disturbed	Grassland / Coastal Sage Scrub*
	Developed	Developed/Disturbed *
	Disturbed Habitat	Developed/Disturbed *
	Natural Floodchannel / Streambed	River Channel*
	Open Water	Pond
	Riparian Forest	Riparian Forest*
	Riparian Forest - Disturbed	Riparian Forest*
	Riparian Scrub	River Wash / Riparian Scrub*
	Riparian Scrub - Disturbed	River Wash / Riparian Scrub*

Summary of landscape change

Since the mid-19th century, nearly all portions of the lower Tijuana River valley have undergone significant changes in land cover. Chief among these changes is habitat loss, a process that has occurred across all major habitat type groups. Estuarine habitat types have together decreased by approximately 40%, the extent of the river corridor has decreased by 75%, and other valley floor habitat types have cumulatively decreased by about two-thirds (64%). Of the eleven individual historical habitat types included in the change analysis, eight have experienced a net loss in area over time. All historical habitat types that occupied more than 60 ha (150 ac) during the early 19th century have experienced net losses of ~40–90%.

The largest single driver of habitat loss has been land conversion for urban development and agriculture, which now covers more than two-thirds of the valley. Notably, the three most extensive habitat types historically (River Wash / Riparian Scrub, Grassland / Coastal Sage Scrub, and Alkali Meadow Complex / High Marsh Transition Zone), which once covered more than 80% of the lower valley, have together declined by 78%, largely as a result of this conversion. Though urban development is most striking in Mexico (where 95% of the valley has been converted), more total acreage has actually been developed in the U.S. (~1,400 ha versus ~1,200 ha). In total, 59% of the U.S. portion of the valley has been developed, most of it on the low mesa at the northern side of the valley where all native land cover types have been lost. Urban lands can provide some benefit for wildlife, but these benefits are generally limited to a relatively small subset of species that have adapted to altered conditions.

For some habitat types, loss has primarily been driven by conversion to other habitat types. This process is particularly pronounced in the estuary. Though 'Subtidal Water & Mudflat/Sandflat' and Salt Flat / Open Water have both declined in total by 40–50%, less than 3% of their decline is directly attributable to urban development. The primary change to 'Subtidal Water & Mudflat/Sandflat' has instead been conversion to Salt Marsh, while the primary change to Salt Marsh has been conversion to higher and drier habitat types like Grassland / Coastal Sage Scrub. These shifts towards higher-elevation habitats in the estuary are thought to be driven by the deposition and accumulation of sediment from the watershed (see pp. 180–81).

Habitat conversion has also been dramatic in the river corridor, which today features one of the largest stands of gallery riparian forest in coastal southern California. Though these forests are a defining feature of the valley today, our findings suggest that riparian forest was not historically present in the lower valley. The river corridor was instead dominated by River Wash and Riparian Scrub. River Wash / Riparian Scrub once covered 92% of the river corridor in the U.S. (with the remainder taken up by the river channel itself); today it covers only 35% of the corridor, with Riparian Forest covering all but 2% of the remainder. The dramatic change in the structure and composition of riparian habitats was likely related to the initiation of perennial stream flows in the early 1980s, as well as accompanying changes in groundwater levels (see p. 184). This

(continued on p. 179)

Key findings

- In the U.S., net losses (42–83%) in the extent of all major mapped historical habitat types (ones that occupied >60 ha during the mid-19th century).
- In Mexico, more pronounced net habitat losses (94–97%) and replacement of braided river channel with a concrete channel.
- Conversion of valley floor Alkali Meadow Complexes (wetlands) to Grassland and Coastal Sage Scrub (drylands).
- Conversion of River Wash and Riparian Scrub to gallery Riparian Forest in the U.S.
- Conversion of estuarine habitat types from lower to higher-elevation features (e.g. intertidal flats to Salt Marsh; Salt Marsh to High Marsh Transition Zone).
- Increase in non-native species cover, especially in Grassland, Coastal Sage Scrub, and Riparian Scrub habitats.

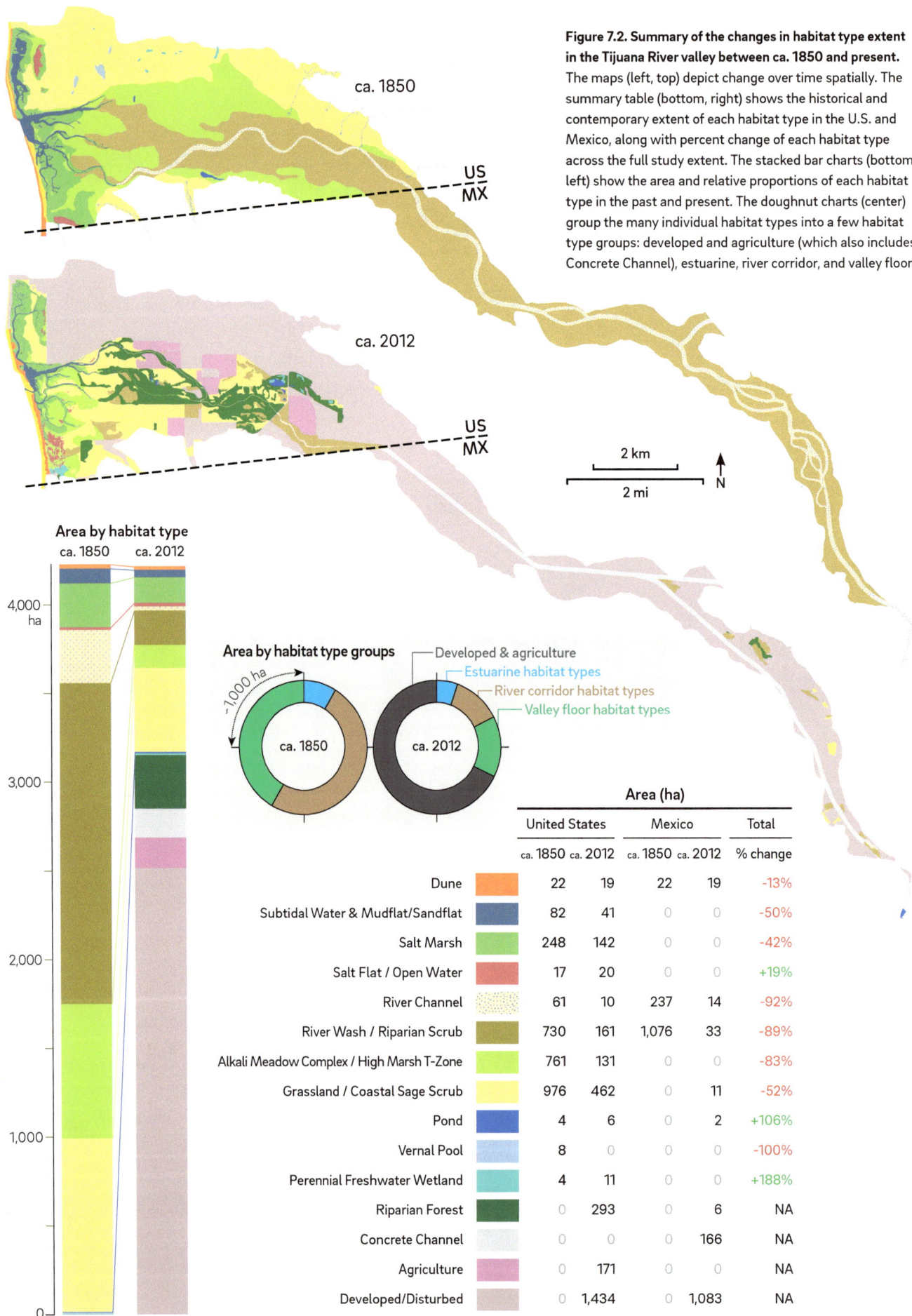

Figure 7.2. Summary of the changes in habitat type extent in the Tijuana River valley between ca. 1850 and present. The maps (left, top) depict change over time spatially. The summary table (bottom, right) shows the historical and contemporary extent of each habitat type in the U.S. and Mexico, along with percent change of each habitat type across the full study extent. The stacked bar charts (bottom, left) show the area and relative proportions of each habitat type in the past and present. The doughnut charts (center) group the many individual habitat types into a few habitat type groups: developed and agriculture (which also includes Concrete Channel), estuarine, river corridor, and valley floor.

ca. 1850

ca. 2012

US / MX

US / MX

2 km

2 mi

N

Area by habitat type
ca. 1850   ca. 2012

4,000 ha

3,000

2,000

1,000

0

Area by habitat type groups

~1,000 ha

ca. 1850

ca. 2012

Developed & agriculture
Estuarine habitat types
River corridor habitat types
Valley floor habitat types

	Area (ha)				
	United States		Mexico		Total
	ca. 1850	ca. 2012	ca. 1850	ca. 2012	% change
Dune	22	19	22	19	-13%
Subtidal Water & Mudflat/Sandflat	82	41	0	0	-50%
Salt Marsh	248	142	0	0	-42%
Salt Flat / Open Water	17	20	0	0	+19%
River Channel	61	10	237	14	-92%
River Wash / Riparian Scrub	730	161	1,076	33	-89%
Alkali Meadow Complex / High Marsh T-Zone	761	131	0	0	-83%
Grassland / Coastal Sage Scrub	976	462	0	11	-52%
Pond	4	6	0	2	+106%
Vernal Pool	8	0	0	0	-100%
Perennial Freshwater Wetland	4	11	0	0	+188%
Riparian Forest	0	293	0	6	NA
Concrete Channel	0	0	0	166	NA
Agriculture	0	171	0	0	NA
Developed/Disturbed	0	1,434	0	1,083	NA

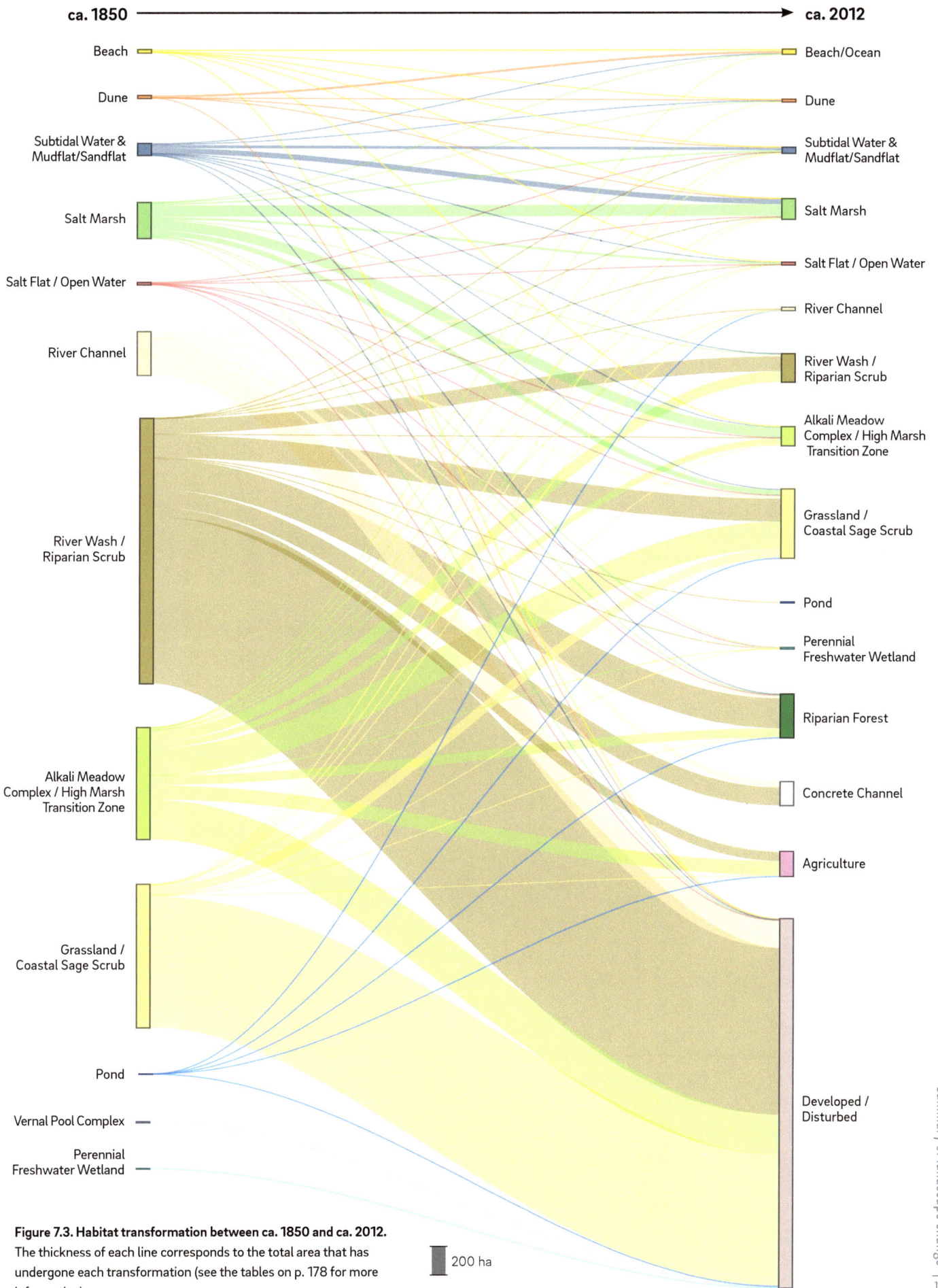

ca. 1850 → **ca. 2012**

ca. 1850 labels (left):
Beach
Dune
Subtidal Water & Mudflat/Sandflat
Salt Marsh
Salt Flat / Open Water
River Channel
River Wash / Riparian Scrub
Alkali Meadow Complex / High Marsh Transition Zone
Grassland / Coastal Sage Scrub
Pond
Vernal Pool Complex
Perennial Freshwater Wetland

ca. 2012 labels (right):
Beach/Ocean
Dune
Subtidal Water & Mudflat/Sandflat
Salt Marsh
Salt Flat / Open Water
River Channel
River Wash / Riparian Scrub
Alkali Meadow Complex / High Marsh Transition Zone
Grassland / Coastal Sage Scrub
Pond
Perennial Freshwater Wetland
Riparian Forest
Concrete Channel
Agriculture
Developed / Disturbed

200 ha

Figure 7.3. Habitat transformation between ca. 1850 and ca. 2012.
The thickness of each line corresponds to the total area that has undergone each transformation (see the tables on p. 178 for more information).

**Areas with the same land cover
ca. 1850 & ca. 2012**

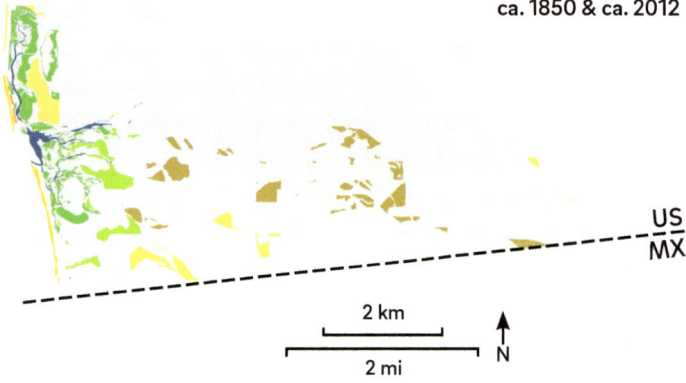

Figure 7.4. Areas that have the same land cover type today as ca. 1850. These "persistent" areas account for 8% of the total study extent. Some of the persistent areas are true remnant habitat patches, which often have unique values for restoration efforts. Others were converted at one point in time but have since recovered (like the Model Marsh restoration project in the southern part of the estuary).

US
MX

2 km
2 mi
N

Table 7.2. Land cover change matrix. Read across rows to find the percentage of each historical habitat type that has been converted to each contemporary habitat type. Cells with grey backgrounds quantify "persistent" areas that have the same land cover today as they did ca. 1850. These are the areas shown in Figure 7.4 (above).

ca. 2012 →
ca. 1850 ↓

ca. 1850 \ ca. 2012	Beach/Ocean	Dune	Subtidal Water & Mudflat/Sandflat	Salt Marsh	Salt Flat / Open Water	River Channel	River Wash / Riparian Scrub	Alkali Meadow Complex / High Marsh T-Zone	Grassland / Coastal Sage Scrub	Pond	Vernal Pool	Perennial Freshwater Wetland	Riparian Forest	Concrete Channel	Agriculture	Developed/Disturbed
Beach	52%	29%	<1%	<1%	<1%	0%	0%	1%	0%	0%	0%	0%	0%	0%	0%	17%
Dune	75%	6%	1%	2%	0%	0%	0%	0%	0%	0%	0%	0%	0%	0%	0%	17%
Subtidal Water & Mudflat/Sandflat	7%	5%	26%	46%	<1%	0%	1%	5%	9%	0%	0%	0%	<1%	0%	0%	<1%
Salt Marsh	<1%	2%	5%	33%	7%	0%	2%	28%	15%	0%	0%	1%	3%	0%	0%	2%
Salt Flat / Open Water	0%	0%	3%	50%	<1%	0%	0%	17%	2%	0%	0%	18%	1%	0%	0%	8%
River Channel	0%	0%	0%	0%	0%	4%	4%	0%	8%	1%	0%	<1%	6%	14%	1%	62%
River Wash / Riparian Scrub	0%	0%	<1%	<1%	<1%	<1%	6%	<1%	9%	<1%	0%	<1%	11%	7%	4%	63%
Alkali Meadow Complex / High Marsh T-Zone	<1%	<1%	1%	1%	<1%	1%	10%	6%	24%	0%	0%	<1%	8%	0%	13%	36%
Grassland / Coastal Sage Scrub	0%	0%	<1%	1%	<1%	<1%	1%	1%	6%	0%	0%	<1%	<1%	0%	0%	92%
Pond	0%	0%	0%	0%	0%	1%	0%	0%	7%	0%	0%	6%	0%	0%	<1%	86%
Vernal Pool	0%	0%	0%	0%	0%	0%	0%	0%	0%	0%	0%	0%	0%	0%	0%	100%
Perennial Freshwater Wetland	0%	0%	0%	0%	0%	0%	0%	0%	0%	0%	0%	0%	0%	0%	0%	100%

Due to 20th-century **conservation efforts**, only a very small percentage of the estuary's tidal habitats have been developed. The primary change to these habitat types has instead been conversion to other native habitat types.

Urban development of the low mesa on the north side of the valley in the U.S. resulted in complete loss of historical Vernal Pools and Perennial Freshwater Wetlands.

Nearly half of the original area of Subtidal Water & Mudflat/Sandflat now sits higher in the tidal frame and has converted to Salt Marsh, probably from **sediment accumulation** in the estuary.

More than 40% of the estuary's original Salt Marsh has also shifted higher in the tidal frame and now is occupied by High Marsh Transition Zone and Grassland / Coastal Sage Scrub, probably due to **sediment accumulation** and **early land reclamation**.

Apart from urban development, the largest change to the Alkali Meadow Complex wetlands was conversion to more xeric grasslands and shrublands. **Groundwater depletion** is thought to have been a major contributing factor.

A substantial percentage of former Riparian Scrub habitat has converted into Riparian Forest, likely due to the **perennialization of streamflow** during the 1980s and accompanying changes in groundwater levels.

(continued from p. 175)

finding raises a number of important management questions, including whether or not riparian forests are sustainable under the current intermittent streamflow regime and how to manage potential trade-offs between riparian species that might favor one riparian habitat type over the other (see pp. 128–29).

Lastly, Alkali Meadow Complex (a facultative wetland type dependent on high ground water levels that once dominated the valley floor) has largely been converted to Grassland / Coastal Sage Scrub (a more xeric terrestrial habitat type). This wetland to dryland conversion is likely related to early land use changes (the expansion of agriculture during the late 19th and early 20th centuries) and to decreases in groundwater levels during the early to mid-20th century (see p. 189). The change is reflective of an overall decrease in the ratio of wetlands to drylands from more than 3:1 to less than 2:1 (these calculations exclude developed and agricultural areas; including them as dryland habitats would make the decrease in the ratio much more dramatic).

As a result of habitat loss and conversion, only 8% of the lower river valley's surface area is occupied by the same habitat today as during the mid-19th century (Fig. 7.4). Mapping these "persistent" areas is a first step towards identifying, preserving, and building on important habitat remnants. Persistent patches are concentrated in the estuary, but there are number of interesting areas. The existence, for example, of persistent patches of Riparian Scrub (despite the agricultural legacy of most of these sites) points to the resilience of this particular habitat type and the endurance of key physical and ecological processes (e.g., flooding and seed dispersal). There are also intriguing patches of persistent Grassland / Coastal Sage Scrub, some of which seem to be true remnants of 19th century land cover. Remnant habitats can be important parts of a region's natural and cultural heritage and can be useful for habitat restoration efforts.

This said, even areas mapped with the same habitat type classification in ca. 1850 and ca. 2012 have experienced at least some degree of change. Species composition and habitat quality have been impacted by the introduction of non-native species, environmental contaminants, and a variety of other factors (changes that are not well captured by the land cover analysis). Non-native species now make up a significant component of several habitat types, including Grassland / Coastal Sage Scrub (where they dominate roughly 45% of the mapped area) and River Wash / Riparian Scrub (where they dominate roughly 70% of the mapped area; see pp. 190–91). Since changes in species composition and habitat quality can affect the functions provided by natural areas, these changes must be kept in mind when making comparisons between different areas over time.

The following sections provide more detail about land cover changes in the valley and the primary drivers behind these trends. Some of the changes to the river valley are essentially irreversible, such as widespread urban development of the low mesa on the northern side of the valley and of the river corridor in Mexico. In many areas of the valley, however, land cover changes have been much less drastic, providing a potential opportunity for restoration of natural habitat types and associated ecological functions (the Tijuana River valley constitutes the largest contiguous area of undeveloped alluvial valley floors in coastal southern California). Assessing the relative permanence of land cover changes will be important for prioritizing future restoration efforts within the valley.

Change in the Tijuana Estuary

Intertidal mudflats have converted to salt marsh, likely due to sediment accumulation

Of the 75 ha (180 ac) of intertidal flats (Mudflat/Sandflat) historically present in the estuary, approximately 35 ha (90 ac) have been converted to higher-elevation Salt Marsh. This change has been most pronounced in the northern arm of the estuary, where cordgrass, pickleweed, and other salt marsh plants have colonized much of the area along Oneonta Slough formerly occupied by unvegetated mudflats. Some marsh progradation occurred by 1925, though sizeable areas of mudflat were still present. By the late 20th century, however, the mudflats in northern part of the estuary had virtually disappeared (Fig. 7.5, Swanson 1987a).

Because subtidal water and intertidal flats were classified together in the contemporary mapping, it was not possible to quantify the absolute change in Mudflat/Sandflat area, though comparison of historical maps and aerial photographs with modern aerial photos suggests a significant overall loss in Mudflat area over time (Fig. 7.5). In addition, the approximate area occupied historically by Mudflat/Sandflat alone (75 ha [180 ac]) exceeds substantially the combined area occupied by Subtidal Water and Mudflat/Sandflat today (approximately 40 ha [100 ac]).

Though dam construction has reduced overall suspended sediment delivery to the mouth of the estuary by an estimated 50% (Brownlie and Taylor 1981), sediment transported from the watershed appears to be a primary driver of the historical conversion of intertidal flats to salt marsh. Vertical accretion of up to 10 cm on the marsh plain during the 1980 flood (Zedler 1983), 1.9–8.5 cm (0.8–3.3 in) in low marsh areas following winter storms in 1992–3 (Cahoon et al. 1996), and up to 12.7 cm (5 in) in low marsh and mudflats following floods in 1997–8 (Ward et al. 2003) suggest sediment deposition during episodic floods can still be quite significant (c.f. Wallace et al. 2005). On average, vertical accretion rates along Oneonta Slough between 1963 and 1998 significantly exceeded the rate of sea level rise (0.71–1.23 cm/yr [0.28–0.48 in] of accretion versus 1–3 mm/yr [0.04–0.1 in/yr] of sea-level rise; Weis et al. 2001). A net increase in land surface elevations relative to sea level would be expected to cause a gradual transition to habitat types situated higher in the tidal frame.

1852   1928   2014

Figure 7.5. Sedimentation has resulted in the conversion of intertidal mudflat to salt marsh in Oneonta Slough, as seen from a comparison of the 1852 T-sheet, the 1928 aerials, and the 2014 aerials. *(Harrison 1852, courtesy NOAA; San Diego County 1928; NAIP 2014)*

Salt marsh to transition-zone conversion has caused a westward shift in the inland extent of the estuary

Wetlands in the central and southern portions of the estuary have also been converted to habitat types situated higher in the tidal frame. This change has primarily been driven by the conversion of approximately 70 ha (170 ac) of Salt Marsh to High Marsh Transition Zone (classified as Alkali Meadow Complex / High Marsh Transition Zone because it was not possible to differentiate these two habitat types in the historical mapping). Sediment accumulation appears to have been the primary driver of this change. The high rates of sediment delivery are likely the result of increased erosion caused by rapid urbanization within Goat Canyon: Webber (2010) estimates that Goat Canyon can deliver up to 79,000 tons of sediment to the estuary annually, and Callaway and Zedler (2004) found that up to 30 cm (12 in) of sediment were deposited in the southern portion of estuary during a single (1994–5) storm season. Overall, Salt Marsh area within the estuary has decreased by 42%, from approximately 250 ha (610 ac) to 140 ha (350 ac; Fig. 7.6). In areas that remained Salt Marsh, sedimentation during the late 19th century still contributed to increases in soil salinity and decreases in species diversity (Zedler and West 2008).

Because High Marsh Transition Zone was not differentiated from Alkali Meadow Complex in the historical mapping, it is not possible to also quantify changes in the extent of this habitat type, though its western boundary has clearly shifted seaward (further west) over time. Though the displacement of salt marsh by high marsh transition zones and other upland habitat types is a significant conservation and management concern, it is also important to note that high marsh transition zone is itself a threatened habitat type with a range of ecological values (e.g., Desmond et al. 2001; see p. 160). Throughout California, estuarine-terrestrial transition zones have experienced major declines due to development, but the Tijuana Estuary retains an unusually undeveloped transition zone that has a high degree of connectivity both with the marsh and with drier habitat types further up the valley.

The seaward shift in the inland extent of salt marsh, combined with a landward shift in the beach-dune system (see p. 183) has resulted in an apparent "compression" of the estuary since the mid-19th century. It is possible that this trajectory has reached an inflection point— accelerated sea-level rise could push the landward boundary of the estuary inland over the next century (Naughton 2013), which would counteract at least one major component of the historical trend.

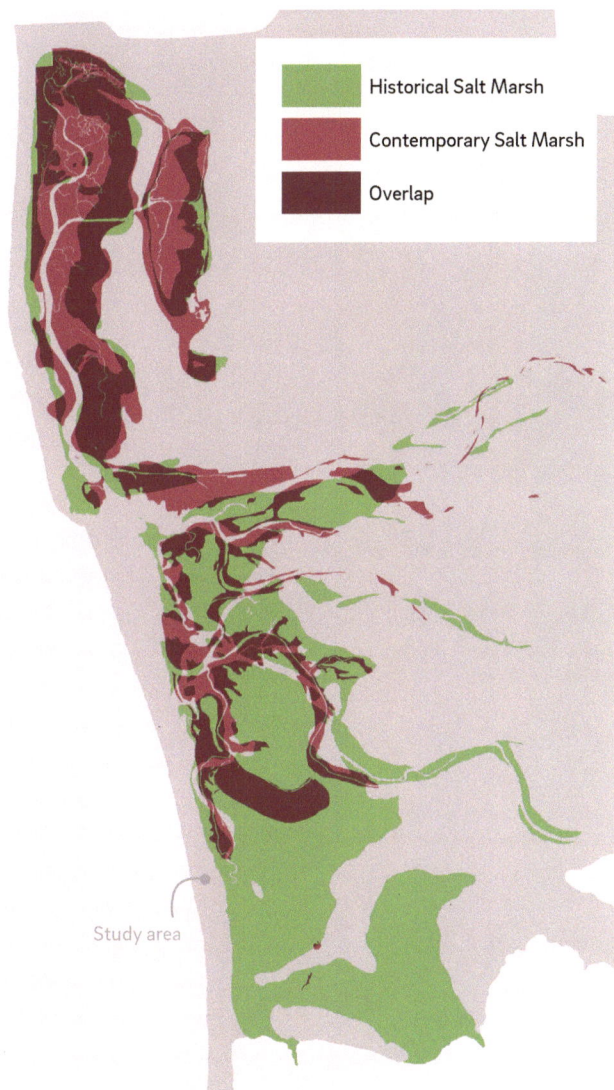

Historical Salt Marsh

Contemporary Salt Marsh

Overlap

Study area

Figure 7.6. Distribution of Salt Marsh ca. 1850 and today. The most extensive losses have been at the south end of the estuary.

Figure 7.7. Land cover changes at the south end of the estuary. The historical salt flat has been lost; new salt flats have established north of the old location. A large freshwater perennial wetland has formed at the mouth of Yogurt Canyon. *(Base map: NAIP 2014)*

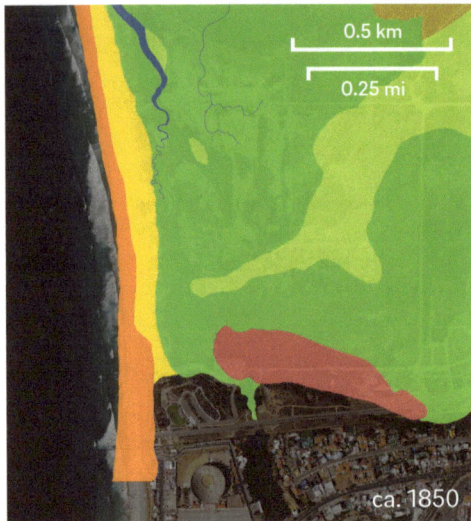

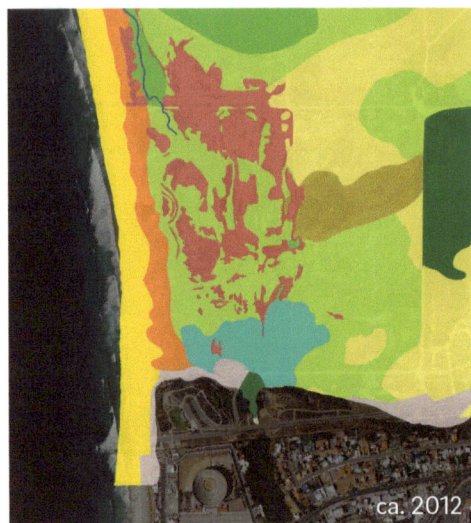

Legend:
- Salt Marsh
- Salt Flat / Open Water
- Perennial Freshwater Wetland
- Alkali Meadow Complex / High Marsh Transition Zone

A perennial freshwater wetland has established in the southern part of the estuary

Freshwater discharges from Yogurt Canyon have decreased salinity levels on the southern side of the estuary, resulting in the establishment of an approximately 6 ha (15 ac) Perennial Freshwater Wetland in an area historically occupied by Salt Marsh and Salt Flat / Open Water (Figs. 7.7–7.8; J. Crooks, personal communication). Though there is historical evidence for occasional brackish (but potable) seepage from the banks of the estuary's sloughs (Hatherley 1936b), there is no data to suggest the historical presence of Perennial Freshwater Wetlands in the estuary. Despite the loss of Perennial Freshwater Wetlands on the low mesa on the north side of the valley (see p. 190), the development of the feature in the estuary has contributed to net increase in the mapped extent of the habitat over time (from 4 ha [10 ac] ca. 1850 to 11 ha [27 ac] ca. 2012).

Salt flat area has remained stable, but the distribution and character of salt flats has changed

Though the total area of Salt Flat / Open Water today (approximately 20 ha [49 ac]) is comparable to the historical area (17 ha/42 ac), the distribution of the habitat type has changed: salt flats today comprise a network of mostly small patches occurring primarily on the southern side of the estuary. These flats are likely areas with compacted soils from 20th century military activity that now resist re-vegetation (Fig. 7.7, Figs. 7.9–7.10; J. Zedler, personal communication). As a result, contemporary salt flats are not necessarily analogous to historical flats. The large salt flat on the northern side of the estuary, known historically as the "Salt Works" (Pascoe 1869, see p. 155), became the site of sewage disposal ponds during the mid-20th century (USGS 1953, 1967). In the 1980s , these ponds were opened to tidal circulation, resulting in the establishment of salt marsh vegetation (D'Elgin et al. n.d.).

Figure 7.8. A new perennial freshwater wetland supported by freshwater discharges from Yogurt Canyon. *(Photo by Samuel Safran, April 2015)*

The dune system is simpler and has migrated inland

The dune system that fringes the estuary on the west has migrated inland by up to 200 m (650 ft) since the mid-19th century (Fig. 7.11). Along the central-northern portion of the estuary, where the dunes historically formed a double ridge fronting the estuary (see pp. 156–59), the system has become simplified to a single-ridge dune. Along the northernmost part of the estuary, dunes were eliminated by beach-side development. Overall, however, total Dune area has remained fairly constant at approximately 20 ha (50 ac). Possible causes for barrier beach retreat include sea-level rise, decreases in sediment delivery from the Tijuana River to the littoral zone (see p. 40), and the retreat of the headlands at Imperial Beach and Playa de Tijuana (c.f. Swanson 1987b). Any changes in longshore sand transport or the frequency and magnitude of storm events, such as those that might occur with large-scale climate cycles, would also be expected to influence the position and stability of the barrier beach (Swanson 1987b, Orme et al. 2011).

Dunes
ca. 1850

Dunes
ca. 2012

200 m

1,000 ft

N

Figure 7.9. (above, top) Salt flats on the southern side of the estuary, likely the result of 20th century soil compaction associated with human activity in the area (see Fig. 7.10). Although not evident in this particular photograph, these features do concentrate salts and frequently form a visible white crust.

Figure 7.10. (above, bottom) Soil compaction and other disturbances from military activities at Border Field Auxiliary Landing Field are visible in this April 1945 aerial photo of the southern portion of Tijuana Estuary. The U.S. armed forces operated bases here from 1916–1961 (Carter 2011).

Figure 7.11. (right) Inland movement of the dune system between ca. 1850 and ca. 2012.

(Salt flats: photo by Samuel Safran, April 2015; 1945 aerial photograph: USDA 1945, courtesy Brian Rehwinkel; Beach movement base map: NAIP 2014)

Change in the Tijuana River and valley

The river corridor is now dominated by riparian forest, not riparian scrub (U.S.)

Today, the lower Tijuana River valley features one of the largest stands of gallery riparian forest in coastal southern California (Fig. 7.12). Though these forests support a wide variety of riparian species and serve as critical habitat for the federally endangered least Bell's vireo (*Vireo bellii pusillus*; USFWS 1994b, Unitt et al. 2004), our findings suggest that riparian forest was not historically present in the lower valley. The river corridor was instead dominated by River Wash and Riparian Scrub, which historically covered 92% of the river corridor in the U.S. (with the rest taken up by the river channel itself). Today, Riparian Scrub covers only 35% of the corridor, with Riparian Forest covering all but 2% of the remainder. Apart from urban development, the single largest driver of River Wash / Riparian Scrub habitat loss has been from conversion to Riparian Forest. In total, the extent of River Wash / Riparian Scrub in the U.S. has decreased by nearly 80%, from 730 to 161 ha (1,800 to 398 ac). Riparian Forests now cover 293 ha (724 ac). The change in riparian vegetation structure and composition is also thought to have had an impact on a variety of riparian wildlife (see pp. 128–29).

The drivers of the conversion from riparian scrub to riparian forest are complicated and not yet understood with certainty. Riparian forest first established in the lower valley during the late 20th century, but the river corridor had already been altered in other ways by that time. Clearing for agriculture and grazing in the late 19th and early 20th centuries severely reduced the extent of riparian scrub along the river corridor (Rempel 1992), as did activities such as sand and gravel mining and extensive clearing downstream of the international boundary (Wyman 1937a, Rempel 1992). The dramatic shift from riparian scrub habitat to gallery riparian forest began in January and February of 1980, when major floods scoured much of the existing vegetation from the river corridor and were quickly followed by the establishment of riparian tree species such as arroyo willow (*Salix lasiolepis*) and Goodding's black willow (*S. gooddingii*; Boland 2014).

Large floods like the 1980 event support the establishment of willow forests because they (1) clear ground-cover and canopy-forming vegetation, which provides the seedlings with necessary space and light; (2) deposit or expose fine sediments, which is the substrate needed by seeds; and (3) often lead to surface wetness through the spring, which is required for germination (J. Boland, personal communication). But large flood events that satisfied each of these requirements (see pp. 132, 111, and 105, respectively) were a regular occurrence during the historical period and yet did not lead to the establishment of forests. This begs the question: what has changed?

We hypothesize that the dramatic change in the structure and composition of the valley's riparian habitats was related to the initiation of perennial streamflow in the 1979 and associated changes in groundwater levels (Fig. 7.13). Though through the 19th and much of the 20th century the Tijuana River flowed only intermittently or ephemerally (see pp. 96–99), by 1979 (after the completion of the Tijuana River Flood Control Project, in which the river was channelized through Mexico) the river was flowing year-round due to urban runoff and sewage releases (PWA 1987; Fig. 7.13). Research from other dryland rivers relating streamflow permanence to riparian vegetation cover has shown that perennial conditions favoring riparian forests over scrub (Lite and Stromberg 2005). Flow perennialization could impact riparian vegetation directly by inducing higher local humidity levels and lower leaf-to-air vapor pressure deficits (which would be expected to increase the photosynthesis rates) and indirectly in association with changes in ground-water levels. As noted by Lite and Stromberg (2005):

> Sites with perennial flow tend to be situated in gaining reaches, where inflowing ground water would sustain stable, shallow ground-water levels across the flood plain even during times of extended drought. At the highly intermittent

Fig. 7.12. Gallery riparian forest, dominated by arroyo willow and Gooding's black willow, lines the river channel in many areas historically occupied by river wash and riparian scrub. *(Photo by Samuel Safran, April 2015)*

sites, which typically are in losing reaches, ground-water depths and fluctuations likely have periodically exceeded… survivorship tolerance ranges for *P. fremontii* [Fremont cottonwood] and *S. gooddingii* [Goodding's willow].

Though groundwater levels were seasonally shallow in the Tijuana River Valley (as little as 2 ft [<1 m] below the surface in some years across much of the valley), there was still significant intra-annual variability, perhaps, we hypothesize, enough to preclude the establishment of riparian forests. The annual range of fluctuation across nine wells in the valley during the 1915 water year averaged 2.16 m (7.08 ft), with a minimum range of 1.13 m (3.72 ft) and a maximum range of 4.17 m (13.69 ft; but note that local ground-water pumping could have exacerbated seasonal fluctuations, even during the early 1900s; Ellis and Lee 1919). These fluctuations created maximum seasonal groundwater depths (~2–5 m [7–15 ft]; Ellis and Lee 1919) that met or exceeded reported thresholds for the survival of Goodding's willow saplings (~2–3 m [7–10 ft]; Lite and Stromberg 2005). It is also worth noting that the only location known to have supported mature Fremont cottonwoods during the historical period was adjacent to the Agua Caliente hot springs, a reach of the river that had perennial flows and artesian groundwater conditions (see p. 131).

Streamflow conditions changed again in the early 1990s, when wastewater treatment and management resumed intermittent flows in most years (Fig. 7.13). Since the riparian forests persisted, it seems likely that most of the trees that established during the period of perennial flow ca. 1980 were able to reach sufficient size to allow year-round access to groundwater. Despite resumed intermittent flows, additional areas of riparian forest established following floods in 1993 and 2005 (Boland 2014). More information on local groundwater conditions during those years would be useful for understanding how these later events fit into the hypothesis for the initial establishment of riparian forest in 1980 outlined above (in the simplified version, intermittent flows and associated seasonal fluctuations in groundwater after 1990 would be expected to preclude the development of riparian forests). Flow permanence was relatively high after the 1993 flood (the river subsequently flowed for nearly 20 straight months), which could help explain the apparent contradiction. Groundwater data should also be analyzed to confirm that seasonal groundwater levels were shallow and relatively steady during the 1980s (as hypothesized/depicted in Fig. 7.13).

Despite their relatively recent establishment, the future of the riparian forests in the Tijuana River valley is uncertain. The return to intermittent flow conditions, and the possibility of future changes to the hydrologic regime, could potentially limit forest regeneration and persistence over the long term. In addition, an outbreak during the summer of 2015 of Kuroshio shot hole borer beetle, which "farms" a fungus that can damage native plants, impacted tens of thousands of trees in the valley (Fig. 7.13). Overall, an estimated 71% of willow trees were infested, with half of these trees showing signs of damage (Boland 2016). The outbreak is expected to cause significant changes in riparian forest structure and function (Boland 2016). The management implications of the changes detailed here are discussed in more detail on page 195.

Dec. 1979   Feb. 1980

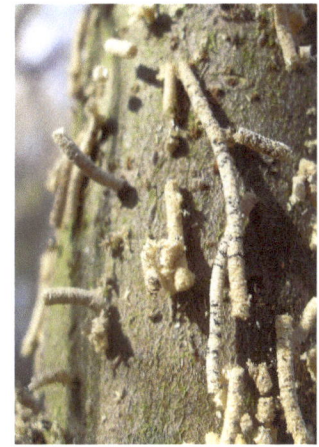

Figure 7.13. Changes in riparian habitats over time. The chart (below) plots the dominant riparian habitat type in the U.S. on a timeline along with possible drivers of change. Large flood events (top line) are required for the establishment of riparian trees; indeed, floods in 1980 marked a shift from riparian scrub to forest, but it is clear that events of this size also occurred earlier and did not result in forests. One change just prior to 1980 was a shift from intermittent/ephemeral to perennial streamflow (second line), as reflected in the chart of percent of days with flow (third line). We hypothesize that this change, as well as related changes in the level and stability of groundwater (bottom line) contributed to the shift in riparian vegetation structure and composition. The introduction of a non-native shot hole borer beetle in 2015 is noted at the top-right of the chart. This species has damaged tens of thousands of trees in the valley and is expected to alter the forests' structure and function. A photo of "frass" coming out of beetle holes in an infested tree is shown at right. The beetle's effects are also seen in the photos along the bottom of the page, all taken from the Hollister St. Bridge, looking east. The 1979 photograph (facing page, bottom-left) shows a scrub-dominated river corridor one month prior to a major flood. The 1980 photograph (facing page, bottom right), taken just after the flood, shows the scoured river bed. Riparian forests soon established at the location and persisted through most of 2015 (below left), but recently suffered severe beetle-related damage (bottom right). (*Flood data: see p. 110; Days with flow: see p. 99; Measured groundwater data: Rempel 1992; Riparian plants drawings: Jen Natali; Frass photo: John Boland, 2016, PeerJ, 10.7717/PeerJ.2141; 1979 and 1980 photos: John Boland; 2015 and 2016 photos: Google*)

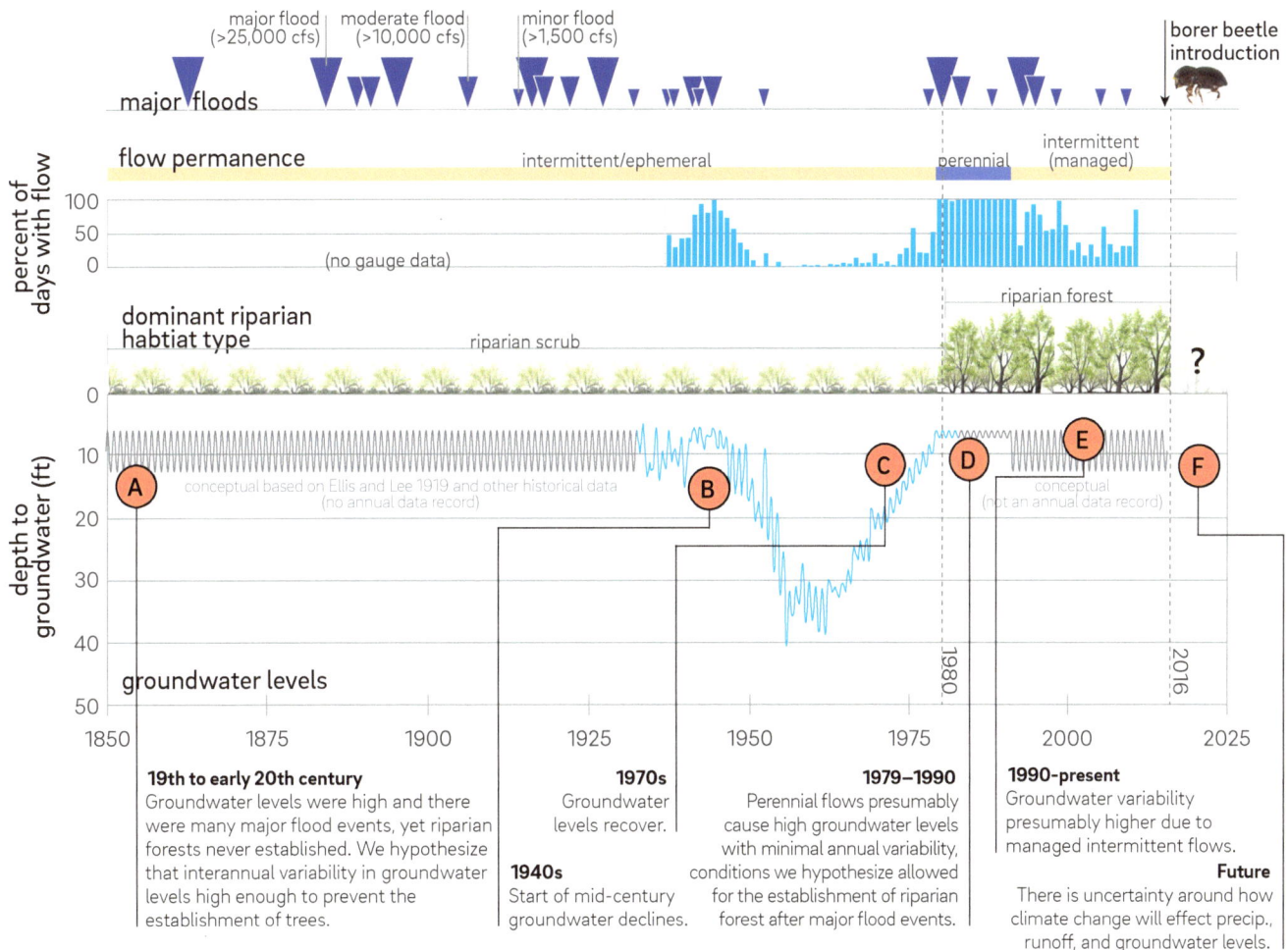

19th to early 20th century
Groundwater levels were high and there were many major flood events, yet riparian forests never established. We hypothesize that interannual variability in groundwater levels high enough to prevent the establishment of trees.

1940s
Start of mid-century groundwater declines.

1970s
Groundwater levels recover.

1979–1990
Perennial flows presumably cause high groundwater levels with minimal annual variability, conditions we hypothesize allowed for the establishment of riparian forest after major flood events.

1990-present
Groundwater variability presumably higher due to managed intermittent flows.

Future
There is uncertainty around how climate change will effect precip., runoff, and groundwater levels.

Apr. **2015**

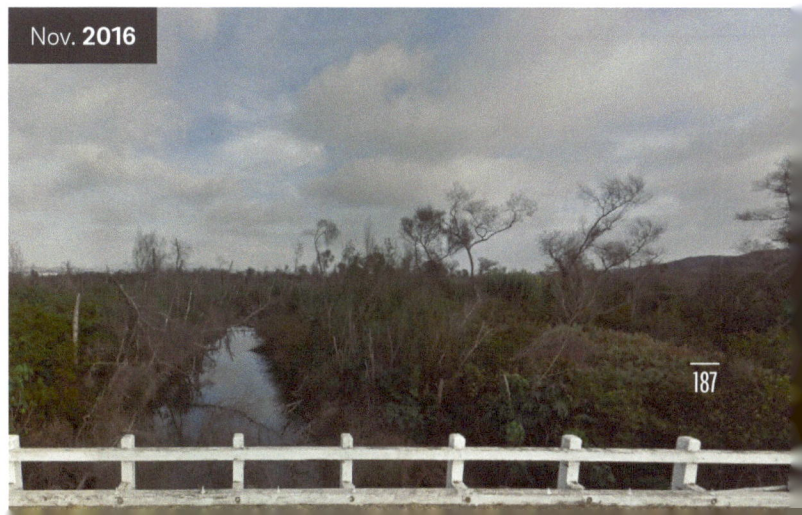
Nov. **2016**

187

The river channel in Mexico has been straightened and lined with concrete

From the international border south to Matanuco Canyon (approximately 18 km [11 mi]), the compound river channel has been replaced by a straightened, concrete, trapezoidal flood control channel (Fig. 7.14). As a result, the river corridor, which historically ranged from 70–1,400 m (200–4,600 ft) wide, is now uniformly 100 m (330 ft) wide. The Tijuana River concrete channel was completed in 1979 as part of the Tijuana River Flood Control Project. Since 2008, 7.5 km (4.7 mi) of Rio Alamar (Cottonwood Creek) were also converted to a concrete channel (see pp. 94–95). In both cases, the channelization process has largely eliminated the alluvial streambed and riparian vegetation, as well as attendant physical and ecological processes (e.g., hyporheic exchange).

The vast majority of the river corridor in Mexico has been developed

Ninety seven percent of the River Wash / Riparian Scrub habitat and 94% of the River Channel habitat in Mexico has been converted to urban development. Several small patches of Riparian Scrub still exist, totaling approximately 30 ha (70 ac), but for the most part these are hydrologically disconnected from the river channel. In addition, the corridor supports several small patches of Riparian Forest and Grassland / Coastal Sage Scrub. A small area (approximately 14 ha or 35 ac) of undeveloped River Channel still exists in Matanuco Canyon, just downstream of Rodríguez Dam (Fig. 7.15, also see Fig. 5.37).

Figure 7.14. A concrete trapezoidal flood control channel, completed in 1979, carries flows from downstream of Matanuco Canyon to the international border. It is seen here beneath the railroad bridge near the former site of the Agua Caliente hot springs. *(Photo by Samuel Safran, April 2015)*

Figure 7.15. A small section of vegetated river channel still exists in Matanuco Canyon, just downstream of Rodríguez Dam. Between 94 and 97% of the native riparian habitat in Mexico has been lost to urban development. *(Photo by Samuel Safran, April 2015)*

Alkali meadow complex has been eliminated, likely due to both urban/agricultural development and groundwater withdrawals

The vast alkali meadow complex that historically occupied the valley floor and was supported by seasonally high groundwater levels has been almost entirely converted to non-wetland habitat types and land uses. Though 131 ha (320 ac) are still classified as Alkali Meadow Complex / High Marsh Transition Zone in the contemporary mapping, nearly all of these areas are actually just High Marsh Transition Zone adjacent to the estuary (Fig. 7.16). Approximately 25% of the former Alkali Meadow Complex / High Marsh Transition Zone area is now occupied by Grassland / Coastal Sage Scrub, while another 50% approximately is occupied by Agriculture or Developed/Disturbed areas. In addition to urban and agricultural development, groundwater extraction in the valley during the early to mid-20th century likely contributed to the loss of the formerly extensive alkali meadow complex, a process well-documented in other systems (Elmore et al. 2006).

The conversion of this facultative wetland type to more xeric habitat types has contributed to an overall decrease in the ratio of wetlands to drylands across the lower valley, from more than 3:1 to less than 2:1 (a 45% decrease in the ratio). If urban and agricultural areas are included in this calculation as dryland habitat types, the change in ratio is even more dramatic (from 3:1 to 1:4 or -92%). This said, the wetland to dryland ratio has decreased by much less in the U.S. alone (-12% without urban and agricultural areas and −80% with them) where the relative loss of wetland area was lower (Mexico lost 96% of its wetland area but the U.S. only lost 57%).

ca. 1850

ca. 2012

Fig. 7.16. Distribution of Alkali Meadow Complex / High Marsh Transition zone ca. 1850 and ca. 2012. All of the remaining area mapped as this habitat type is now High Marsh Transition Zone.

The low mesa has been almost entirely developed

The low mesa north of the valley floor in the U.S. has been extensively developed, with a near complete loss of the Grassland / Coastal Sage Scrub, Vernal Pool, and Perennial Freshwater Wetland habitat types that once occupied the area (Fig. 7.17). Vernal Pools now have a mapped extent of zero, while some Perennial Freshwater Wetlands can be found within the river corridor and estuary (see p. 182). The loss of the Vernal Pool complex on the low mesa is in line with an extensive loss of this sensitive habitat type across the region, which once covered approximately 51,800 ha (200 mi^2) in San Diego County, but is now quite limited (Bauder and McMillan 1998). Though they covered only a small portion of the lower river valley, vernal pools in the region have high rates of species endemism (Bauder and McMillan 1998), so the contributions of these features to the biodiversity of the valley were likely disproportionate with their size (as were losses in species diversity when the pools were developed). Though most Grassland / Coastal Sage Scrub is now found below the low mesa in areas where it did not exist historically, nearly 60 ha (150 acres) of land formerly occupied by Grassland / Coastal Sage Scrub are also occupied by it today (see p. 178). At least one of these persistent areas is likely a true remnant patch; a review of historical maps and photographs suggests that large portions of the Grassland / Coastal Sage Scrub area east of Oneonta Slough have existed continuously since the 19th century, without major urban or agricultural development during the intervening period.

Habitat composition and quality has been altered by non-native vegetation

Non-native vegetation now dominates approximately 315 ha (775 ac) on the U.S. side of the valley (as determined from vegetation mapping by SANDAG [2012]; Fig. 7.18). Roughly 45% of the contemporary Grassland / Coastal Sage Scrub and 70% of the River Wash / Riparian Scrub is dominated by non-native vegetation. The primary non-native species within the valley include tamarisk (*Tamarix* spp.), arundo (*Arundo donax*), mustard (*Brassica nigra*), chrysanthemum (*Glebionis coronaria*), and grasses such *Avena* spp., *Bromus* spp., and *Lolium* spp. (SANDAG 2012). Non-native species have surely altered the ecological functions provided by different habitat types, both in ways we might view as both negative and positive. When their structural characteristics are similar, for example, *Tamarix* forests can have similar ecological functions to native *P. fremontii* forests, and can provide habitat in places where broad-leafed deciduous species are absent, reduced, or can no longer be sustained due to abiotic changes (Lite and Stromberg 2005 and references therein). At the same time, shifts in community composition to Tamarix can also negatively affect animal species richness, diversity, and abundance (Lite and Stromberg 2005 and references therein) and radically alter the structure and function of certain ecosystems, including the estuary's salt marshes (Crooks 2002, Whitcraft et al. 2007). Ultimately, a detailed accounting of the impacts of non-native species in the lower river valley is beyond the scope of this study, but these impacts should be recognized when considering changes in land cover over time.

Figure 7.17. Development on the low mesa over time.
This set of photographs shows development of the low mesa between 1928 and 2012. A handful of vernal pools are visible in the 1928 photograph at the modern day site of the Naval Outlying Landing Field in Imperial Beach.
(*Left: San Diego County 1928; Right: NAIP 2012*)

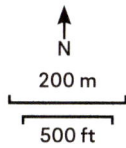

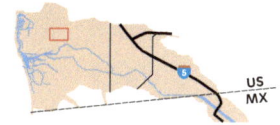

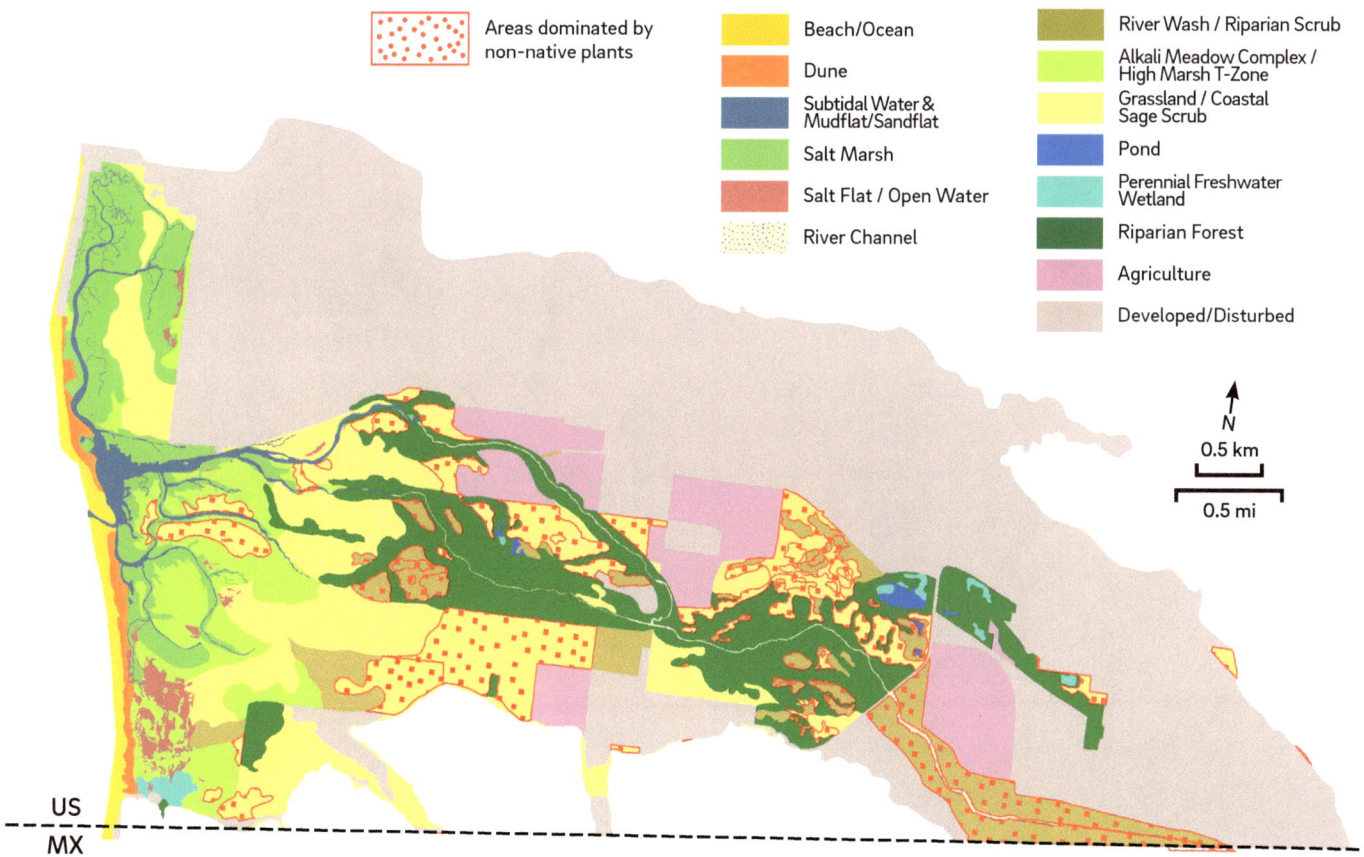

N
200 m
500 ft

Areas dominated by non-native plants

Beach/Ocean

Dune

Subtidal Water & Mudflat/Sandflat

Salt Marsh

Salt Flat / Open Water

River Channel

River Wash / Riparian Scrub

Alkali Meadow Complex / High Marsh T-Zone

Grassland / Coastal Sage Scrub

Pond

Perennial Freshwater Wetland

Riparian Forest

Agriculture

Developed/Disturbed

N
0.5 km
0.5 mi

Figure 7.18. Areas dominated by non-native plants cover approximately 315 ha (775 ac) on the U.S. side of the valley, including large areas of Grassland / Coastal Sage Scrub and River Wash / Riparian Scrub habitat types.

8
CONCLUSIONS

The research presented in this report provides an in-depth look at the Tijuana River valley as it existed prior to major modifications of the landscape, and of the changes that have occurred over the intervening years. In many ways, the valley is a radically different environment today than it was in the mid-19th century. Hydrology, sediment dynamics, tidal influence, channel stability, and land cover have all been significantly impacted by urban and agricultural development and other modifications. Despite these changes, however, the river valley still supports extensive areas of undeveloped habitat and provides immense value both for people and nature.

As we look toward the next century and beyond, there are many uncertainties about how the valley will respond to climate change and continued land use changes. How will population growth trends and urban development influence habitat quality and distribution? How will climate change alter physical processes like streamflow and sediment flux? Will flood risk increase? Are there biological thresholds likely to be exceeded under a changing climate?

The historical reconstruction presented in this report does not provide a simple blueprint for the future valley. It will not be possible—or necessarily even desirable—to recreate the landscape that once existed. Further research, including the work described on the following page, will help address these and other pressing questions. However, understanding what the landscape looked like and how it functioned historically can inform ongoing restoration planning and provide a starting point for envisioning a more resilient landscape for the future. In combination with contemporary research and scenario modeling, the historical ecology research presented here provides a tool for identifying restoration opportunities, setting restoration targets, and understanding how physical setting and processes continue to shape the landscape of the Tijuana River valley.

In considering how to apply the historical ecology research to restoration planning, a number of fundamental questions arise. How do we identify restoration targets that are grounded in history yet resilient to future changes? How can we most effectively restore some of the historical extent and diversity of wetland habitat types given current constraints on hydrology and land use? How do we navigate the trade-offs in ecological function inherent in different restoration visions? How do we integrate regional conservation planning priorities into the development of local restoration goals? While these questions have no easy answers, the findings discussed in this report do suggest a number of management implications. The following section describes a partial list of considerations for management.

(Photo by Samuel Safran, October 2015)

193

Future Historical Ecology Research

Further research and analysis in key areas will help to address some of the outstanding questions and uncertainties highlighted in this report. A full account of potential future research directions for the Tijuana River valley is beyond the scope of this report; we focus here on recommended research avenues that would broaden and deepen our understanding of the historical landscape and the processes that shaped it.

- *Conduct additional archival data collection focused on the Mexican side of the river valley.* While we endeavored to assemble a rich dataset for the entire study area and visited numerous archives within Mexico, logistical constraints inevitably limited our ability to pursue all promising leads and potential data sources for the Mexican side of the river valley.

- *Develop conceptual models to illustrate the connections* between physical processes (e.g., streamflow, sediment flux, tidal cycles, groundwater fluctuation) and the ecological patterns found within the river valley historically.

- *Situate information on the historical Tijuana River valley within a regional context* by comparing findings from this report with reconstructions of other estuaries and river systems in the area.

- *Further analyze 20th century landscape trajectories and drivers of change.* Our research focused primarily on understanding mid-19th century landscape conditions in comparison with the modern-day landscape. More in-depth research into how the landscape evolved between these two endpoints could provide many insights regarding causes of change and future landscape potential.

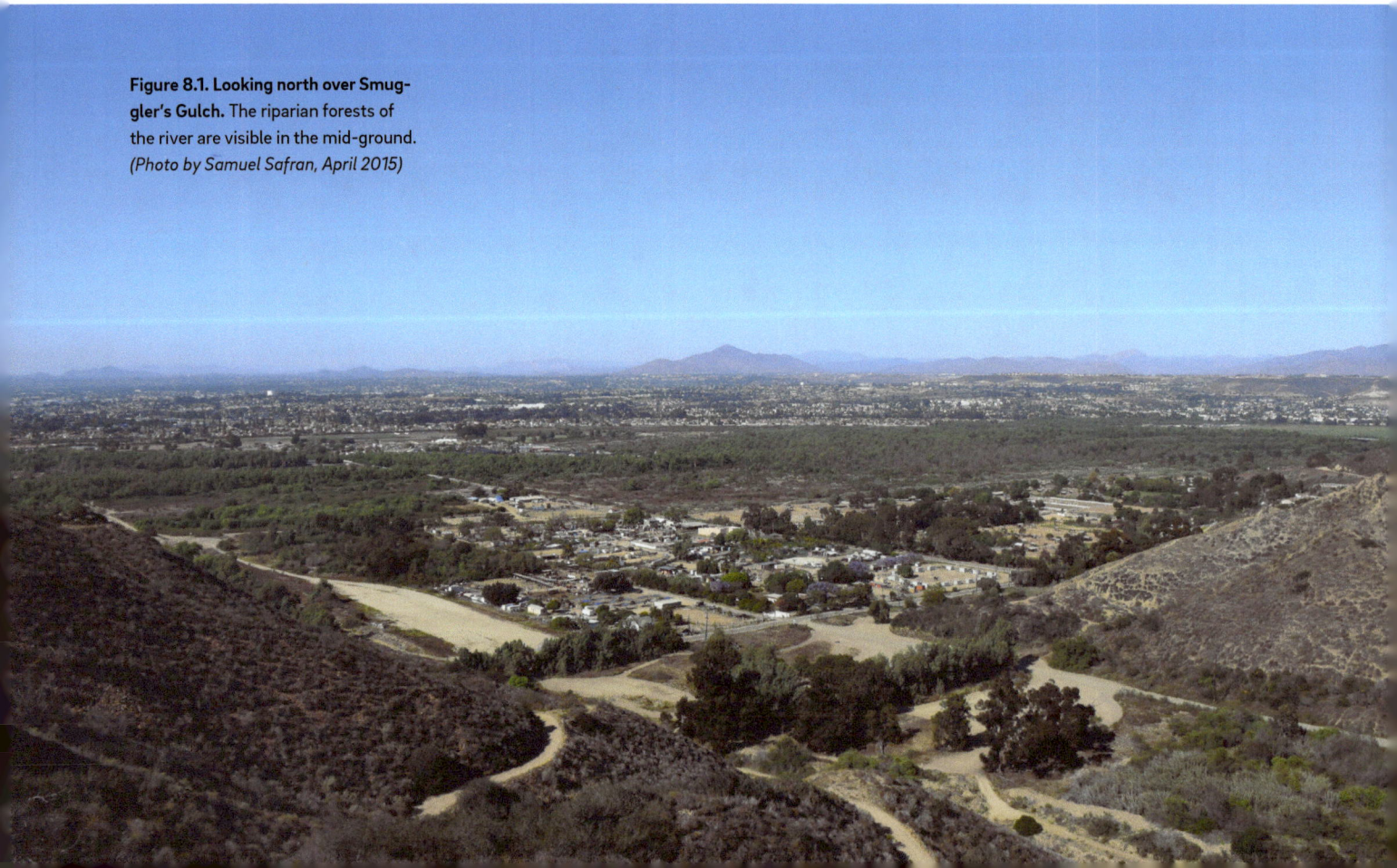

Figure 8.1. Looking north over Smuggler's Gulch. The riparian forests of the river are visible in the mid-ground. *(Photo by Samuel Safran, April 2015)*

Management Implications

- *Assess the long-term feasibility of, and trade-offs associated with, different riparian habitat types.* The transition to perennial streamflow in the 1980s, accompanied by rising groundwater levels and several large flood events that scoured the river channel and floodplain, were likely key factors enabling the establishment of willow-dominated riparian forests along the Tijuana River (see pp. 184–86). However, with streamflow managed to once again be intermittent and the recent invasion of the non-native shot hole borer beetle, it is unclear whether riparian forests will be able to survive and regenerate in the long term. The conversion from river wash and riparian scrub to riparian forest was accompanied by a major change in ecological function (p. 128–29): species such as the endangered Pacific pocket mouse (once present, now absent) depended on the presence of a more open scrub-dominated riparian corridor, while species such as the endangered southwestern willow flycatcher (once absent, now present) benefit from the continued presence of today's riparian forests. The endangered least Bell's vireo (which nests in low, dense riparian vegetation but forages in the canopies of mature trees) was historically present during the breeding season prior to the establishment of riparian forests, so it is possible that a future conversion to riparian scrub would not preclude this sensitive species. At present, shot hole borer beetle infestation rates appear to be lower in riparian scrub than in riparian forest (Boland 2016), a trend that should be further monitored and studied given the implications for riparian habitats in the future. Though the riparian forests are a novel habitat in the lower river valley and have suffered from the beetle outbreak, they still constitute one of the largest contiguous areas of riparian forest in San Diego County, where this habitat type has declined steeply over time (Zedler et al. 1992). Setting appropriate restoration targets for the riparian corridor will require careful analysis—at both the local and regional scale—of the feasibility of maintaining riparian forest over the long term, as well as of the trade-offs between the ecological functions provided by different riparian habitat types.

- *Consider accommodating natural channel movements.* Dramatic channel avulsions have ample historical precedence: the course of the river changed frequently over the past 150 years, alternately flowing through the southern, middle, and northern parts of the valley, with movement driven by high flow events (see p. 112). The Tijuana River has continued to move across the valley; as recently as 1993, floodwaters cut a new channel north of the river's previous course. Since then, there has been continuing and often-challenging efforts to constrain flows through berm construction and dredging of a "pilot channel." Given that the river is naturally dynamic and often shifted course during historical flood events and that these events were critical for maintaining habitat heterogeneity, it may be both practical and beneficial to manage the river to allow for more natural channel movement.

- *Explore feasibility of re-establishing groundwater-dependent wetlands.* The alkali meadow complex and other wetland habitat types that historically occupied much of the lower valley floor were reliant on seasonally high groundwater levels (see p. 70). Groundwater pumping and clearing for agriculture resulted in the complete loss of these habitat types, likely as early as the mid-20th century. However, groundwater levels have substantially rebounded since their historical low point in the 1960s, and are now close to their former levels (see pp. 52–57). It may now be possible to reestablish "missing" wetland types, such as alkali meadow complex, that were dependent on high groundwater. Since fires are known to help facilitate the recovery of alkali meadows in other areas with high groundwater that have experienced a type-conversion from meadow to xeric shrubland (Pritchett and Manning 2009), any efforts to recover meadows in the Tijuana River valley might also require careful management of the local fire regime.

- ***Prioritize restoration of salt marsh and other intertidal habitats.*** Estuarine habitats have undergone both loss (approximately 40% decrease in total area) and large-scale conversion (see pp. 180–83). The most significant loss of salt marsh has occurred in the southern part of the estuary (i.e., south of Tijuana River Slough), related to elevation increases due to excess sedimentation from hillside erosion in Tijuana canyons and decreases in tidal prism since the mid-19th century. These findings point toward the need for continued efforts to restore intertidal habitats, particularly in the heavily impacted southern arm of the estuary, to maintain desired functions. Restoration efforts may be aided by sediment management approaches (e.g., sediment catch basins and source control in the communities of Tijuana) and managing the tidal regime to increase tidal prism, which is estimated to have decreased by 55–85% over time. However, quickening rates of sea level rise in the future may necessitate shifting approaches to managing sediment delivery to the estuary, as sediment loads could become a resource that contributes to the ability of salt marshes to keep pace with rising water levels.

- ***Facilitate adaptive management by continuing to track changes through mapping and monitoring.*** Transformations in habitats and processes from the past to the present represent one part of change trajectories that will continue into the future. Being able to effectively respond to these changes will require sustained tracking of both drivers and responses. Continued habitat mapping, conducted in a manner that can crosswalk with the results of this study, will allow for assessment of evolving natural processes, anthropogenic stressors, and management interventions. Monitoring of focal species, especially those that have responded to past changes in the system, will provide another tool to indicate underlying change in supporting ecosystems. Monitoring of key drivers, such as sea level, tidal prism, salinity, and sediment dynamics, will provide a more mechanistic understanding of key processes and can help trigger potential shifts in management approaches, such as sediment management.

- ***Share the story of the valley through communication and outreach.*** The maps, photographs, and stories highlighted here provide a glimpse into what the Tijuana River Valley looked like in the not-so-distant past. Using these materials to tell the story of valley to various audiences, including students, educators, residents, decision-makers, and other stakeholders on both sides of the border has the potential to capture the collective imagination and motivate continued protection and restoration of this system.

(Photo by Erin Beller, May 2013)

Adams F. 1913. *Irrigation resources of California and their utilization.* Washington: Government Printing Office.

Adams F. 1928. *Tia Juana River Irrigation District.* Frank Adams Papers, MS ADAMS-1, Box 34. *Courtesy of Water Resources Collections and Archives (WRCA), UC Riverside.*

Adams F, Favela JL. 1929. Map of the watershed of Tia Juana River. International Water Commission, United States and Mexico. *Courtesy of the International Boundary and Water Commission Records Office.*

Aguado E. 2005. Temperature & precipitation. In *Tijuana River watershed atlas.* San Diego State University, Dept. of Geography. San Diego State University Press, Institute for Regional Studies of the Californias.

Alexander WE. ca. 1912. *Plat book of San Diego County, California.* Los Angeles, CA: Pacific Plat Book Co.

Alverson CS. 1914. *Report of hydraulic studies on the Tijuana River of U.S. and Mexico.* Bureau of Water Development, City of San Diego.

Andrews. 1853. 33D Congress, 1st Session, Monday, December 12. William Gwin, editor. The Congressional Globe.

Andrews E, Antweiler R, Neiman PJ, et al. 2004. Influence of ENSO on flood frequency along the California coast. *Journal of Climate* 17(2):337–348.

Armour CL, Duff DA, Elmore W. 1991. The effects of livestock grazing on riparian and stream ecosystems. American Fisheries Society.

Atwood J, Bontrager D. 2001. California Gnatcatcher (*Polioptila californica*). In *The birds of North America, no. 574,* ed. A Poole and F Gill. Philadelphia, PA.

Automobile Club of Southern California. 1917. Automobile roadmap touring San Diego County California. Los Angeles. *Courtesy of Earth Science and Map Library, UC Berkeley.*

Automobile Club of Southern California. ca. 1930. Automobile road map of San Diego County, California. Los Angeles, California. *Courtesy of Earth Sciences & Map Library, UC Berkeley.*

Bailey V. 1939. The solitary lives of two little pocket mice. *Journal of Mammalogy* 20(3).

Baker VB. 1977. Stream-channel response to floods, with examples from central Texas. *Geological Society of America Bulletin* 88(8).

Bancroft HH. 1888. *History of California, Volume II.* San Francisco: A.L. Bancroft & Co.

Barbour MG, Solomeshch AI, Buck JJ, et al. 2007. *Classification, ecological characterization, and presence of listed plant taxa of vernal pool associations in California.* United States Fish and Wildlife Service Agreement/Study.

Barnes GW. 1879. The hillocks or mound-formations of San Diego, California. *The American Naturalist* 13(9):565–571.

Barreto G. Departmento de Obras Publicas, Del Gobierno del Territorio Norte de la Baja California. 1937. Mapa del Territorio Norte de la Baja California. *Courtesy of Mapoteca Manuel Orozco y Berra.*

Bartlett JR. 1963. *Personal narrative of explorations and incidents in Texas, New Mexico, California, Sonora, and Chihuahua.* New York: D. Appleton & Co.

Battalio B, Danmeier D, Williams P. 2006. *Predicting closure and breaching frequencies of small tidal inlets - A quantified conceptual model.* 30th International Coatal Engineering Conference, San Diego, CA, USA.

Bauder ET, Bohonak AJ, Hecht B, et al. 2011. *A draft regional guidebook for applying the Hydrogeomorphic approach to assessing wetland functions of vernal pool depressional wetlands in Southern California.* San Diego, CA: San Diego State University.

Bauder ET, McMillan S. 1998. Current distribution and historical extent of vernal pools in southern California and northern Baja California, Mexico. *Ecology conservation and management of vernal pool ecosystems. California Native Plant Society, Sacramento*:56–70.

Beasley TD, Schuyler JD. 1889. Official map of San Diego County, California: compiled from latest official maps of U.S. surveys, railroad and irrigation surveys, county records and other reliable sources. California.

Bedford TA, Cromwell G. 1910. Map of San Diego County California. Rodney Stokes Co. Inc. Publishers. San Diego. *Courtesy of Earth Science and Map Library, UC Berkeley.*

Beller E, Baumgarten S, Grossinger R, et al. 2014. *Northern San Diego county lagoons: historical ecology investigation: regional patterns, local diversity, and landscape trajectories. SFEI Publication #722.* San Francisco Estuary Institute, Richmond, CA.

Beller EE, Downs PW, Grossinger RM, et al. 2016. From past patterns to future potential: using historical ecology to inform river restoration on an intermittent California river. *Landscape Ecology* 31(3).

Beller EE, Grossinger RM, Salomon MN, et al. 2011. *Historical ecology of the lower Santa Clara River, Ventura River, and Oxnard Plain: an analysis of terrestrial, riverine, and coastal habitats. SFEI contribution #641.* San Francisco Estuary Institute, Oakland, CA.

Belsky AJ, Matzke A, Uselman S. 1999. Survey of livestock influences on stream and riparian ecosystems in the western United States. *Journal of Soil and Water Conservation* 54(1):419–431.

Berstein B, Merkel K, Chesney B, et al. 2011. *Recommendations for a southern California regional eelgrass monitoring program.* National Marine Fisheries Service, Technical Report 632.

Biondi F, Gershunov A, Cayan DR. 2001. North Pacific decadal climate variability since 1661. *Journal of Climate* 14(Letter):5–10.

Black SF. 1913. *San Diego county, California: a record of settlement, organization, progress, and achievement.* Chicago: S. J. Clarke Publishing.

Black SF, Smythe WE. 1913. *San Diego and Imperial Counties, California, Volume I & II.* Chicago: S. J. Clarke Publishing.

Blackburn O. 1931. Blackburn's map of San Diego County. Los Angeles. *Courtesy of Earth Science and Map Library, UC Berkeley.*

Blanco J. 1901. *Memoria de la Sección Mexicana de la Comisión Internacional de Límites entre México y los Estados Unidos que restableció los monumentos de El Paso al Pacífico.* John Polhemus y Compañía.

Boland JM. 2014. Factors determining the establishment of plant zonation in a Southern Californian riparian woodland. *Madroño* 61(1):48–63.

Boland JM. 2016. The impact of an invasive ambrosia beetle on the riparian habitats of the Tijuana River Valley, California. *PeerJ.*

Bolla JR. n.d. *Memories of living at the estuary in Imperial Beach.*

Bonillas YS, Urbina F. 1912. Informe acerca de los recursos naturales de la parte Norte de la Baja California, especialmente del Delta del Río Colorado. In *Memoria de la Comision del Instituto Geologico de Mexico que exploro la region norte de la Baja California. Parergones del Instituto Geologico de Mexico, Tomo IV, Numeros 2 a 10.,* ed. Instituto Geologico de Mexico.

Bonin CL, Zedler JB. 2008. Southern California salt marsh dominance relates to plant traits and plasticity. *Estuaries and Coasts* 31(4).

Böse E, Wittich E. 1912. Informe relativo a la exploracion de la Region Norte de la Costa Occidental de la Baja California. In *Notas preliminares relativas a un reconocimiento geologico por el curso del Atoyac (Rio Verde) de Oaxaca. Parergones del Instituto Geologico de Mexico, Tomo IV, Numeros 2 a 10.,* ed. Paul Waitz, 307–529. Mexico: Instituto Geologico de Mexico.

Boudini DJ. 1854. *Milijo land grant case files.* 73–75. *Courtesy of The Bancroft Library, UC Berkeley.*

Bowman JN. 1947. *The area of the mission lands. Courtesy of The Bancroft Library, UC Berkeley.*

Brinkhoff T. 2014. Mexico: Baja California. City Population. http://citypopulation.de/Mexico-BajaCalifornia.html.

Brown AK. 2005. *Reconstructing early historical landscapes in the Northern Santa Clara Valley.* Russell K. Skowronek. Santa Clara, CA: Santa Clara University.

Brown CM, Pallamary MJ. 1988. *History of San Diego County land surveying experiences.* Privately published.

Brownlie WR, Taylor BD. 1981. *Sediment management for southern California mountains, coastal plains and shoreline. Part C. Coastal sediment delivery by major rivers in southern California.* California Institute of Technology, Pasadena, CA.

Bryant WE. 1889. A catalogue of the birds of Lower California, Mexico. *Proceedings of the California Academy of Sciences* 2:237–320.

Brylski P, Hays L, Avery J. 1998. *Pacific Pocket Mouse (*Perognathus longimembris pacificus*) recovery plan.* U.S. Fish and Wildlife Service Region 1.

BSI Consultants Inc. et al. 1994. *Tijuana River Valley, two alternatives report flood control and infrastructure study.*

Buordo EA. 1956. A review of the General Land Office survey and of its use in quantitative studies of former forests. *Ecology* 37:754–768.

Burbeck EM. ca. 1900. Map of the City of San Diego California. San Diego, CA. *Courtesy of Coronado Public Library.*

Cabrillo JR, Bolton HE. 1916. The Cabrillo-Ferrelo Expedition. In *Spanish exploration in the Southwest 1542–1706*, ed. Herbert Eugene Bolton: C. Scribner's Sons.

Cahoon DR, Lynch JC, Powell AN. 1996. Marsh vertical accretion in a Southern California estuary, U.S.A. *Estuarine, Coastal and Shelf Science* 43:19–32.

California Department of Finance. 2012. Historical census populations of California, Counties, and Incorporated Cities, 1850–2010. California State Data Center, Demographic Research Unit, Department of Finance. http://www.dof.ca.gov/Reports/Demographic_Reports/documents/2010-1850_STCO_IncCities-FINAL.xls.

California Department of Parks and Recreation and Department of Fish and Game. 1972. *A review of land uses in the Tijuana River Valley.* California Department of Parks and Recreation and Department of Fish and Game.

California State Legislature. 1907. *Appendix to the journals of the senate and assembly of the thirty-seventh session of the legislature of the state of California.*

Callaway JC. 2001. Hydrology and Substrate. In *Handbook for restoring tidal wetlands*, ed. Joy B. Zedler, 89–118. Boca Raton: CRC Press.

Callaway RM, Jones S, Wayne FRJ, et al. 1990. Ecology of a mediterranean-climate estuarine wetland at Carpinteria, California: plant distributions and soil salinity in the upper marsh. *Canadian Journal of Botany* 68:1139–1146.

Callaway JC, Zedler JB. 2004. Restoration of urban salt marshes: lessons from southern California. *Urban Ecosystems* 7:107–124.

Cañizares Jd, Thickens VE, Mollins M. 1952. Putting a lid on California: an unpublished diary of the Portola expedition. *California Historical Society Quarterly* 31(4):343–354.

Capen L. 1831. Salt hay. *New England Farmer* 9(35):273.

Carruthers WM. 1912. A few suggestions to the alfalfa diaryman. *Pacific Rural Press.* February 10. *Courtesy of California Digital Newspaper Collection.*

Carter NC. 2011. Border Field state park and its monument. *Eden: Journal of the California Garden & Landscape History Society* 14(4).

CDWR (California Division of Water Resources). 1942. Map of Tia Juana River in Mexico showing Rodriquez irrigation product and other features. *Courtesy of Water Resources Collections and Archives (WRCA), UC Riverside.*

CESAR (Center for Earth Systems Analysis Research). 2000. Tijuana River watershed digital vegetation file. Center for Earth Systems Analysis Research, El Colegio de la Frontera Norte, San Diego, CA.

Chabreck RH. 1968. *The relation of cattle and cattle grazing to marsh wildlife and plants in Louisiana.* Louisiana Wild Life and Fisheries Commission.

Chase, McLean. 1928. *Tia Juana River irrigation district communiation to state engineer regarding proposed Mexian development on Tijuana River.* San Diego. *Courtesy of Water Resources Collections and Archives (WRCA), UC Riverside.*

Chase MK, Guepel GR. 2005. *The use of avian focal species for conservation planning in California.* USDA Forest Service.

Chin EH, Aldridge BN, Longfield RJ. 1991. *Floods of February 1980 in southern California and central Arizona.* U.S. Geological Survey, National Oceanic and Atmospheric Administration,, Washington D.C.

Christenson LN, Sweet EL. 2008. *Ranchos of San Diego County.* Arcadia Publishing.

Cicourel MB. 1921. *Oposición a la solicitud de concesión de las aguas del manantial "Agua Caliente".* Mexico D.F.

City of Imperial Beach. n.d. *City History.* Imperial Beach, California. http://www.imperialbeachca.gov.

City of San Diego. 1935. Water Resources, City of San Diego, California, Water Department, Division of Development & Conservation. [San Diego, Calif.]. *Courtesy of Earth Science and Map Library, UC Berkeley.*

City of San Diego. 1973. *Tia Juana River Valley land use and flood control alternatives. A joint study by the Office of the City Manager and Planning Department.*

City of San Diego. 1976. *Tia Juana River valley plan and environmental impact report.*

City of San Diego, Page & Turnbull. 2010. *San Ysidro Historic Context Statement.* City of San Diego: City Planning and Community Investment.

Clark. U.S. Coast Survey. 1886. Survey of the coast of Lower California, register no. 54343. *Courtesy of NOAA (National Oceanic and Atmospheric Administration).*

Cleisz S, Currie G, Ehrenkrantz R, et al. 1989. *A management framework for the Tijuana River valley.* Department of Landscape Architecture: California State Polytechnic University, Pomona. *Courtesy of TJNERR.*

Coats RN, Williams PB, Cuffe CK, et al. 1995. *Design guidelines for tidal channels in coastal wetlands.* Philip Williams & Associates, Ltd. #934, San Francisco, CA.

Collins BD, Montgomery DR. 2001. Importance of archival and process studies to characterizing pre-settlement riverine geomorphic processes and habitat in the Puget Lowland. *Water Science and Application* 4:227–243.

Collins JN, Grenier JL, Didonato J, et al. 2007. *Ecological connections between baylands and uplands: examples from Marin County.* San Francisco Estuary Institute.

Collins JN, Grossinger RM. 2004. *Synthesis of scientific knowledge concerning estuarine landscapes and related habitats of the South Bay Ecosystem.* Technical report of the South Bay salt pond restoration project. San Francisco Estuary Institute, Oakland.

Colson LB. 1896. Across the border. In *The land of sunshine: a southewestern magazine*, ed. Charles F. Lummis. Volume VI. Los Angeles, CA: Land of Sunshine Publishing Co.

Conagua-DGE. 2014. [Precipitation data from Presa Rodriguez Dam] CNA-SMN-SCDI, Climatología estadística, Datos contenidos en la base de datos climatológica, Estación 02038, Presa Rodriguez. *http://smn.conagua.gob.mx/tools/resources/diarios/2038.txt* Accessed April, 2015.

Cooper WS. 1967. *Coastal dunes of California.* Boulder, CO: Geological Society of America.

Corona AP. n.d. La presa Abelardo L. Rodríguz, modelo de ingeniería hidráulica. Ayuntamiento de Tijuana.

Corona AP. n.d. The Rancho Tía Juana (Tijuana) Grant. *The Journal of San Diego History* 50.

County of San Diego. 1970. *The coastal lagoons of San Diego County.* Prepared by The Environmental Task Force, County of San Diego. *Courtesy of San Elijo Lagoon Conservancy.*

Cowardin LM, Carter V, Golet FC, et al. 1979. *Classification of wetlands and deepwater habitats of the United States.* Washington: Fish and Wildlife Service, Biological Services Program, U.S. Department of the Interior.

Cox GW. 1984. The distribution and origin of mima mound grasslands in San Diego County, California. *Ecology* 65(5):1397–1405.

Cox GW, Zedler JB. 1986. The influence of Mima Mounds on vegetation patterns in the Tijuana Estuary Salt Marsh, San Diego County, California. *The Bulletin of the Southern California Academy of Sciences* 85(3).

Craig D, Williams PL. 1998. Willow Flycatcher (*Empidonax traillii*). In *The riparian bird conservation plan: a strategy for reversing the decline of riparian-associated birds in California*, ed. California Partners in Flight.

Crespí J, Bolton HE. 1927. *Fray Juan Crespí, missionary explorer on the Pacific coast, 1769–1774.* Berkeley, CA: University of California Press.

Crespí J, Brown AK. 2001. *A description of distant roads: original journals of the first expedition into California, 1769–1770.* San Diego, CA: San Diego State University Press.

Crooks J. 2002. Characterizing the consequences of invasions: the role of introduced ecosystem engineers. *Oikos* 97:153–166.

Croswell M. 1870. At San Diego and the gold mines. *Overland monthly and Out West Magazine* 5(4). San Francisco.

Crowell. 1906. Map of the City of San Diego, Cal. and vicinity. *Courtesy of Special Collections & Archives, UC San Diego Library.*

Cruse RE. 1937. Areas subject to overflow in lower reaches of Tia Juana River California and Mexico. San Diego County Flood Control. Los Angeles, California. *Courtesy of Water Resources Collections and Archives (WRCA), UC Riverside.*

Cuero D, Shipek FC. 1968. *The autobiography of Delfina Cuero, a Diegueño Indian.* Los Angeles: Dawson's Book Shop.

Cummings LS, Puseman K, Varney RA. 2004. *Pollen and macrofloral analysis at site CA-SDI-16,047 in Goat Canyon, California.* Paleo Research Institute, San Diego, CA.

Daily Alta California. 1868. Letter from southern California: the San Pasqual battlefield, San Bernardino Rancho, a ruined town, the fat of the land, indian boomerang, among the Spanish, American settlers, other notes. November 13. *Courtesy of California Digital Newspaper Collection.*

Daily Alta California. 1873. The region of the Colorado. September 29. *Courtesy of California Digital Newspaper Collection.*

Daily Alta California. 1886. Coast news. February 13. *Courtesy of California Digital Newspaper Collection.*

Daily Alta California. 1890a. Trouble at Tia Juana: an American arrested for shooting at a Mexican guard. June 19. *Courtesy of California Digital Newspaper Collection.*

Daily Alta California. 1890b. Evading the exclusion act. April 26. *Courtesy of California Digital Newspaper Collection.*

Daily Alta California. 1891. After the deluge: death and destrction caused by the flood at Tia Juana; two men swept down by the resistless torrent; a survivor's thrilling experience-the southern overland roads blocked by masses of earth. February 27. *Courtesy of California Digital Newspaper Collection.*

Dark S, Stein ED, Bram D, et al. 2011. *Historical ecology of the Ballona Creek watershed. Technical Report #671.* Southern California Coastal Water Research Project, Costa Mesa, CA.

Das T, Dettinger MD, Cayan DR. 2010. *Potential impacts of global climate change on Tijuana River watershed hydrology - an initial analysis.* Sustainability Solutions Institute, University of California San Diego.

Day S. 1870. [Plat map] Fractional Township No 19 South, Range No 2 West, San Bernardino Meridian. *Courtesy of Bureau of Land Management.*

D'Elgin T, Krautheim V, Leonard B, et al. n.d. Tijuana River National Estuarine Research Reserve, high school teachers' guide. Challenge Cost Share Program, National Park Service, and the Friends of San Diego Wildlife Refuges.

Dedina SL. 1991. The political ecology of transboundary development: land use, flood control, and politics in the Tijuana River valley, 1920–1990. Master of Science, Geography, University of Wisconsin-Madison, Madison, WI.

Demére TA. 2005. Geology. In *Tijuana River watershed atlas.* San Diego State University, Dept. of Geography. San Diego State University Press, Institute for Regional Studies of the Californias.

Denton. 1900. Baja California Mapa de la region comprendida entre la latutyd 32 norte y la linea divisoria de Mexico y Los Estados Unidos de America. Mexico. *Courtesy of Mapoteca Manuel Orozco y Berra.*

Denton G, Lauteren. 1875. Mapa de la region septentrional de la Baja California incluyendo los terrenos de la ex. mision de Santa Catarina. *Courtesy of Mapoteca Manuel Orozco y Berra.*

Department of Public Works [Division of Water Resources, State of California]. 1929. Irrigated lands, Tia Juana Valley- San Diego Co., not included in irrigation districts. *Courtesy of San Diego History Center.*

Department of Public Works [State of California]. 1935. *San Diego County Investigation.* Bulletin No. 48. Division of Water Resources.

Desmond J, Williams G, Vivian-Smith G, et al. 2001. The diversity of habitats in southern California coastal wetlands. In *Handbook for restoring tidal wetlands*, ed. Zedler JB. Boca Raton, LA: CRC Press.

Dingler JR, Clifton HE. 1994. Barrier systems of California, Oregon, and Washington. In *Geology of Holocene barrier island systems*, ed. Richard A Davis Jr. Berlin, New York: Springer-Verlag Berlin Heidelberg.

Dugan JE, Hubbard DM. 2010. Loss of coastal strand habitat in Southern California: the role of beach grooming. *Estuaries and Coasts* 33(1):67–77.

Durán IG. 1989. Tijuana y Tia Juana dos poblados fronterizos. In *Historia de Tijuana*, ed. Jésus Ortiz Figueroa and Davud Piñera Ramírez. Tijuana: Universidad Autonoma de Baja California Centro de Investigaciones Históricas UNAM-UABC.

Ecarg N. 1912. The located line. *Overland Monthly* 59(3):219.

Elliott M. 1998. *Intertidal sand and mudflats & subtidal mobile sandbanks: An overview of dynamic and sensitivity characteristics for conservation management of marine SACs.* UK Marine SACs Project.

Ellis AJ, Lee CH. 1919. *Geology and ground waters of the western part of San Diego County, California.* Water-supply paper (Washington, D.C.), no. 446. Washington, DC: Government Printing Office.

Elmore AJ, Manning SJ, Mustard JF, et al. 2006. Decline in alkali meadow vegetation cover in California: the effects of groundwater extraction and drought. *Journal of Applied Ecology* 43:770–779.

Emory WH. 1857. *Report of the United States and Mexican Boundary survey, made under the direction of the Secretary of the Interior* 1. United States House of Representatives.

EPA (U.S. Environmental Protection Agency). 2009. *Environmental Assessment (EA) for the expansion of the water distribution and wastewater collection system for unserved areas in Tijuana, B.C.* U.S. Environmental Protection Agency.

Ervast A. 1921. Map showing location of wells and cultivated land as of May 10th 1921 in Tia Juana Valley. Coronado Water Company. *Courtesy of San Diego History Center.*

Fetzer L. 2005. *San Diego County place names A to Z.* San Diego, CA: Sunbelt Publications, Inc.

Fink HT. 1891. *The Pacific Coast scenic tour from Southern California to Alaska, the Canadian Pacific Railway, Yellowstone Park, and the Grand Cañon.* Boston: Berwick & Smith, Printers.

Fisher R, Case T. 2000. Final report on herpetofauna monitoring in the Tijuana Estuary.

Flick R. 2005. Dana Point to the international border. In *Living with the changing California coast*, ed. Gary B. Griggs, Kiki Patsch, and Lauret E. Savoy. Berkeley, Calif.: University of California Press.

Foster ZC, Moran WJ. 1930. *Soil Survey of Calveston County, Texas.* 31, Bureau of Chemistry and Soils.

Fox CJ, Willey HJ. 1883. Map of the County of San Diego California. San Diego, California. *Courtesy of San Diego Natural History Museum.*

Freeman JE. 1854. *Field notes of the base lines in Townships 17, 18, and 19 South, Ranges 1, 2, and 3 West, San Bernardino Meridian.* Book 167–37. *Courtesy of Bureau of Land Management, Sacramento, CA.*

Friedman JM, Lee VJ. 2002. Extreme floods, channel change, and riparian forests along ephemeral streams. *Ecological Monographs* 72(3):409–425.

Gallegos. 2002. Southern California in transition: Late Holocene occupation of southern San Diego County, California. In *Catalysts to complexity: Late Holocene of the California coast.* ed. J. Erlandson and T.L. Jones: Cotsen Institute of Archaeology, University of California, Los Angeles

Gamble LH, Wilken-Robertson M, Johnson K, et al. 2004. *Cultural ecology and the indigenous landscape of the Tijuana River watershed.*

Gander FF. 1936. Consortium of California Herbaria, record for *Suaeda taxifolia* from "Bluff at north side of mouth of San Dieguito Creek". San Diego Natural History Museum Herbarium.

Gayman W. 1971. *Environmental impact study for the proposed Tijuana River flood control channel.* Department of the Army Los Angeles District Corps of Engineers, Science Division Ocean Science and Engineering, Inc.

Gayman W. 1978. *Estimation of present and past tidal prisms in Batiquitos Lagoon. Courtesy of Scripps College.*

Goals Project. 1999. *Baylands Ecosystem Habitat Goals. A report of habitat recommendations prepared by the San Francisco Bay Area Wetlands Ecosystem Goals Project.* U.S. Environmental Protection Agency and S.F. Bay Regional Water Quality Control Board, San Francisco and Oakland, CA.

Goals Project. 2015. *The Baylands and Climate Change: What We Can Do. Baylands Ecosystem Habitat Goals Science Update 2015.* California State Coastal Conservancy, Oakland, CA.

Goodwin P, Kamman RZ. 2001. Mixing and circulation in tidal wetlands. *Journal of Coastal Research* Vol. SI(27):109–120.

Graf WL. 1981. Channel instability in a braided, sand bed river. *Water Resources Research* 17(4):1087–1094.

Graf WL. 1983. Flood-related channel change in an arid-region river. *Earth Surface Processes and Landforms* 8(2):125–139.

Graf WL. 1988. *Fluvial processes in dryland rivers.* Caldwell, NJ: The Blackburn Press.

Graf WL. 2000. Locational Probability for a Dammed, Urbanizing Stream: Salt River, Arizona, USA. *Environmental Management* 25(3):321.

Graham M. 2016. Shot borer beetle gets hot in Tijuana River Valley. March 15. *San Diego Reader.*

Gran KB, Tal M, Wartman ED. 2015. Co-evolution of riparian vegetation and channel dynamics in an aggrading braided river system, Mount Pinatubo, Philippines. *Earth Surface Processes and Landforms* 40(8):1101–1115.

Gray A. 1886. *Synoptical flora of North America: the gamopetalae.* Smithsonian Institution.

Gray AB. 1849. Topographical sketch of the port of San Diego, and measurement of the marine league for determining initial point of boundary between the United States and Mexican Republic, as surveyed by the United States Commission. *Courtesy of Coronado Public Library.*

Greer K. 2014. Spatial distribution and habitat assessment of *Panoquina errans* (Lepidoptera: *Hesperiidae*) in San Diego County, California. *The Journal* 47:17–27.

Grewell BJ, Callaway JC, Ferren Jr. WR. 2007. Estuarine wetlands. In *Terrestrial vegetation of California*, ed. Michael G. Barbour, Todd Keeler-Wolf, and Allan A. Schoenherr, 124–154. Berkeley, Los Angeles, London: University of California Press.

Gribovszki Z, Szilágyi J, Kalicz P. 2010. Diurnal fluctuations in shallow groundwater levels and streamflow rates and their interpretation—a review. *Journal of Hydrology* 385(1):371–383.

Grossinger RM. 2005. Documenting local landscape change: the San Francisco Bay area historical ecology project. In *The historical ecology handbook: a restorationist's guide to reference ecosystems*, ed. Dave Egan and Evelyn A. Howell, 425–442. Washington, D.C.: Island Press.

Grossinger RM, Askevold RA. 2005. *Historical analysis of California Coastal landscapes: methods for the reliable acquisition, interpretation, and synthesis of archival data. Report to the U.S. Fish and Wildlife Service San Francisco Bay Program, the Santa Clara University Environmental Studies Institute, and the Southern California Coastal Water Research Project, SFEI Contribution 396*. San Francisco Estuary Institute, Oakland. 48.

Grossinger RM, Stein ED, Cayce K, et al. 2011. *Historical wetlands of the Southern California coast: an atlas of U.S. Coast Survey t-sheets, 1851–1889. SFEI contribution #586, SCCWRP technical report #589*. San Francisco Estuary Institute, Oakland, CA.

Grossinger RM, Striplen CJ, Askevold RA, et al. 2007. Historical landscape ecology of an urbanized California valley: wetlands and woodlands in the Santa Clara Valley. *Landscape Ecology* 22:103–120.

Guldbaum D. 1917. Mapa general del Distrito Norte de la Baja California. 1:400,000 *Courtesy of Mapoteca Manuel Orozco y Berra*.

Haas WE, Unitt P. 2004. Willow Flycatcher *Empidonax traillii*. In *San Diego County Bird Atlas*, ed. Philip Unitt: Sunbelt Publications.

Hall WH. 1888. *Irrigation in California (Southern) : the field, water-supply and works, organization and operation in San Diego, San Bernardino and Los Angeles counties* Sacramento, CA State Office, J.D. Young, Supt. State Printing.

Haltiner J, Swanson M. 1987. Technical appendix A.2: Geomorphology and hydrology of the Tijuana River. In *Technical appendices: Tijuana estuary enhancement hydrologic analysis*, ed. Philip Williams & Associates. San Francisco, CA.

Hanley N, Ready R, Colombo S, et al. 2009. *The impacts of knowledge of the past on preferences for future landscape change.* Journal of Environmental Management 90.

Hardcastle ELF, Gray AB. 1850. Boundary between the United States & Mexico.

Harley JB. 1989. Historical geography and cartographic illusion. *Journal of Historical Cartography* 15:80–91.

Harris, Cromwell. 1915. Mapa de la Baja California y Golfo de California segun datos oficiales y privados. *Courtesy of Mapoteca Manuel Orozco y Berra*.

Harris JA, Hobbs RJ, Higgs E, et al. 2006. *Ecological restoration and global climate change*. Restoration Ecology 14.

Harrison AM. 1852. Map of the coast of California from San Diego Bay to the boundary. U.S. Coast Survey.

Harwood. 1931. *The Tia Juana Estuaries and the plant and animal Life in them.*

Hatherley AE. 1936a. *Memorandum of conference with Mr. A. E. Hatherley and others in my office Friday evening, May 29th, 1936. Courtesy of Water Resources Collections and Archives (WRCA), UC Riverside.*

Hatherley AE. 1936b. *Memorandum of statement of Mr. Hatherley. Courtesy of Water Resources Collections and Archives (WRCA), UC Riverside.*

Hauser SA. 2006. *Distichlis spicata*. U.S. Department of Agriculture, Forest Service, Rocky Mountain Research Station, Fire Sciences Laboratory. http://www.fs.fed.us/database/feis/.

Hertlein LG, Grant US. 1944. *The Geology and Paleontology of the Marine Pliocene of San Diego, California*. Memoirs of the San Diego Society of Natural History, Volume II.

Herzog LA. 1990. *Where North meets South: cities, space, and politics on the U.S.-Mexican Border*. Center for Mexican American Studies, University of Texas at Austin.

Higgins EB. 1949. Annotated distributional list of the ferns and flowering plants of San Diego County, California. *Occasional Papers of the San Diego Society of Natural History* (8).

Higgins HC, Coleman RW, Brown GM, et al. 1994. *Archaeological investigations at South Bay International Wastewater Treatment Plant site and outfall facilities,cultural resource identification and geotechnical test monitoring.* Submitted to the International Boundary and Water Commission.

Higgs E. 2012. History, novelty, and virtue in ecological restoration. In *Ethical adaptation to climate change: human virtues of the future*, ed. Allen Thompson and Jeremy Bendik-Keymer. Cambridge, MA: MIT Press.

Holland RF. 1986. *Preliminary descriptions of the terrestrial natural communities of California. Unpublished report.* California Department of Fish and Game, Natural Heritage Division, Sacramento, CA.

Holmes LC, Pendleton RL. 1918. *Reconnoissance soil survey of the San Diego region, California*. U.S. Department of Agriculture, Washington. U.S. Department of Agriculture.

Huey LM. 1931. Notes on two birds from San Diego County, California. *The Auk*:620–621.

Hull JS. 1937. *Widths of submerged channels of Tia Juana River.* Charles H. Lee Papers, MS 76/1, Folder 1440. *Courtesy of Water Resources Collections and Archives (WRCA), UC Riverside.*

Hull JS. John S. Hull to Dr. Roger R. Revelle, 1 March 1941. Roger Randall Revelle Papers (1929–1991), Box 145, Folder 6. UC San Diego Special Collections & Archives. *Courtesy of UC San Diego Special Collections & Archives.*

Humple D, Geupel GR. 2004. Song sparrow (*Melospiza melodia*). In *The Riparian Bird Conservation Plan: a strategy for reversing the decline of riparian-associated birds in California. California Partners in Flight.*, ed.

IBWC (International Boundary and Water Commission). 1960. *Flow of the Colorado River and other western boundary streams and related data.* Western Water Bulletin.

IBWC (International Boundary and Water Commission). 1976. Tijuana River flood control project, San Diego County, California. Information brochure. *Courtesy of University of California, San Diego.*

Ilarregui JS. 1850. Plano de la parte austral del Puerto de S. Diego, y del terreno comprendido entre dicha parte, el punto inicial en la costa del Pacífico y la sesta estación hecha en la dirección de la línea que divide las repúblicas de México y de los Estados-Unidos. *Courtesy of the Bancroft Library.*

Ilarregui JS, de Chavero FM. 1850. Linea divisoria entre Mexico y los Estados Unidos, conforme al tratado de Guadalupe Higaldo. *Courtesy of Mapoteca Manuel Orozco y Berra.*

INEGI (Instituto Nacional de Estadística y Geografía). 2012. *Volumen y crecimiento. Población total por entidad federative, 1895 a 2010.* http://www3.inegi.org.mx/sistemas/sisept/Default.aspx?t=mdem0148&s=est&c=29192.

INEGI (Instituto Nacional de Estadística y Geografía). 2014. *X Censo general de población y vivienda 1980. Censos y conteos de población y vivienda.* http://www3.inegi.org.mx/sistemas/tabuladosbasicos/LeerArchivo.aspx?ct=46601&c=16762&s=est&f=1.

Inman DL, Masters PM. 1991. *Budget of sediment and prediction of the future state of the coast.* U.S. Army Corps of Engineers, Los Angeles District.

International Boundary Commission. 1901. Boundary between the United States and Mexico by the International Boundary Survey under the convention of July 29th 1882, Revived February 18th 1889. Engraved by R.F. Bartle & Co., Washington, D.C., U.S.A. No. 18. New York. 1:60,000 *Courtesy of David Rumsey Map Collection.*

International Water Commission. 1930. *Report of the American section of the International Water Commission, United States and Mexico. Message from the President of the United States transmitting report submitted by the American Section of the International Water Commission, United States and Mexico.* [United States] 71st Cong., 2d sess. House. Doc. 359. Washington, D.C.

Irelan W. 1888. *Eighth annual report of the state mineralogist: for the year ending October 1, 1888.* California State Mining Bureau, Sacramento.

Irlandi E, Macia S, Serafy J. 1997. Salinity reduction from freshwater canal discharge: effects on mortality and feeding of an urchin (*Lytechinus variegatus*) and a gastropod (*Lithopoma tectum*). *Bulletin of Marine Science* 61(3):869–879.

Izbicki JA. 1985. *Evaluation of the Mission, Santee, and Tijuana hydrologic subareas for relaimed-water use, San Diego County, California.* U.S. Geological Survey. *Courtesy of (USGS) U.S. Geological Survey.*

Jackson ST, Hobbs RJ. 2009. *Ecological restoration in the light of ecological history.* Science 325(5940).

Jacobs D, Stein ED, Longcore T. 2011. *Classification of California estuaries based on natural closure patterns: templates for restoration and management.* Southern California Coastal Water Research Project. Technical Report 619.

Jepson WL. 1907. Jepson field books. *Courtesy of the Jepson Herbarium, University of California, Berkeley.*

Johannessen C. 1964. Marshes prograding in Oregon: aerial photographs. *Science* 146.

Johnson DL, Burnham JLH. 2012. Introduction: overview of concepts, definitions, and principles of soil mound studies. *The Geological Society of America Special Papers* (490):1–19.

Joint Committee on Water Problems of the California Legislature. 1953. *Allocation of the waters of the Tijuana River under the treaty between United States and Mexico of 1944.* Fifth partial report on water problems of the State of California. Senate of the State of California.

Jones J. 1892. *Mark Patty v. E.P. Colgan, State Controller, Transcript on Appeal.* Supreme Court of the State of California. Courtesy of the California State Archives.

Katz GL, Denslow MW, Stromberg JC. 2012. The Goldilocks effect: intermittent streams sustain more plant species than those with perennial or ephemeral flow. *Freshwater Biology* 57:467–480.

Keeley JE, Zedler PH. 1996. Characterization and global distribution of vernal pools. *Ecology, Conservation, and Management of Vernal Pool Ecosystems.*

Keeler-Wolf T, Elam DR, Lewis K, et al. 1998. California vernal pool assessment preliminary report. State of California Department of Fish and Game.

Kennedy MP, Tan SS. 2007. Geologic map of the oceanside 30' x 60' quadrangle, California. USGS, editor. California Department of Conservation California Geological Survey.

Kennedy MP, Tan SS. 2008. *Geologic map of the San Diego 30' x 60' quadrangle, California.* California Department of Conservation, California Geological Survey. 23.

Klare RE. 1935. Automobile Road Map of San Diego County, California. Route and Map Service Department, Automobile Club of Southern California. Los Angeles, California. *Courtesy of Earth Sciences & Map Library, UC Berkeley.*

Knight EE. ca. 1900. South San Diego and Imperial Beach. *Courtesy of Bancroft Library, UC Berkeley.*

Knox RW. 1933. Mexcian Boundary to Chula Vista, register no. 5371. U.S. Coast and Geodetic Survey. *Courtesy of NOAA (National Oceanic and Atmospheric Administration).*

Knox RW. 1934. *T-5679 descriptive report: Southern California coast Mexican boundary to San Diego bay.* U.S. Coast and Geodetic Survey & Department of Commerce. *Courtesy of NOAA (National Oceanic and Atmospheric Administration).*

Kondolf GM, Downs PW. 1996. Catchment approach to planning channel restoration. In *River channel restoration: guiding principles for sustainable projects,* ed. Andrew Brookes and F. Douglas Shields. Chichester, New York: J. Wiley.

Kondolf GM, Smeltzer MW, Railsback SF. 2001. Design and performance of a channel reconstruction project in a coastal California gravel-bed stream. *Environmental Management* 28(6):761-776.

Kovalchik BL, Elmore W. 1992. *Effects of cattle grazing systems on willow-dominated plant associations in central Oregon.* McArthur, D. Bedunah, and CL Wambolt, compilers. Proceedings-Symposium on Ecology and Management of Riparian Shrub Communities. USDA Forest Service General Technical Report INT-289. Intermountain Research Station, Ogden, UT.

Ku T-L, Kern JP. 1974. Uranium-series age of the upper Pleistocene Nestor Terrace, San Diego, California. *Geological Society of America Bulletin* 85(11):1713–1716.

Kus B. 2002. Least Bell's Vireo (*Vireo bellii pusillus*). In *The riparian bird conservation plan: a strategy for reversing the decline of riparian-associated birds in California,* ed.: California Partners in Flight.

Kus B, Hopp SL, Johnson RR, et al. 2010. Bell's Vireo (*Vireo bellii*). The Birds of North America Online (P.G. Rodewald, Ed.). Cornell Lab of Ornithology, Ithaca.

Laliberte AS, Ripple WJ. 2004. Range contractions of North American carnivores and ungulates. *BioScience* 54(2):123–138.

Largier JL, Hollibaugh JT, Smith SV. 1997. Seasonally hypersaline estuaries in Mediterranean-climate regions. *Estuarine, Coastal, and Shelf Science* 45:789–797.

Lautz LK. 2008. Estimating groundwater evapotranspiration rates using diurnal water-table fluctuations in a semi-arid riparian zone. *Hydrogeology Journal* 16(3):483–497.

Lee CH. 1937. Well Map, Tia Juana Valley. *Courtesy of San Diego History Center.*

Lee CH. 1940. *Report to valley water committee on ground water supply of Tia Juana Valley in San Diego County, California.* Institute of Marine Resources.

Lee M, *San Diego Union-Tribune.* 2012. Two million gallon of sewage hit TJ River - again.

Lightner J. 2013. *San Diego native plants in the 1830s.* San Diego Flora, San Diego, CA.

Limbaugh C. 1955. *Fish life in the kelp beds and the effects of kelp harvesting.* University of California, La Jolla, California.

Lite SJ, Stromberg JC. 2005. Surface water and ground-water thresholds for maintaining *Populus-Salix* forests, San Pedro River, Arizona. *Biological Conservation* 125:153–167.

Longcore T, Osborne KH. 2015. Butterflies are not grizzly bears: Lepidoptera conservation in practice. In *Butterfly Conservation in North America: efforts to help save our charismatic microfauna,* ed. C. Jaret Daniels. Dordrecht: Springer Netherlands.

Los Angeles Daily Herald. 1889. Hippolyte Goujon sentenced for smuggling cigars over the Mexican line. March 5. *Courtesy of Library of Congress.*

Los Angeles Herald. 1878. Untitled. February 24. *Courtesy of California Digital Newspaper Collection.*

Los Angeles Herald. 1888. Track and tie. April 11. *Courtesy of California Digital Newspaper Collection.*

Los Angeles Herald. 1897. January 16. *Courtesy of Library of Congress Newspapers.*

Los Angeles Herald. 1899. A great farm map. May 21. *Courtesy of California Digital Newspaper Collection.*

Los Angeles Herald. 1904. A day in old Mexico. April 10. *Courtesy of California Digital Newspaper Collection.*

Los Angeles Herald. 1905. Heaviest weather in eight years off coast at San Diego. March 14. *Courtesy of California Digital Newspaper Collection.*

Los Angeles Lithographic Co. ca. 1889. Map of subdivision of town of Oneonta. *Courtesy of Huntington Art Collections, San Marino, California.*

Lundquist JD, Cayan DR. 2002. Seasonal and spatial patterns in diurnal cycles in streamflow in the western United States. *Journal of Hydrometeorology* 3(5):591–603.

Luomala K. 1978. Tipai-Ipai. In *Handbook of North American Indians, 8 (California)*, ed. Robert F. Heizer, 592–609. Washington, DC: Smithsonian Institution.

Lynch HB. 1931. *Rainfall and stream run-off in Southern California since 1769*. Metropolitan Water District of Southern California, Los Angeles, CA.

Manies KL. 1997. Evaluation of General Land Office survey records for analysis of the northern Great Lakes hemlock-hardwood forests. M.Sc., University of Wisconsin, Madison.

Mansfield HB. 1889. Hydrography off coast of California from Boundary Monument to Sand Ridge, register no. 1889. U.S. Coast and Geodetic Survey. *Courtesy of NOAA (National Oceanic and Atmospheric Administration)*.

Mantua N, Hare S. 2002. The Pacific Decadal Oscillation. *Journal of Oceanography* 58(1):35–44.

Marshall JT. 1948. Ecologic races of song sparrows in the San Francisco Bay Region: Part I. Habitat and abundance. *The Condor* 50(5):193–215.

Mathias J. 2011. Mammalogy at the intersection of mercy and truth. Field Book Project. Smithsonian National Museum of Natural History.

Mattoni R, Longcore TR. 1997. The Los Angeles coastal prairie: a vanished community. *Crossosoma* 23(2):71-102.

Mayer JA. 1987. "Soil morphology of the Tijuana River National Estuarine Sanctuary: northwest portion." Master's thesis, San Diego State University.

MBC Applied Environmental Sciences. 2015. *Status of the kelp beds in 2014: kelp bed surveys—Ventura, Los Angeles, Orange, and San Diego Counties*. Costa Mesa, California.

McDonough O, Hosen J, Palmer M. 2011. *Temporary streams: the hydrology, geography, and ecology of non-perennially flowing waters*. Nova Science Publishers, Hauppauge, New York. 259–290.

McGlashan H, Ebert F. 1918. *Southern California floods of January, 1916*. Washington: Government Printing Office.

Mearns EA. 1898. *Descriptions of three new forms of pocket-mice from the Mexican border of the United States*. 299–302.

Meigs P. 1925. Baja California research materials 1925–1979. University of California, San Diego. Journal I, 1925, January 1—August 3, Baja California.

Meinzer OE. 1923. *Outline of ground-water hydrology with definitions. Geological Survey Water-Supply Paper 494*. U.S. Department of the Interior and U.S. Geological Survey, U.S. Government Printing Office, Washington D.C. 71. *Courtesy of U.S. Geological Survey*.

Mendenhall TC. 1890. *Report of the Superintendent of the U.S. Coast and Geodetic Survey showing the progress of the work during the fiscal year ending with June, 1889*. U.S. Coast and Geodetic Survey, Washington, D.C. 61. *Courtesy of National Oceanic and Atmospheric Association*.

Metsker Maps. ca. 1950. Metsker's map of San Diego County California. Seattle, Washington. *Courtesy of University of California San Diego*.

Michel SM. 2000. Defining hydrocommons governance along the border of the Californias: a case study of transbasin diversions and water quality in the Tijuana-San Diego metropolitan region. *Nat. Resources J.* 40:931.

Milliman JD, Farnsworth KL. 2011. *River discharge to the coastal ocean: a global synthesis*. Cambridge; New York: Cambridge University Press.

Minan JH. 2002. Recent developments in wastewater management in the Coastal Region at the United States-Mexico Border. *San Diego International Law Journal* 3:51.

Minnich RA, Vizcaíno EF. 1998. *Land of chamise and pines: historical accounts and current status of Northern Baja California's vegetation*. University of California Press.

Mock PJ. 2004. California gnatcatcher *(Polioptila californica)*. In *The riparian bird conservation plan: a strategy for reversing the decline of riparian-associated birds in California*, ed. California Partners in Flight.

Moffatt and Nichol Engineers. 1987. *Silver Strand littoral cell preliminary sediment budget report: Coast of California Storm and Tidal Waves Study*. U.S. Army Corps of Engineers, Los Angeles District, Planning Division, Coastal Resources Branch, Los Angeles, Calif.

Moffatt and Nichol, Inc. 1957. *Report on the possible development of Oneonta Lagoon as a small craft harbor to the honorable Board of Supervisors and the Planning Commission, County of San Diego, California*. Long Beach, CA. *Courtesy of Water Resources Collections and Archives (WRCA), UC Riverside*.

Mora J. 1928. City of San Diego, this whimsical map and history of a city in California. The Marston Company. San Diego, California. *Courtesy of University of California San Diego*.

Morin HE. 1916. Directing relief work by aeroplane during the recent San Diego Flood. *Flying* 5(2).

Morning Oregonian. 1916. Looters busy in flood wake. January 31. *Courtesy of Historic Oregon Newspapers*.

Nagano CD. 1982. *The population status of seven species of insects inhabiting Tijuana Estuary National Wildlife Refuge, San Diego County, California*. Office of Endangered Species.

NAIP (National Agriculture Imagery Program). 2012. [Natural color aerial photos of San Diego County]. Ground resolution: 1m. National Agriculture Imagery Program (NAIP). U.S. Department of Agriculture (USDA), Washington, D.C.

NAIP (National Agriculture Imagery Program). 2014. [Natural color aerial photos of San Diego County]. Ground resolution: 1m. National Agriculture Imagery Program (NAIP). U.S. Department of Agriculture (USDA), Washington, D.C.

Naughton MP. 2013. "SLAMM (Sea Level Affecting Marshes Model) modeling of the effects of sea level rise on coastal wetland habitats of San Diego County." Master's thesis, San Diego State University.

Nationwide Environmental Title Research. 2016. Historic aerials viewer [Aerial photograph mosaics of the lower Tijuana River Valley from 1953, 1964, 1966, 1968, 1971, 2002, 2003, 2005]. http://www.historicaerials.com/. Accessed December, 2015.

Nelson EW. 1922. Lower California and its natural resources. *Memoirs of the National Academy of Sciences* XVI:11.

Newell FH. 1901. *Report of progress of stream measurements for the calendar year 1899*. Twenty-First Annual Report of the United States Geological Survey, Part 4. Washington: Government Printing Office. *Courtesy of Internet Archive.*

Nichols MM. 1953. *Tijuana River Slough*. Scripps Institution of Oceanography, Marine Foraminifera Laboratory. *Courtesy of Tijuana River National Estuarine Research Reserve.*

NLCD (National Land Cover Database). 2011. Multi-resolution Land Characteristics Consortium. Retrieved on November 11, 2014 from: http://www.mrlc.gov/.

NOAA (National Oceanic and Atmospheric Administration). 2016. NOAA Tide Predictions, Imperial Beach, California, 2017. National Oceanic and Atmospheric Administration.

Noe GB, Zedler JB. 2001a. Spatio-temporal variation of salt marsh seedling establishment in relation to the abiotic and biotic environment. *Journal of Vegetation Science* 12(1):61-74.

Noe GB, Zedler JB. 2001b. Variable rainfall limits the germination of upper intertidal marsh plants in southern California. *Estuaries* 24(1):30–40.

North WJ, James DE, Jones LG. 1993. *History of kelp beds (*Macrocystis*) in Orange and San Diego Counties, California.* Fourteenth International Seaweed Symposium, 277–283.

North CP, Warwick GL. 2007. Fluvial fans: myths, misconceptions, and the end of the terminal-fan model. *Journal of Sedimentary Research* 77:693–701.

NWS (National Weather Service). 2013. Monthly precipitation for San Diego. National Oceanic and Atmospheric Administration. http://www.wrh.noaa.gov/sgx/climate/san-pcpn.htm.

Ojeda Revah L. 2000. Land use and the conservation of natural resources in the Tijuana River Basin. In *Shared space: rethinking the U.S.-Mexico border environment*, ed. Lawrence Herzog. La Jolla, California: Center for U.S.-Mexican Studies, University of California, San Diego.

O'Leary J. 2005. Vegetation. In *Tijuana River watershed atlas*. San Diego State University, Dept. of Geography. San Diego State University Press, Institute for Regional Studies of the Californias.

Orcutt CR. 1886a. Botanical notes. *The West American Scientist* 2(13):15. San Diego Society of Natural History.

Orcutt CR. 1886b. Northern lower California. *The West American Scientist* 2(16):37-41. San Diego Society of Natural History.

Orcutt CR. 1887. Aquatic plants of the vicinity of San Diego. *The West American Scientist* 3(25):23–125.

Orme AR, Griggs G, Revell D, et al. 2011. Beaches changes along the southern California coast during the 20th century: A comparison of natural and human forcing factors. *Shore & Beach* 79(4):38-50.

Orozco R. 1889. Plano topografico de Tijuana. Tijuana de Zaragoza. *Courtesy of Mapoteca Manuel Orozco y Berra.*

Orr BK, Diggory ZE, Coffman GC, et al. 2011. Riparian vegetation classification and mapping: important tools for large-scale river corridor restoration in a semi-arid landscape. In *Proceedings of the CNPS Conservation Conference: strategies and solutions*, 17–19 January 2009, California Native Plant Society, ed. B.K. Orr J. W. Willoughby, K. Schierenbeck, and N. Jensen [eds.]. Sacramento, CA.

Pacific Rural Press. 1872. Agricultural notes: California. February 17. *Courtesy of California Digital Newspaper Collection.*

Pacific Rural Press. 1873. Agricultural notes. December 20. *Courtesy of California Digital Newspaper Collection.*

Pacific Rural Press. 1877a. Agricultural notes: California. July 14. *Courtesy of California Digital Newspaper Collection.*

Pacific Rural Press. 1877b. Agricultural notes: California. November 24. *Courtesy of California Digital Newspaper Collection.*

Pacific Rural Press. 1885. Agricultural notes: California. September 12. *Courtesy of California Digital Newspaper Collection.*

Pacific Rural Press. 1896. Latest statements about Canaigre. August 15. *Courtesy of California Digital Newspaper Collection.*

Page JM. 1955. Water Problems of the Tia Juana River, San Diego, California. *Courtesy of Water Resources Collections and Archives (WRCA), UC Riverside.*

Papenfuss TJ, Parham JF. 2013. Four new species of California legless lizards *(Anniella). Breviora: Museum of Comparative Zoology* (536):1–17.

Parkinson ES. 1894. *Wonderland; or, twelve weeks in and out of the United States. Brief account of a trip across the continent—short run into Mexico—ride to the Yosemite Valley—steamer voyage to Alaska, the land of glaciers—visit to the Great Shoshone Falls and a stage ride through the Yellowstone National Park*. Trenton, N.J.: MacCrellish & Quigley.

Parry CC. 1849. Correspondence : Parry and Torrey, 1846–1868. *Courtesy of the Internet Archive.*

Parry CC. [1849]2014. *Parry's California notebooks*. James Lightner. San Diego: San Diego Flora.

Pascoe J. 1869. *Field notes of the subdivision lines of Township 18 and 19 South Range 2 West, San Bernardino Meridian, California*. Books 168-34, 168-35, 168-37. *Courtesy of Bureau of Land Management, Sacramento, California.*

Patsch K, Griggs G. 2007. *Development of sand budgets for California's major littoral cells: Eureka, Santa Cruz, Southern Monterey Bay, Santa Barbara, Santa Monica (including Zuma), San Pedro, Laguna, Oceanside, Mission Bay, and Silver Strand littoral cells*. Institute of Marine Sciences, University of California, Santa Cruz California Department of Boating and Waterways, California Coastal Sediment Management Workgroup. 115.

Peel MC, Finlayson BL, McMahon TA. 2007. Updated world map of the Köppen-Geiger climate classification. *Hydrology and earth system sciences discussions* 4(2):439–473.

Pennings SC, Bertness MD. 1999. Using latitudinal variation to examine effects of climate on coastal salt marsh pattern and process. *Current topics in wetland biogeochemistry* 3:100–111.

Pennings SC, Callaway RM. 1992. Salt march plant zonation: the relative importance of competition and physical factors. *The Ecological Society of America* 73(2).

Pennings SC, Grant MB, Bertness MD. 2005. Plant zonation in low-latitude salt marshes: disentangling the roles of flooding, salinity and competition. *Journal of Ecology* 93.

Perry H. 1936. *Marvin L. Allen, et al. (Plaintiffs) vs. California Water and Tel. Company (Defendants)*. Superior Court of the State of California: County of San Diego. *Courtesy of Water Resources Collections and Archives (WRCA), UC Riverside.*

Peterson GD, Cumming GS, Carpenter SR. 2003. Scenario planning: a tool for conservation in an uncertain world. *Conservation biology* 17(2):358–366.

Pethick JS. 1992. Saltmarsh geomorphology. In *Saltmarshes: morphodynamics, conservation and engineering significance*, ed. J.R.L. Allen and K. Pye, 41–62. Cambridge: Cambridge University Press.

Pickart AJ, Barbour MG. 2007. Beach and dune. In *Terrestrial vegetation of California*, ed. Michael G. Barbour, Todd Keeler-Wolf, and Allan A. Schoenherr. Berkeley: University of California Press.

Plasencia Novarro A, El Mexicano Gran Diario Regional. 2011. Historia de las inundaciones en Tijuana: PRIMERA PARTE. January 23.

Poole CH. U.S. District Court California, Southern District. 1854. Plan of the Rancho of Melijo. Land Case Map E-1074. *Courtesy of Tijuana River 2.*

Pourade RF. 1964. *The glory years*. San Diego: Union-Tribune Pub. Co.

Pourade RF. 1965. *Gold in the sun*. San Diego: Union-Tribune Pub. Co.

Pritchett DW, Manning SJ. 2009. Effects of fire and groundwater extraction on alkali meadow habitat in Owens Valley, California. *Madroño* 56(2):89-98.

Proffitt TD. 1994. *Tijuana: the history of a Mexican metropolis*. San Diego, Calif.; Lanham, MD: San Diego State University Press.

Pryde PR. 2004. *San Diego: an introduction to the region*. Fourth Edition edition. San Diego: Pearson & Sunbelt.

Purer EA. 1936. Studies of certain coastal sand dune plants of southern California. *Ecological Monographs* 6(1).

Purer EA. 1939. Ecological study of vernal pools, San Diego County. *Ecology* 20(2):217–229.

Purer EA. 1942. Plant ecology of the coastal salt marshlands of San Diego County, California. *Ecological Monographs* 12(1):81–111.

PWA (Philip Williams & Associates). 1995. *A preliminary assessment of a salinity model for the Tijuana estuary*. National Biological Service, San Diego. *Courtesy of TJNERR.*

PWA (Philip Williams & Associates). 1987. Technical Appendices: Tijuana Estuary enhancement hydrologic analysis.

Pyrde PR. 1976. Chapter 8: water supply for the county. In *San Diego : an introduction to the region*, ed. P.R. Pyrde. Dubuque, Iowa: Kendall/Hunt Pub. Co.

Quayle EH. 1944. Physiographic block diagram of the San Diego and La Jolla quadrangles and some of the mountainous region to the East. *Courtesy of Standford University Library.*

Radeloff VC, Mladenoff DJ, He HS, et al. 1999. Forest landscape change in the northwestern Wisconsin Pine Barrens from pre-European settlement to the present. *Canadian Journal of Forest Research* 29:1649–1659.

Radeloff VC, Mladenoff DJ, Manies KL, et al. 1998. Analyzing forest landscape restoration potential: presettlement and current distribution of oak in the northwest Wisconsin Pine Barrens. *Transcriptions of the Wisconsin Academy of Sciences Arts and Letters* 86:189–205.

Ramírez DP. 1985. *Historia de Tijuana.* et al. Roberto Moreno de los Arcos. Tijuana, Baja California: Centro de Investigaciones Historicas UNAM-UABC.

Rand McNally. 1935. Standard map of California. *Courtesy of Coronado Public Library.*

Randall JM. 2000. *Mesembryanthemum crystallinum.* In *Invasive plants of California's wildlands*, ed. Carla C. Bossard, John M. Randall, and Marc C. Hoshovsky, 244–246. Berkeley: University of California Press.

Rankin CR. 1909. Tijuana to Mantuco Cañon. Southern Pacific Railroad Company records. 1889–1997. Record Group 2, Operating and Maintenance Department records, 1889–1997, Tube 78. *Courtesy Department of Special Collections, Stanford University Libraries.*

Rebert P. 2001. *La Gran Línea: mapping the United States-Mexico boundary, 1849–1857.* Austin: University of Texas Press.

Regional Water Management Group. 2013. *2013 San Diego Integrated Regional Water Management Plan.* 1–112.

Rempel R. 1992. *Hydrogeological assessment of the Tijuana River valley.* California State Water Resources Control Board.

Rhemtulla JM, Mladenoff DJ. 2007. Why history matters in landscape ecology. *Landscape ecology* 22(1–3).

RHJV (Riparian Habitat Joint Venture). 2004. *The riparian bird conservation plan: a strategy for reversing the decline of riparian associated birds in California. California Partners in Flight.* http://www.prbo.org/calpif/pdfs/riparian_v-2.pdf.

Richards ML. 2002. *Sand in my shoes: life in the Tijuana Sloughs 1931–1944.*

Riedman M, Estes J. 1988. A review of the history, distribution and foraging ecology of sea otters. In *The community ecology of sea otters*, ed., 4–21: Springer.

Rintoul DT, Lowe JR, Hutchens JR, et al. 1936. *Bridge report. Courtesy of the California State Railroad Museum.*

Rodney Stokes Co. 1920. Map of San Diego County, California. 1:125,000. *Courtesy of Earth Science and Map Library, UC Berkeley.*

Rodriguez Galeana, M. 1920. *Secretario de Fomento, Cuarta Division, 4(19)335R, Caja-R-2. p. 2. Districto Norte de la B.C. Courtesy of El Archivo Historico del Agua.*

Rovirosa AS. 1963. *Geografía del estado de Baja California.* México, D.F.: Ediciones Económicas.

Ruhlen G, Jr. 1900. Untitled map. *Courtesy of South Bay Historical Society.*

Ryan HL, Humphreys T. 1889. Map of San Diego County, California: compiled from the latest and most reliable official sources and special surveys for Dodge & Burbeck. Lith. H.S. Crocker & Co. San Francisco, CA. *Courtesy of Bancroft Library, UC Berkeley.*

Sacramento Daily Union. 1891. Destruction by water. February 28. *Courtesy of California Digital Newspaper Collection.*

Safford H, North M, Meyer M. 2012. Climate change and the relevance of historical forest conditions. In *Managing Sierra Nevada forests. General Technical Report PSW-GTR-237*, ed. Malcolm North. Albany, CA: U.S. Department of Agriculture, Forest Service, Pacific Southwest Research Station.

Safford H, Wiens JA, Hayward GD. 2012. The growing importance of the past in managing ecosystems of the future. In *Historical environmental variation in conservation and natural resource management*, ed. Gregory Hayward John Wiens, Hugh Safford, and Catherine Giffen. John Wiley & Sons.

SANDAG (San Diego Association of Governments). 2012. Western San Diego County Vegetation. Retrieved on October 13, 2014 from: http://www.sangis.org/.

Sanders TA, Edge WD. 1998. Breeding bird community composition in relation to riparian vegetation structure in the western United States. *The Journal of Wildlife Management*:461–473.

San Diego and Arizona Eastern Railway. ca. 1900. Preliminary Lines National City to Mex-US line South. Southern Pacific Railroad Records, M1010, RG 2, Tube 0033. *Courtesy Department of Special Collections, Stanford University Libraries.*

San Diego and Arizona Railway. 1910. Sketch showing Pile Trestles as constructed between San Diego and Tijuana, Engineer's Office, E-4-684. Southern Pacific Railroad Records, M1010, RG 2, Flat Box 184, Folder D-74. *Courtesy Department of Special Collections, Stanford University Libraries.*

San Diego and Arizona Railway. 1922. Lines of the San Diego & Arizona RY. in California and the Tijuana & Tecate RY. in Mexico. *Courtesy of California State Railroad Museum.*

San Diego County. 1928. [Aerial photographs, flown between November 1928 & March 1929] *Courtesy of Department of Public Works, San Diego County.*

San Diego County Flood Control. 1937. *Preliminary examination report San Diego, San Luis Rey and Tia Juana Rivers.* San Diego.

San Diego Land and Town Company. ca. 1880. Map of the south western portion of the County of San Diego. San Diego, California. *Courtesy of Bancroft Library, UC Berkeley.*

San Diego Land and Town Company. 1887. Birds-eye view of National City San Diego County, California. Frank Burgess and Co. Publishers. National City, California. *Courtesy of David Rumsey Map Collection.*

San Diego Regional Water Quality Control Board. 1967. *Water quality control policy for the Tijuana River Basin in California.* *Courtesy of UC San Diego.*

San Diego Sun. 1891. Sad cry for help: never too late to do good in any charitable cause. *Courtesy of TJNERR.*

San Diego Union. 1887. Tia Juana: a trip to the new town on the Mexican border. September 4. *Courtesy of Universidad Autonoma de Baja California.*

San Francisco Call. 1895. Floods in the south. January 20. *Courtesy of California Digital Newspaper Collection.*

San Francisco Call. 1905. Cool weather follows storm: falling temperatures are reported from southern portion of the state. February 7. *Courtesy of California Digital Newspaper Collection.*

San Ysidro Border Press. 1931. October 9. *Courtesy of San Diego Public Library.*

San Ysidro Chamber of Commerce. 1915. San Ysidro, California. *Courtesy of Special Collections & Archives, UC San Diego Library.*

Sausalito News. 1916. Victims of flood washed out to sea; 60 lives are lost: bodies of seventeen victims taken from San Diego Bay; Tijuana is flooded. February 5. *Courtesy of California Digital Newspaper Collection.*

Savage HN. 1918. City of San Diego California water resources impounding and carrying system. San Diego, California. *Courtesy of Water Resources Collections and Archives (WRCA), UC Riverside.*

Savage HN. 1920. *Tia Juana River (Garcia) Dam Site.* Hiram Newton Savage Photographs, 1905–1933, Accession no. 385, WRCA MS 76/16, Box 2, Folders 44–45. *Courtesy of Water Resources Collections and Archives (WRCA), UC Riverside.*

Sawyer JO, Keeler-Wolf T, Evens J. 2009. *A manual of California vegetation, second edition.* Sacramento, CA: California Native Plant Society.

Schneider C. 1919. Notes on American willows IV species and varieties of section Longfoliae. *Botanical Gazette* 67(4).

Schoenherr S. 2015. The Tijuana River Valley historic sites. *South Bay Historical Society Bulletin* (9).

SCMWC (Southern California Mountain Water Co.). n.d. Map Showing: Lines surveyed for pipes, flumes, ditches &; location of reservoirs & other information relating to the company's Morena Water System. San Diego County, California. *Courtesy of International Boundary and Water Commission.*

Scolam S. Mapoteca Manuel Orozco y Berra. 1919. Parte de una region del estado de Baja Callifornia entre Santa Rosa y Tijuana. Baja, California. *Courtesy of Mapoteca Manuel Orozco y Berra.*

Scripps Institution of Oceanography. 2005. [Silver Strand littoral cell.] Living with coastal change then and now. University of California, San Diego. http://coastalchange.ucsd.edu/st1_thenandnow/silver.html. Accessed January 5, 2017.

SCWMP (Southern California Wetlands Mapping Project). 2012. [Wetland and riparian habitat mapping for southern California coastal wetlands] Region 8 of Southern California Coastal Watershed Wetland Mapping Project Area (CONUS_wet_poly). http://www.socalwetlands.com Accessed April 16, 2013. California State University Southern California Coastal Water Research Project and The Center for Geographic Studies, Northridge, editor.

SDSU (San Diego State University) Dept. of Geography. 2005. *Tijuana River watershed atlas.* San Diego State University Press, Institute for Regional Studies of the Californias.

Seamans P. 1988. Wastewater creates a border problem. *Water Pollution Control Federation* 60(10):1799–1804.

Sebold KR, Record HABSHAE. 1992. *From marsh to farm: the landscape transformation of coastal New Jersey.* Washington, D.C.: Historic American Buildings Survey/Historic American Engineering Record, National Park Service, U.S. Department of the Interior.

Serra J, Tibesar A. 1955. Diary of the expedition from Loreto to San Diego. In *Writings of Junípero Serra*, ed. Antonine Tibesar. Washington, D.C.: Academy of American Franciscan History. Volume I.

Shalowitz AL. 1964. *Shore and sea boundaries, with special reference to the interpretation and use of Coast and Geodetic Survey data, United States.* U.S. Department of Commerce, Coast and Geodetic Survey. [Washington]: Government Printing Office.

Shipek FC. 1982. Kumeyaay socio-political structure. *Journal of California and Great Basin Anthropology* 4(2).

Shipek FC. 1989. An example of intensive plant husbandry: the Kumeyaay of southern California. In *Foraging and farming: the evolution of plant exploitation*, ed. David R Harris and Gordon C Hillman: Unwin Hyman.

Shipek FC. 1993. Kumeyaay plant husbandry: Fire, water, and erosion management systems. In *Before the wilderness: Environmental management by native Californians*, ed. TC Blackburn and K Anderson. Menlo Park, CA: Ballena Press.

Shuirman G, Slosson JE. 1992. *Forensic Engineering: environmental case histories for civil engineers and geologists.* San Diego: Academic Press, Inc.

Sickley TA, Mladenoff DJ, Radeloff VC, et al. 2000. A Pre-European Settlement Vegetation Database for Wisconsin. http://gis.esri.com/library/userconf/proc00/professional/papers/PAP576/p576.htm.

Sipe RJ, McBean K. 1935. Mexican Boundary to San Diego Bay. Southern California Coast, California. U.S. Coast and Geodetic Survey. *Courtesy of NOAA (National Oceanic and Atmospheric Administration).*

Skaradek W. 2010. Saltgrass *distichlis spicata (l.) greene.* USDA NRCS. http://plants.usda.gov/factsheet/pdf/fs_disp.pdf.

Slingerland R, Smith ND. 2004. River avulsions and their deposits. *Annual Review of Earth and Planetary Sciences* 32:257-285.

Smith RL. 1980. Alluvial scrub vegetation of the San Gabriel River Floodplain, California. *Madroño* 27(3):126–138.

Sogge MK, Marshall RM, Sferra SJ, et al. 1997. A Southwestern Willow Flycatcher Natural History Summary and Survey Protocol. Technical Report NPS/NAUCPRS/NRTR- 97/12. USGS Colorado Plateau Research Station, Northern Arizona University, Flagstaff, Arizona

Spalding MJ, Ganster P, University of California San Diego, et al. 1999. *Sustainable development in San Diego-Tijuana: environmental, social, and economic implications of interdependence.* La Jolla: Center for U.S.-Mexican Studies, University of California, San Diego.

Spencer WD. 2005. Recovery research for the endangered Pacific pocket mouse: An overview of collaborative studies. *USDA Forest Service Gen. Tech. Rep.*

Stanford B, Grossinger R, Beagle J, et al. 2013. *Alameda Creek Watershed historical ecology study.* San Francisco Estuary Institute-Aquatic Science Center, Richmond, CA.

Stearns, Abel. 1852. *Melijo (also called "Milijo") [San Diego County and Baja California, Mexico] BANC MSS Land Case Files 91 SD.* p. 5. U.S. District Court, Southern District. *Courtesy of The Bancroft Library, UC Berkeley.*

Stein ED, Cayce K, Salomon M, et al. 2014. *Wetlands of the Southern California Coast - historical extent and change over time.* SCCWRP Technical Report 826, SFEI Report 720, Southern California Costal Water Research Project, San Francisco Estuary Institute, California State University Northridge Center for Geographical Studies.

Stein ED, Dark S, Longcore T, et al. 2007. *Historical Ecology and Landscape Change of the San Gabriel River and Floodplain.* SCCWRP Technical Report #499.

Stein ED, Dark S, Longcore T, et al. 2010. Historical ecology as a tool for assessing landscape change and informing wetland restoration priorities. *Wetlands* 30:589–601.

Steneck RS, Graham MH, Bourque BJ, et al. 2002. Kelp forest ecosystems: biodiversity, stability, resilience and future. *Environmental Conservation* 29(4):436–459.

Stephens F. 1908. MVZ Archival Field Notebooks: Stephens F. 1908 (Part 2), Section 1: Journal and catalog: San Diego County, California 1908 (Part 2). *Courtesy of Museum of Vertebrate Zoology, UC Berkeley.*

Stephens F. 1912. MVZ Archival Field Notebooks: Stephens F. 1909 (Part 2), 1910, 1912, Section 4: Journal and catalog: San Diego County, California 1912. *Courtesy of Museum of Vertebrate Zoology, UC Berkeley.*

Stephens F, Fenn WJ. 1906. *California mammals.* San Diego, California: West Coast Pub. Co.

Storie RE, Carpenter EJ. 1923. *Soil survey of the El Cajon area, California.* U.S. Department of Agriculture: Bureau of Chemistry and Soils. *Courtesy of University of Michigan.*

Storie RE, Carpenter EJ. 1930. Soil map: el Cajon area: California. U.S. Department of Agriculture: Bureau of Chemistry and Soils. *Courtesy of the University of Alabama.*

Stromberg JC, Bagstad KJ, Leenhouts JM, et al. 2005. Effects of stream flow intermittency on riparian vegetation of a semiarid region river (San Pedro River, Arizona). *River Research and Applications* 21:925–938.

Stromberg JC, Beauchamp VB, Dixon MD, et al. 2007. Importance of low-flow and high-flow characteristics to restoration of riparian vegetation along rivers in arid south-western United States. *Freshwater Biology* 52(4):651–679.

Sullivan G, Noe GB. 2001. Distribution of plant species in coastal wetlands of San Diego County. In *Handbook for restoring tidal wetlands*, ed. Joy B. Zedler. Boca Raton: CRC Press.

Swanson M. 1987a. Technical appendix A.1: Geomorphology of the Tijuana Estuary. In *Technical appendices: Tijuana estuary enhancement hydrologic analysis*, ed. Philip Williams & Associates. San Francisco, CA.

Swanson M. 1987b. Technical appendix A.3: Geomorphology of the barrier beach. In *Technical appendices: Tijuana estuary enhancement hydrologic analysis*, ed. Philip Williams & Associates. San Francisco, CA.

SWCA Environmental Consultants. 2004. *Cultural and paleontological resources study for the Tijuana River Valley Regional Park trails and habitat restoration enhancement project*. San Diego, CA.

Swetnam TW, Allen CD, Betancourt JL. 1999. Applied Historical Ecology: Using the Past to Manage for the Future. *Ecological Applications* 9(4):1189–1206.

Taggart CP. 1869. *Pueblo lands of San Diego. Exceptions. Argument in favor of the Fitch Map and Survey*. San Diego Union Book and Job Printing Office.

Takekawa JY, Thorne KM, Buffington KJ, et al. 2013. *Downscaling climate change models to local site conditions: San Diego National Wildlife Refuge Complex*. U.S. Geological Survey, Western Ecological Research Center, Vallejo, CA.

Tanaka H, Smith TE, Huang CH. 1984. The Santiago Peak volcanic rocks of the peninsular ranges batholith, Southern California: volcanic rocks associated with coeval gabbros. *Bulletin of Volcanology* 47(1):153–171.

Teng C. 1994. *The effects of the international boundary treatment plant on the riparian habitat of the Tijuana River valley*. United States Fish and Wildlife, San Diego. *Courtesy of TJNERR.*

The Arizona Sentinel. 1874. High Water. February 21. *Courtesy of Library of Congress Newspapers.*

Thomas C. 1950. Map of San Diego. *Courtesy Department of Special Collections, Stanford University Libraries.*

Thorne KM, MacDonald GM, Ambrose RF, et al. 2016. *Effects of climate change on tidal marshes along a latitudinal gradient in California*. 2331–1258, US Geological Survey.

Tipton AM. 2008. Life with the military at Border Field. California Department of Parks and Recreation.

Tooth S. 2000. Process, form and change in dryland rivers: a review of recent research. *Earth-Science Reviews* 51(1–4):67–107.

TRNERR (Tijuana River National Estuarine Research Reserve). 2014. Temporal investigations of marsh ecosystems. http://trnerr.org/time/.

TRNERR (Tijuana River National Estuarine Research Reserve). 2015. Tijuana River National Estuarine Research Reserve: History. Tijuana River National Estuarine Research Reserve. http://trnerr.org/about/history/.

TRNERR (Tijuana River National Esturarine Research Reserve). 2010. *Comprehensive management plan: Tijuana slough national wildlife refuge & border field state park*. Tijuana River National Estuarine Research Reserve (TRNERR).

Trujillo Muñoz G. 2010. *Gente de frontera : personajes memorables de Baja California*. Tijuana, B.C.: Centro Cultural Tijuana, CONACULTA.

Trussell R. 1891. The Tijuana Flood of 1891. In. *The San Pasqual Trussells: A brief description of the origin and history of the Trussell family which settled in San Pasqual Valley in the late 19th century*. R. Rhodes Trussell.

TRVRP (Tijuana River Valley Regional Park). 2007. *Area specific management directives*. Tijuana River Valley Regional Park (TRVRP).

TRVRT (Tijuana River Valley Recovery Team). 2012. *Recovery strategy: Living with the water*. State Water Resources Control Board. 1–32.

Underwood J, York A. 2004. *Archaeological data recovery excavations in Goat Canyon, sites CA-SDI-13485 and CA-SDI-16047 Locus B, Border Field State Park, San Diego County, California*.

United Nations Department of Economic and Social Affairs/Population Division. 2005. Table A.12: Population of urban agglomerations with 750,000 inhabitants or more in 2005, by country, 1950–2015. *World Urbanization Prospects: The 2005 Revision*. http://www.un.org/esa/population/publications/WUP2005/2005WUP_DataTables12.pdf.

Unitt P. 1987. *Empidonax traillii extimus*: an endangered subspecies. *Western Birds* 18(3):137–162.

Unitt P, Klovstad AE, Haas WE, et al. 2004. *San Diego County bird atlas*. San Diego, California: San Diego Natural History Museum.

Unknown. 1881. Map showing the location of "Los Coronados" Islands. San Diego, California. *Courtesy of Mapoteca Manuel Orozco y Berra.*

Unknown. 1890. Came in time. *The Great Southwest.* 11(1).

Unknown. 1910. [Photograph] First train S.D. and A. Ry. chamber of commerce excursion July 29, 1910. *Courtesy Coleccionista de Tijuana.*

Unknown. 1917. Water Supply. *Municipal Journal* XLII(10):344–345.

Unknown. 1921. *Manantial de agua caliente sobre la margen izquierda y dentro delcauce del Rio de Tijuana.* Correspondencia relativa a concesion de derechos de aguas para el establecimiento de baños medicinales. *Courtesy of Archivo Historico de Agua.*

Unknown. 1946. Mosaico fotografico diversas regiones del Norte de Baja California. *Courtesy of Mapoteca Manuel Orozco y Berra.*

Unknown. 1955. Imágenes verticles: Rio Tijuana en Baja California Norte, Obra 1178, Caja 1/1, Escala/Altura 1:8000, Fecha 1955. *Courtesy of Fundación ICA.*

Unknown. 1965. [Photograph showing "proposed floodway channel" and "present overflow area"] Tijuana, Baja California—San Diego, California. *Courtesy International Boundary and Water Commission Records Office.*

Unknown. 1976. *A history of Imperial Beach. Courtesy of SDSU University Archives.*

URS Corporation. 2012. *Analysis of extreme peak flows for the main Tijuana River, San Diego, California.* San Diego.

U.S. Army Engineer District. 1974. *Tijuana River flood control project.* San Diego.

U.S. Census. 1882. *Report on the productions of agriculture as recorded at the Tenth Census (June 1, 1880).* U.S. Census Office, Washington, D.C.: Government Printing Office.

U.S. Census. 1895. *Report on the statistics of agriculture in the United States at the Eleventh Census: 1890.* U.S. Census Office, Washington, D.C.: Government Printing Office.

U.S. District Court California, Southern District ca. 1840. [Diseño del Rancho Milijo : San Diego County, Calif. and Baja California, Mexico] Land Case Map B-1073. *Courtesy of Bancroft Library, UC Berkeley.*

U.S. Engineer Office. 1937. Areas subject to overflow in lower reaches of Tiajuana River. Los Angeles, CA. *Courtesy of Water Resources Collections and Archives (WRCA), UC Riverside.*

USAED (U.S. Army Engineer District). 1963. [Map] Overflow area and recommended plan of improvement. *Courtesy of Water Resources Collections and Archives (WRCA), UC Riverside.*

USAED (U.S. Army Engineer District). 1964. Review report for flood control: Tijuana River basin, California: overflow area and recommended plan of improvement. Los Angeles, CA. *Courtesy of Water Resources Collections and Archives (WRCA), UC Riverside.*

USAED (U.S. Army Engineer District). 1974. *Tijuana River Flood Control Project: Draft Environmental Statement. Courtesy of Northwestern University.*

USCGS (U.S. Coast and Geodetic Survey). 1895. San Diego Bay, California, plate no. 2395, register no. 5106. Washington, D.C. 1:40000. *Courtesy of NOAA (National Oceanic and Atmospheric Administration).*

USCGS (U.S. Coast and Geodetic Survey). 1974. Revised Shoreline Map T-11893 (2) California Pacific Ocean Coastline Tia Juana River. *Courtesy of NOAA (National Oceanic and Atmospheric Administration).*

USFWS (U.S. Fish and Wildlife Service). 1994a. Endangered and threatened wildlife and plants; determination of endangered status for the Pacific Pocket Mouse. *Department of the Interior Federal Register* 59(188):49752–49764.

USFWS (U.S. Fish and Wildlife Service). 1994b. Designation of critical habitat for the least Bell's vireo. *Federal Register* 59(22):4845–4867.

USFWS (U.S. Fish and Wildlife Service). 2010. Pacific pocket mouse (*Perognathus longimembris pacificus*) 5-year review: summary and evaluation.

USGS (U.S. Geological Survey). 1903. Topography, Cuyamaca Quadrangle. San Diego County, California. *Courtesy of U.S. Geological Survey (USGS).*

USGS (U.S. Geological Survey). 1904. California: San Diego quadrangle. *Courtesy of U.S. Geological Survey (USGS).*

USGS (U.S. Geological Survey). 1913. *Topographic instructions of the United States Geological Survey.* Washington, D.C.: Government Printing Office.

USGS (U.S. Geological Survey). 1930. San Diego quadrangle. *Courtesy of U.S. Geological Survey (USGS).*

USGS (U.S. Geological Survey). 1943. California: San Ysidro quadrangle. *Courtesy of U.S. Geological Survey (USGS).*

USGS (U.S. Geological Survey). 1953. San Ysidro quadrangle. *Courtesy of U.S. Geological Survey (USGS).*

USGS (U.S. Geological Survey). 1967. Imperial Beach quadrangle. *Courtesy of U.S. Geological Survey (USGS).*

USGS (U.S. Geological Survey). 1980a. [Aerial photograph] AR1VEZT00050037. *Courtesy of U.S. Geological Survey (USGS).*

USGS (U.S. Geological Survey). 1980b. [Aerial photograph] AR1VEZT00050039. *Courtesy of U.S. Geological Survey (USGS).*

USGS (U.S. Geological Survey). 1989. NP0NAPP001836047. San Diego, California. *Courtesy of U.S. Geological Survey (USGS).*

USGS (U.S. Geological Survey). 1990. [Aerial photograph] NP0NAPP001854030. San Diego, California. *Courtesy of U.S. Geological Survey (USGS).*

USGS (U.S. Geological Survey). 1994. Imperial Beach. California. Western Mapping Center. *Courtesy of U.S. Geological Survey (USGS).*

USGS (U.S. Geological Survey). 1996. Imperial Beach Quadrangle. *Courtesy of USGS NMD Historical Map Archives.*

USGS (U.S. Geological Survey). 2003. *Part 6 Publication Symbols: Standards for 1:24,000- and 1:25,000-Scale Quadrangle Maps.* U.S. Department of the Interior and U.S. Geological Survey.

USGS (U.S. Geological Survey). 2009. Pacific pocket mouse monitoring plan for Marine Corps Base Camp Pendleton: short term studies and long term goals. Western Ecological Research Center.

USGS (U.S. Geological Survey). 2012. Imperial Beach Quadrangle. *Courtesy of U.S. Geological Survey (USGS).*

USGS (U.S. Geological Survey). 2014. National Hydrography Dataset (NHD).

Uyeda KA, Deutschman DH, Crooks JA. 2013. Abiotic limitation of non-native plants in the high salt marsh transition zone. *Estuaries and coasts* 36(6):1125–1136.

Van Dyke TS, Leberthon TT, Taylor A. 1888. *The city and county of San Diego: illustrated and containing biographical sketches of prominent men and pioneers.* San Diego, CA Leberthon & Taylor.

Van Etten. 1935. *San Diego County investigation.* State of California Department of Public Works. Publication of Water Resources, Edward Hyatt, State Engineer. *Courtesy of the Library of the University of California Davis.*

Van Wormer S. 2005. *A Land Use History of the Tia Juana River Valley.*

Varty AK, Zedler JB. 2008. How waterlogged microsites help an annual plant persist among salt marsh perennials. *Estuaries and Coasts* 31(2):300–312.

Vizcaíno S, Bolton HE. 1916. The Vizcaíno Expedition. In *Spanish exploration in the Southwest 1542–1706*, ed. Herbert Eugene Bolton: C. Scribner's Sons.

von Bloeker JC, Jr. 1931. *Perognathus pacificus* from the type locality. *Journal of Mammalogy* 12(4):369–372.

Vyverberg K. 2010. *A Review of Stream Processes and Forms in Dryland Watersheds.* California Department of Fish and Game, Sacramento, CA.

Waitz DP. 1928. *Brief report about geological conditions of Rodriquez No. 1 dam-site, at lower California.* Comission Nacional de Irrigacion. *Courtesy of Archivo Historico de Agua.*

Wallace KJ, Callaway JC, Zedler JB. 2005. Evolution of tidal creek networks in a high sedimentation environment: a 5-year experiment at Tijuana Estuary, California. *Estuaries* 28(6):795–811.

Ward FB. 1889. Baja California: the southwestern corner of our country, an overland journey from San Diego to Ensenada-how the country looks. *Sacramento Daily Union.* May 25. *Courtesy of California Digital Newspaper Collection.*

Ward KM, Callaway JC, Zedler JB. 2003. Episodic colonization of an intertidal mudflat by native cordgrass (*Spartina foliosa*) at Tijuana Estuary. *Estuaries* 26(1):116–130.

Warner CD. 1891. *Our Italy.* New York, NY: Harper & Brothers, Franklin Square.

Webber SR. 2010. *The role of local watersheds on sediment accumulation in the Tijuana Estuary Reserve.* A thesis presented to the faculty of San Diego State University.

Webster EB. 1913. *Report on the northern district of lower California.* San Diego, CA.

Webster WS. 1903. Trip to Tia Juana, Mexico. *Coronado Tent City Daily Program. Courtesy of California Digital Newspaper Collection.*

Weis DA, Callaway JC, Gersberg RM. 2001. Vertical accretion rates and heavy metal chronologies in wetland sediments of the Tijuana Estuary. *Estuaries* 24(6A):840–850.

West JM. 2001. Tijuana Estuary. In *Handbook for restoring tidal wetlands*, ed. Joy B. Zedler, 32-33. Boca Raton: CRC Press.

Wetmore PH, Herzig C, Alsleben H, et al. 2003. Mesozoic tectonic evolution of the Peninsular Ranges of southern and Baja California. *Geological Society of America Special Papers* (374):93–116.

Whipple AA, Grossinger RM, Davis FW. 2011. Shifting baselines in a California oak savanna: nineteenth century data to inform restoration scenarios. *Restoration Ecology* 19(101):88–101.

Whipple AA, Grossinger RM, Rankin D, et al. 2012. *Sacramento-San Joaquin Delta historical ecology investigation: Exploring pattern and process.* San Francisco Estuary Institute-Aquatic Science Center, Richmond, CA.

Whitcraft CR, Talley DM, Crooks JA, et al. 2007. Invasion of tamarisk (*Tamarix* spp.) in a southern California salt marsh. *Biological Invasions* 9(7):875–879.

White CA. 1983. *A history of the rectangular survey system.* Washington, DC: U.S. Government Printing Office.

White J. 1998. Blue Grosbeak (*Guiraca caerulea*). In *The riparian bird conservation plan: a strategy for reversing the decline of riparian associated birds in California*, ed. Stinson Beach, CA: Point Reyes Bird Observatory.

Williams CP. 1933. Unusual features of the Rodriguez Dam construction. *American Water Works Association* 25(3).

Williams PB, Swanson ML. 1987. *Tijuana Estuary enhancement hydrologic analysis.* Philip Williams & Associates. San Francisco, CA.

Wilson W. 1883. *History of San Diego County, California, with illustrations, descriptive of its scenery, farms, residences, public buildings ... from original drawings, with biographical sketches.* San Francisco, CA: W.W. Elliott & Co.

WRCC (Western Regional Climate Center). 2006. San Diego WSO Airport, California: monthly total precipitation (inches). Western Regional Climate Center.

Wright RD. 2005. Topography, Population, and Hydrography. In *Tijuana River watershed atlas.* San Diego State University, Dept. of Geography. San Diego State University Press, Institute for Regional Studies of the Californias.

Wright WS. 1908. Annotated list of the diurnal Lepidoptera of San Diego County, California, Based on Collections during 1906 and 1907. *Journal of the New York Entomological Society* 16(3):153–167.

Wyman T. 1937a. *Report on preliminary examination: San Diego, San Luis Rey, and Tia Juana rivers.* United States Engineer Office, Los Angeles, CA. *Courtesy of Water Resources Collections and Archives (WRCA), UC Riverside.*

Wyman T, Jr. 1937b. Public hearing on flood control of Tia Juana River, Palm City, California. In *Preliminary examination report: San Diego, San Luis Rey and Tia Juana Rivers*, ed., 1–56. San Diego County, CA: San Diego County Flood Control.

Zaragoza B. 2014. *San Ysidro and the Tijuana River Valley.* Charleston, SC: Arcadia Publishing.

Zedler JB. 1983. Freshwater impacts in normally hypersaline marshes. *Estuaries* 6(4):346–355.

Zedler JB. 1996. Coastal mitigation in southern California: the need for a regional restoration strategy. *Ecological Applications* 6(1):84–93.

Zedler JB. 2010. How frequent storms affect wetland vegetation: a preview of climate-change impacts. *Frontiers in Ecology and the Environment* 8(10):540–547.

Zedler JB, Callaway JC, Desmond JS, et al. 1999. Californian salt-marsh vegetation: An improved model of spatial pattern. *Ecosystems* 2:19–35.

Zedler JB, Covin J, Nordby C, et al. 1986. Catastrophic events reveal the dynamic nature of salt-marsh vegetation in Southern California. *Estuaries* 9(1).

Zedler JB, Nordby CS, Kus BE. 1992. *The ecology of Tijuana Estuary: a National Estuarine Research Reserve.* NOAA Office of Coastal Resource Management, Sanctuaries and Reserves Division. Washington D.C.

Zedler JB, West JM. 2008. Declining diversity in natural and restored salt marshes: a 30-year study of Tijuana Estuary. *Restoration Ecology* 16(2):249–262.

Zedler PH. 1987. *The ecology of Southern California vernal pools: a community profile.* Biological Report U.S. Fish and Wildlife Service.